Formation and Regeneration
of Nerve Connections

Formation and Regeneration of Nerve Connections

Sansar C. Sharma
James W. Fawcett
Editors

Birkhäuser
Boston • Basel • Berlin

Sansar C. Sharma
Department of Ophthalmology
New York Medical College
Valhalla, NY 10595
USA

James W. Fawcett
The Physiological Laboratory
Cambridge CB2 3EG
England

Library of Congress Cataloging-in-Publication Data

Formation and regeneration of nerve connections / edited by Sansar C.
  Sharma and James W. Fawcett.
      p. cm.
    Includes bibliographical references and index.
    ISBN 0-8176-3563-7 (H : alk. paper).—ISBN 3-7643-3563-7 (alk.
paper)
      1. Developmental neurology.   2. Nervous system--Regeneration.
  3. Nerves--Growth.   4. Visual pathways.   5. Neural Pathways.
  I. Sharma, S. C. (Sansar C.), 1938– . II. Fawcett, James W.,
  1950– .
      [DNLM: 1. Nerve Regeneration.   2. Neurons—physiology.   3. Retina-
  -growth & development.   WW 270 F724]
  QP363.5.F67   1992
  596'.01'88—dc20
  DNLM/DLC
  for Library of Congress   92-49631
                                                                       CIP

ISBN 0-8176-3563-7
ISBN 3-7643-3563-7

Typeset by Atlis Graphics, Mechanicsburg, PA
Printed and bound by Quinn Woodbine, Inc., Woodbine, N.J.
Printed in the United States of America

9 8 7 6 5 4 3 2 1

Contents

Preface

The collection of essays in this book arose out of a meeting held in Edinburgh, Scotland in September 1991, timed to mark the retirement of R.M. Gaze (Mike, as he is known to all his friends), and attended by most of his ex-students and many of his collaborators. We, the organizers of this meeting and the editors of this book, were Mike's second (SCS) and penultimate (JWF) Ph.D. students. The diversity of material covered provides a broad synopsis of some of the work currently being pursued in developmental neurobiology and gives an idea of the breadth of Mike's interests and of the interests he has stimulated in his many students and collaborators.

Mike started his scientific career after qualifying in medicine in Edinburgh, then studying Animal Physiology in Oxford. For his D.Phil. he studied thalamic neurophysiology with George Gordon, presenting his thesis *"The Thalamic Response to Sensory Afferent Stimulation"* in 1954. Mike writes of George Gordon "I had (and have) a great respect for him as a scientist. He was also unfailingly kind, helpful and supportive to me. My ideas on how science should be done derive from him." Mike then went off to do his national service in the Royal Army Medical Corps. After this, he was appointed as lecturer in the Physiology Department in Edinburgh and it was there that he started the studies on the frog retinotectal projection which made his name, and which he has continued for the rest of his career.

Developmental Neurobiology is now such a huge and fashionable field that it is difficult to picture what it must have been like to be a developmental neurobiologist in the mid-fifties. The previous two generations had produced two personalities who had in turn dominated this rather unfashionable area of research: Paul Weiss and Roger Sperry. Mike's decision to work on the retinotectal projection, as with so many turning points in anyone's career, came about almost accidentally. When he started in Edinburgh he was asked by David Whitteridge, then the head of the department, what research he proposed to do. He replied that he thought he would continue with thalamic electrophysiology. Whitteridge then asked him what he thought of an interesting recent paper by Roger Sperry. Mike (whose initial reaction was "who is Sperry?") read it, immediately became a retinotectal enthusiast, and remained so for the rest of his career.

In his first experiment on the retinotectal projection, the pathway which Sperry had used as his experimental model, Mike followed the same lines as Sperry's work. These initial experiments gave results that were apparently entirely consistent with Sperry's view of a "hard-wired" nervous system, in which each axon would grow to a chemically defined target and make a connection there. The turning point in Mike's career, and a turning point in developmental neurobiology, came from a whole series of experiments with a variety of co-workers including one of us (SCS), which challenged this view of how the nervous system was put together. The experimental paradigm was the mismatch experiment (which also originated in Sperry's laboratory), involving mismatches of retinal and tectal sizes, or mismatches in the complement of retinal positions in compound eyes made of two nasal or temporal half eyes.

These experiments were started after Mike had read an abstract by George Székely, who came from Szenthágothai's very active experimental embryology group in Péces, Hungary. Székely had worked out a technique for operating on *Xenopus* embryos so that they would grow up with half eyes or with compound eyes made of two mismatched halves. Mike thought that such preparations would provide an interesting test of the early chemospecificity ideas. He got in touch with Székely and this led to extensive collaboration and lifelong friendship. Electrophysiological analysis of the connections formed between these "compound eyes" and the tectum was then carried out (GJS, 1963). The revolutionary results of these experiments, implying a considerable degree of plasticity in the putting together of connections in the brain, were not particularly well received at the time, challenging as they did the views put forward by Sperry and others. Eventually, however, the evidence for readjustment of connections as a result of these various manipulations became overwhelming, and it was from the results of these experiments that we can trace our current concepts of neuronal plasticity.

Mike's subsequent work with his many co-workers is too extensive to list here in detail, but some highlights deserve mention. In 1973 there was a direct demonstration of shifting connections during growth of the retinotectal projection, a shift required to accommodate the persistence of an orderly retinotectal map during development with the different geometries of growth of retina and tectum. The eighties saw experiments which concentrated on the guidance of retinotectal axons and the role of axon guidance in map formation and now, in the nineties, Mike is busy with neuronal cell death, microglia and axon-glial interactions. Apart from his experimental work, he wrote in 1970 a classic book, *The Formation of Nerve Connections*, which was both a history of developmental neurobiology and a contemporary research text. This book was, for many, the entry into this new and exotic science of developmental neurobiology, and was the shaper of many scientific careers, including one of ours (JWF, who still remembers the excitement of reading this monograph as an undergraduate).

All of us who have had the good fortune to do science with him have strong memories of Mike's approach to research. He is always at his happiest doing experiments; making physiological recordings or looking down the microscope; particularly so if the results seem to be running counter to current theories. One could always tell when something interesting was up because Mike would stride rapidly back and forth puffing clouds of smoke from his pipe. Nowadays he just strides back and forth, having managed to stop smoking some years ago when, as he says, "it was taking over control of my life."

Mike has always particularly enjoyed the aesthetics of science, the beauty of images seen down the microscope, and the elegance of a well organized, or well re-organized, retinotectal map. He taught by example, and perhaps the lesson of most lasting value to many of us stemmed from this aesthetic fascination with the data for its own sake. Mike thinks of theories, even his own, as things transitory and ephemeral, and enjoys seeing the data demolish them. When beginning in science it is very easy to think that one's data is somehow at fault if it does not agree with current dogma, but Mike would always encourage one to believe in the data in front of one. This is not necessarily an attitude that guarantees an easy relationship with the scientific establishment, but it is one that has stood us all in good stead in the long run.

Mike has had many associates, and as the reader will see from the following collection, most have stayed in developmental biology, although many have strayed from the straight and narrow path of retinotectology. We all share an appreciation of Mike's central role in the development of our field, and in our own personal careers, and we wish him well in his retirement.

In an interesting and evocative essay on the historical aspects of neural specificity, Marcus Jacobson has compared the theories of specificity and plasticity as they pertain to the developing visual system. Phil Grant reviews what is now known of cell death in the developing retina and its relevance to retino-tectal connection patterns. There follows a short review by Jeremy Taylor on axon pathfinding and guidance involved in the establishment of connections to the tectum. Cellular interactions during embryonic development, especially the evidence that glia cells influence neuronal pathfindings, are discussed by Gooday.

The past decade has witnessed a surge of interest in the analysis of sensory experience in the formation of ipsilateral visual inputs and the dramatic plasticity of these connections. The essay by Udin describes the role of NMDA receptors in intertectal plasticity and is followed by a summary of binocular connection formation by Grant, Brickley and Keating.

The techniques of experimental manipulation of the lower vertebrate visual system, exploring various questions of neuronal specificity, have been extended to the mammalian visual system with exciting results. In particular, Ray Lund and his colleagues address this topic. A comparative essay by James Fawcett on the refinement of topographic projections explores the similarities of this phenomenon across species.

Alan Peters describes the modular organization of pyramidal cells and the development of such modules in the rat neocortex. Such detailed studies are prerequisite to defining the properties of physiological response and the emergence of specificity of such connections. A similar approach to understanding the architecture of a system extends to the study of spinal cord connectivity. The essay by Mendell and Koerber explores neuronal specificity in the sensory fiber projection in the spinal cord.

The next three papers include Agranoff and Heacock's review of retinal explants as applied to regenerative capability of optic axons by an examination of certain aspects of developing and regenerating retinal ganglion cells by Dunlop,

Fraley and Beazley, and a discussion of the development of retinal cell types by Charles Straznicky. Collectively these essays show some of the ways in which the general principles of developmental neurobiology are now applied to the analysis of growth and regeneration.

John Schmidt summarizes his observations on the mechanisms subserving the sharpening of retinotectal projection during the early phases of regenerating optic axons and discusses the implication of similar rules during normal development of retinal projections.

The next two papers are a departure from the usual analytical and experimental approaches to the problem of neuronal specificity. The essay by John Cronly-Dillon essentially indulges the "global features" of the organization and function of the nervous system and draws attention to non-equilibrium thermodynamic and other self-organization theories of physics and their relevance to biology, especially as they relate to pattern formation in the nervous system. The essay by Tim Horder describes various facets of evolution as they pertain to the eye.

The next paper by David Willshaw involves the computational aspects of neuronal specificity and describes the new "candidate" mechanism for the establishment of ordered connections.

Finally, the volume ends with a short essay by Sansar Sharma regarding current issues and dogmas in the formation of neuronal connection and drawing attention to unresolved issues and the future direction of the investigation of neuronal specificity.

On behalf of all the contributors, we should like to present these essays to Mike, whom we know and admire both as scientist and as a friend.

George Adelman deserves special thanks for his foresight, gentle persistence and encouragement in the early phase of this undertaking. We are grateful to James Doran for his encouragement and forbearance at every stage. We would like to thank the Pathology Department of the University of Edinburgh for allowing us to gather in their facility.

New York, USA   S.C.S.
Cambridge, England   J.W.F.
September, 1992

List of Contributors

Bernard W. Agranoff, Department of Biological Chemistry and Mental Health Research Institute, University of Michigan, Ann Arbor, Michigan, 48104-1687, USA

Lyn Beazley, Department of Psychology, University of Western Australia, Nedlands, Australia

Stephen G. Brickley, Division of Neurophysiology & Neuropharmacology, National Institute for Medical Research, London, England

John Cronly-Dillon, Department of Optometry & Vision Sciences, Developmental Neurobiology Unit, UMIST, Manchester, England

Sarah Dunlop, Department of Psychology, University of Western Australia, Nedlands, Australia

James W. Fawcett, The Physiological Laboratory, Downing Street, Cambridge CB2 3EG, England

Sandra Fraley, Department of Ophthalmology, New York Medical School, Valhalla, New York, 10595, USA

Douglas Gooday, MRC Neural Development & Regeneration Group, University of Edinburgh, Edinburgh, Scotland

Philip Grant, Laboratory of Neurochemistry, NINDS, National Institutes of Health, Bethesda, Maryland, 20892, USA

Simon Grant, Department of Anatomy, Charing Cross & Westminster Medical School, London, England

Anne M. Heacock, Mental Health Research Institute, University of Michigan, Ann Arbor, Michigan, 48104-1687, USA

T. J. Horder, Department of Human Anatomy, South Parks Road, Oxford, England

Marcus Jacobson, Department of Anatomy, University of Utah School of Medicine, Salt Lake City, Utah, 84132, USA

Michael J. Keating, Division of Neurophysiology & Neuropharmacology, National Institute for Medical Research, London, England

H. Richard Koerber, Department of Neurobiology, Anatomy, and Cell Science, School of Medicine, University of Pittsburgh, Pittsburgh, Pennsylvania, 15261, USA

Raymond D. Lund, Professor and Head, Department of Anatomy, University of Cambridge, Cambridge, England

Lorne M. Mendell, Department of Neurobiology and Behavior, State University of New York, Stony Brook, New York, 11794, USA

Alan Peters, Department of Anatomy and Neurobiology, Boston University School of Medicine, Boston, Massachusetts, 02118, USA

Jeffrey D. Radel, Department of Neurobiology, Anatomy, and Cell Science, University of Pittsburgh School of Medicine, Pittsburgh, Pennsylvania, 15261, USA

John T. Schmidt, Department of Biological Sciences, State University of New York at Albany, Albany, New York, 12222, USA

Sansar C. Sharma, Department of Ophthalmology, New York Medical College, Valhalla, New York, 10595, USA

Charles Straznicky, Department of Anatomy and Histology, School of Medicine, The Flinders University of South Australia, Adelaide, Australia

Jeremy S. H. Taylor, Department of Human Anatomy, Oxford University, Oxford, England

Susan B. Udin, Department of Physiology, State University of New York, Buffalo, New York, 14214, USA

David Willshaw, Centre for Cognitive Science, University of Edinburgh, Edinburgh EH8 9LW, Scotland

Kathleen T. Yee, Department of Neurobiology, Anatomy, and Cell Science, University of Pittsburgh School of Medicine, Pittsburgh, Pennsylvania, 15261, USA

CHAPTER 1

Historical Development of the Concept of Neuronal Specificity

MARCUS JACOBSON

The conventional way of starting a lecture of this kind is to give the person who is being honored a few pats on the back and then to dismiss him before getting down to serious business, which normally means talking about one's own achievements. I have no intention of playing that opening gambit. The position occupied by Mike Gaze in the history of neuronal specificity is uniquely important. He entered the field in 1957 at a time when the concept of neuronal specificity had reached canonical form in the chemospecificity theory of Roger Sperry. During the previous 20 or 30 years, largely as the result of Sperry's work, there had been a transformation of the theory of neuronal plasticity by imperceptible stages into its opposite, specificity. Gaze was to plan an important part in the transformation of the theory of specificity of the retinotectal system back to something like its earlier state.

The dichotomy between rigidity and plasticity (Table 1.1), like so many others, originated in religion. The dichotomy between the inherent rigidity and the infinite perfectibility of man's nature, and between predetermination and free will, was well established in the Christian religion by the Middle Ages. The concept of neuronal specificity developed from the rationalist belief, starting with Descartes, that the brain is a neuronal machine, obeying the laws of cause and effect (Table 1.2). Descartes had to introduce the soul; exempt from causal laws, to account for free will, but his followers such as La Mettrie, Holbach, and Helvetius denied the existence of the soul. It follows from this view, which we may call "the mechanization of the brain picture" (Jacobson, 1991, Chap. 11; 1993), that the components of the brain machine must have quite specialized (specific) structures and connections with one another in order to play their parts in the functions of the whole. On the basis of very few facts, Descartes erected a vast theoretical edifice to explain brain function. For example, Descartes thought of nerves as hollow tubes in which minuscule particles are conveyed from the sense organs to the brain and thence to the effector organs. He made those conjectures before Fontana discovered nerve fibers in 1781. Gassendi, Borelli, and others in the seventeenth century constructed hydraulic and pneumatic theoretical models of the nervous system, by analogy with macroscopic systems, only scaled to microscopic dimensions, but with the same physical laws operating in the macro- and the

Formation and Regeneration of Nerve Connections
Sansar C. Sharma and James W. Fawcett, Editors
© 1993 Birkhäuser Boston

TABLE 1.1. Comparison between theories of neuronal specificity
and plasticity

Specificity	Plasticity
1. Things connected by necessity	1. Things connected by contingency
2. Diversity in unity: "splitters"	2. Unity in diversity: "lumpers"
3. Things interact as particles	3. Things interact as fields, waves, gradients
4. Mechanist	4. Vitalist
5. Reductionist	5. Holist
6. Selective mechanisms (cell death, synapse elimination)	6. Instructive agents (trophic agents and growth factors)
7. Structure before function	7. Function before structure
8. Competitive > cooperative	8. Cooperative > competitive
9. Teleonomic	9. Teleological
10. Psychology: nativist	10. Psychology: empiricist
11. Philosophy: rationalist	11. Philosophy: empiricist
12. Politics: authoritarian	12. Politics: libertarian
13. Ethics: absolute	13. Ethics: relative

microsystems. It was conceivable that the brain contained even more tubes, pumps, levers, cogs, and springs than the most complex automaton that could be constructed by man. One could not set a limit to the degree of specificity; thus nerves conveying different modalities of sensation might transport qualitatively different particles, which might even represent different sensory qualities.

Without going further, it is obvious that the rationalists constructed a pyramid of theory standing upside down on a point of evidence. This is the kind of theory to which T. H. Huxley referred when he said that a tragedy is a beautiful hypothesis killed by an ugly fact. Such rationalist theoretical structures are notoriously easy to overthrow by a little well-corroborated counterevidence. By contrast, the empiricists took the view that theories of the nervous system, such as a theory of specificity, would have to be built on the basis of rather limited conjectures supported by a wealth of empirical data: essentially starting with the bottom of the pyramid and working gradually up to the apex. Such a theoretical structure is stable and cannot be toppled by a single revolutionary counterexample. It can survive the replacement of much of its evidential fabric by counterevidence. The advantage of this method of constructing a research program is its resistance to refutation, allowing the program to survive and mature, especially at the early stages when it is more vulnerable to refutation. If scientific progress was only a matter of conjecture and refutation, few conjectures would survive refutation, even those conjectures that could prove correct in the long run. Insurance against easy refutation is provided by the tenacity with which scientists hold to their conjectures in the face of counterevidence, hoping to come up with counter-counterevidence and so on. If that is not forthcoming, it is usually possible to retreat behind *ad hoc* hypotheses. These stratagems have been used often in the history of neuronal specificity.

A strong vein of thought tending toward the concept of neuronal plasticity derives from the British empiricists, especially David Hume. David Hume

TABLE 1.2. History of neuronal specificity and plasticity

Specificity	Plasticity

1. Automaton theory—determination
    Descartes, 1664
    LaMettrie, 1745
	1. Nonmaterial soul—free will
	    Descartes, 1664

2. Things connected by necessity
    Rationalists
	2. Things connected by contingency
	    Empiricists and utilitarians

3. Specificity of nerve energies
    Müller, 1826
	3. Association of ideas
	    Hume, 1739

4. Structure–function specificity
    Sensory and motor nerves
    Magendie, 1822
    Cerebral cortex
    Gall, 1825
    Hughlings Jackson, 1863
    Fritsch and Hitzig, 1870
    Ferrier, 1875–1890
    Brodmann, 1907
	4. Association of nerve actions
	    Hartley, 1749
	    Spencer, 1855
	    Kappers, 1936
	    Hebb, 1949

5. Speciation by selection of fittest
    Darwin, 1854
	5. Effects of use and disuse
	    Lamarck, 1815–1822

6. Physiological division of labor
    (tissue and cellular differentiation)
    Milne-Edwards, 1857
    Spencer, 1876
	6. Functional adaptation
	    Spencer, 1852
	    Roux, 1881

7. Cellular functional specificity
    Sperm-egg
    Pfeiffer, 1884
    Serological reactions
    Landsteiner, 1900
    Ehrlich, 1910
    Neuronal typology
    Koelliker, 1854–1898
    Golgi, 1885
    Cell recognition and aggregation
    Wilson, 1907
    Holtfreter, 1939
	7. Equipotential systems
	    Embryo
	    Driesch, 1898
	    Cerebral cortex
	    Flourens, 1842
	    Goltz, 1888
	    Lashley, 1929
	    Neural systems
	    Bethe, 1905
	    Weiss, 1928
	    Anokhin, 1968

8. Cellular selection
    Roux, 1881
	8. Plasticity of neural maps
	    Campbell, 1905

9. Molecular specificity
    Antibodies
    Ehrlich, 1910
    Cellular receptors
    Langley, 1895
    Proteins
    Loeb, 1916
    Cell adhesion molecules
    Weiss, 1941
    "Chemoaffinity"
    Sperry, 1963
	9. Plasticity of nerve cells
	    Koelliker, 1896
	    Wundt, 1904

(1711–1776) was born and died in Edinburgh, but he wrote *A Treatise of Human Nature* in France in 1734–1737 and had it published in London in 1739. Indeed, Hume failed to secure the chair of philosophy in Edinburgh, which is perhaps just as well or he might have remained a poor drudge rather than becoming independently wealthy through his business ventures on the continent. I'm not making the case that the best man had to leave Edinburgh to find fame and fortune, although that has occurred so often that it merits some consideration. In any event, Hume's theory of the association of ideas is one of the foundations of theories of neuronal plasticity. According to Hume, sense perceptions of given objects result in mental impressions and ideas of the objects. The frequent coincidence of two ideas leads to their mental association. Then the idea of one object alone will arouse in the mind the association of both ideas.

The rationalists such as Descartes, Spinoza, and Leibnitz regarded causality as a necessary connection between things, whereas Hume saw no necessary connection, only the association from habit resulting in an association of ideas. This important distinction between things connected by necessity and things connected by contingency has since been the fundamental difference between theories of brain specificity and brain plasticity, respectively (Tables 1.1 and 1.2).

Hume's denial of the necessity of causal relationships undermines the belief in the necessity of ethical principles. Hume's assertion that causality is merely association from habit leads to ethics based on pursuit of pleasure and avoidance of pain, and to toleration of different ethical beliefs and practices. Hume was a moderate skeptic in the sense that his questioning did not leave him in a chronic state of indecision, but he was unwilling to take things for granted without further enquiry. In that respect Gaze is an intellectual offspring of Hume.

The philosophical view of the nature of the nervous system which Gaze has himself articulated is that the complexity of the nervous system makes it irreducible to mere molecular biology; that out of the accumulation of large numbers of relatively small, structural, quantitative changes, significantly large qualitative functional changes emerge. In other words, there are dialectical materialist relationships between structure and function. Gaze has said that there is more to the nervous system than the sum of its parts, and moreover, that the answers to problems of development of neuronal systems are not to be found only in physics and chemistry. Gaze (1970, p. 262) states:

I would agree with Goodwin and Cohen (1969) that the belief that an understanding of global embryological problems will come with the application of the ideas and techniques of molecular biology to embryonic systems, is ill-founded.

Consistent with this view of the limits of molecular biology is Gaze's (1970, p. 263) critique of the limits of reductionism:

Our reductionist arguments from the logical requirements of function may lead us very far astray here. Such an approach works very well when we are dealing with the mechanical or hydraulic aspects of life. . . . But in the case of nervous system we are . . . dealing with a quite different **kind** of system. . . . The difficulties and inadequacies of the logico-reductionist position are clearly illustrated in the work on regeneration within the visual system.

Actually, Gaze's intellectual lineage may be traced from the empiricism of Locke to Hume and later to the utilitarians Bentham and J. S. Mill. His detour into Marxism can also be comprehended by recognizing that the utilitarians influenced the scientific side of Marxism. Empiricism in philosophy goes with liberalism in politics, relativism in ethics, and an aversion to rigid systems in science. By contrast, rationalism in philosophy goes with authoritarianism in politics, ethical absolutism, and a tendency toward system building in science (Table 1.1). The rationalist builds a vast edifice of theory on a slender foundation of fact, whereas the empiricist erects a relatively modest theoretical structure on a broad basis of observation.

A situation in which there are two opposing positions has the effect of driving people exclusively into support of one of the opposing scientific theories. Usually it is an advantage to support only one of two conflicting theories such as the theories of neuronal rigidity or plasticity—at least one of them is likely to be right. But to subscribe in part to both the theory of neuronal specificity and the theory of neuronal plasticity is to be exposed to the censure of the supporters of both systems. This I know from direct experience. Our hypothesis that there are separate classes of neurons which subserve either plasticity or rigidity met with censure from all sides. The idea was that the projection neurons are invariant and the local circuit neurons are the variable component (Hirsch and Jacobson, 1975).

The point of departure for the work of Gaze and his associates was the doctrine that the development of the vertebrate nervous system is highly determined by genetic and developmental mechanisms that are virtually unmodifiable by experience. Weiss (1941a,b) called the process "self-determination." That nativist and rationalist position had been supported by evidence obtained by Weiss (1941, 1952) and Sperry (1945, 1951a, 1951b, 1963, 1965) from experiments on amphibians and fish in which it was shown that after nerves made incorrect connections, the resulting behavior was permanently maladaptive. For example, the maladaptive behavior is permanent after inversion of the eye and regeneration of the optic nerve (Sperry, 1951a, 1951b, 1963, 1965). Those findings show that regeneration of the optic nerve is governed by invariant genetic and developmental mechanisms and not by the adequacy of the functional effect. Their conclusions were based almost entirely on behavior and required them to make a large inductive leap to the genetic and developmental processes. Weiss and Sperry made an enormous inductive leap from behavior to cells and molecules. Our intention was to bridge that distance with empirical data and small inductive steps—by what is called the hypothetico-deductive method.

A hypothesis is an educated guess, a generalization from particular instances, and it is thus an inductive step: the hypothetico-deductive method should be called the inductive-deductive method. Thus, if a rational inductive theory is represented by a large pyramid standing on its apex, an empirical hypothetico-deductive theory can be represented as a structure made of many small pyramids standing on their apexes, forming a large pyramid standing on its base. In other words, empiricism is merely rationalism chopped into small bits.

In 1959 the evidence was overwhelmingly consistent with the theory that the nervous system is formed by highly specific associations between neurons unaffected by function and experience. Remember that 20 or 30 years earlier the consensus was that large parts of the nervous system of mammals are formed of equipotential networks that become organized as a result of activity (Bethe, 1903; Weiss, 1928a,b; Anokhin, 1968). Koelliker (1896, Vol. 2., p. 810) summed up this view of the equipotentiality and plasticity of nerve cells:

I am finally forced to the conclusion that all nerve cells at first possess the same function, and that their differentiation depends solely and entirely on the various external influences or excitations which affect them, and originates from the various possibilities that are available to them to respond to those contingencies.

That neurons are connected by contingency, not by necessity, continues to be the essential feature of theories of neuronal plasticity.

The change from the theory of neuronal plasticity to the theory of neuronal specificity occurred gradually—one theory did not topple the other in a revolutionary way, but both coexisted over the entire period. Evidence corroborating the theory of neuronal specificity did not falsify the theory of plasticity, especially when the validity of the evidence was found to be limited to one genus or even to a single species.

A movement away from theories of neuronal specificity toward a theory of plasticity became quite obvious around 1965, when I left Edinburgh. The movement started several years earlier as the result of three important discoveries of Gaze and his associates. The first was the discovery of adjustments of the pattern of the retinotectal projection following surgical removal of part of the retina or tectum (Gaze et al., 1963, 1965; Gaze and Sharma, 1970). The second was the formulation of Gaze's theory of sliding connections to explain the changes in the retinotectal maps that appeared to be necessary to compensate for disparity between the growth patterns of the retina and the tectum (Straznicky and Gaze, 1971, 1972). The third followed the discovery of the binocular visuotectal projection (Gaze, 1959; Gaze and Jacobson, 1962, 1963), when it was found that functional interactions between the visuotectal projections from the two eyes are necessary for bringing the projections into binocular correspondence in the tectum in Xenopus (Keating, 1968). Techniques play an essential role in all research programs. This one could not have progressed without large advances in the technique of microelectrode recording of visually evoked action potentials in the frog's tectum that were made in the early 1960s (Maturana et al., 1960; Jacobson, 1961, 1962; Jacobson and Gaze, 1964).

Among the most far-reaching discoveries made by Gaze and his associates are the adaptations made by tectal afferents to occupy the available tectal space. These experiments are all of fundamentally the same design, involving disparities in the size of the retina or tectum. Either a reduced retina (actually two identical half-retinas forming a "compound eyed") projects to a whole tectum (Gaze et al.,

1963, 1965), or a whole retina projects to a surgically reduced tectum (Gaze and Sharma, 1970). There are many variations and permutations of this type of experiment, but in all cases the retina projects to the whole available tectum. If we think in terms of retinal and tectal place markers (locus specificities) and in terms of a cellular and molecular map, we are forced to the conclusion that the retinal markers, or tectal markers, or both, change so that they always form matching or complementary sets within given boundaries.

Yoon (1976) showed that at first only the optic axons from the nasal hemiretina project to their former positions in a residual rostral half-tectum in the goldfish, but after a few weeks the optic fibers from the entire retina become compressed into the residual rostral half-tectum. We do not understand what sort of changes occur during those few weeks. That changes occur in the tectum is shown by Yoon's finding that the compression of the projections in a related tectal graft occurs separately from the compression in the surrounding tectum.

If the optic nerve is cut in goldfish or frogs, and a part of the tectum is excised and reimplanted at a new orientation, regenerating optic axons return to their former positions in the tectal graft, demonstrating a very precise selectivity between the tectal graft and the surrounding tectum (Levine and Jacobson, 1974). This is good evidence that the axons do not array themselves in the tectum only by interactions among themselves, or that such interactions are weaker than those between axons and tectum. When we first proposed that the formation of the retinotectal map involves weak axon-to-axon interactions and stronger axon-to-tectum interactions (Levine and Jacobson, 1974), we were not thinking of cell adhesion molecules, but that is now regarded as one of the significant molecular mechanisms (Fraser et al., 1984; Jacobson, 1988).

Moving on to the second important contribution of the Gaze school, Straznicky and Gaze (1971, 1972) reported that there is a large disparity in the patterns of histogenesis of the retina and the tectum—the retina grows by addition of concentric rings, the tectum by addition of crescents at the dorsomedial tectal margin. Gaze then proposed that during the period of retinal and tectal histogenesis, while the retinotectal projection is forming, the optic nerve terminals move continually to adapt to the retinotectal size disparity. This theory of sliding connections was an inductive leap quite uncharacteristic of the skeptical empiricist. John Locke said almost three centuries ago that induction from experience "may provide us convenience, not science" (Locke, 1706, Book IV, Chap. 12, Sec. 10). I don't agree with that—science is impossible without induction, namely generalization from the particular instance to the universal rule. However, the risk of error increases with the height of the inductive leap. Gaze made a risky leap from patterns of retinal and tectal histogenesis to patterns of synaptic connections.

The heuristic effects of Gaze's sliding connections theory can hardly be overestimated. It stimulated efforts to verify and to refute it. There are people who believe that the theory has been verified and others who believe that it has been refuted. I shall not attempt to arrive at a consensus, but it should be remembered that verification and refutation are not symmetrical—a theory can be refuted by some well-corroborated counterevidence, but it cannot be conclusively verified by any amount of confirmatory evidence. The counterevidence has been largely ignored or minimized, so it may be worth recounting briefly. First, continuous radioactive isotope labeling of the newly born retinal and tectal neurons and of the projecting optic axons did not show the disparity between tectal cells and optic nerve terminals predicted by the theory (Jacobson, 1977). Second, the predicted disparity between retinal and tectal histogenesis was not observed. Rather than symmetrical annular growth of the retina, Beach and Jacobson (1978a, 1978b) found that 10 times as many cells are added to the ventral retina as to the dorsal retina in *Xenopus* tadpoles. Thus, both retina and tectum have essentially matching growth patterns. The theory of sliding connections was formulated in 1970, before the significance of the extensive retinal cell death first became known (Rager and Rager, 1978; Beazley, 1981; Beazley et al. 1986). The amount of retinotectal disparity that has actually been demonstrated may be compensated by death of retinal axons that form incorrect connections.

Has all this counterevidence resulted in abandonment of the theory? Not in the least! Opposing theories can coexist for decades. The reasons are that the evidence is inconclusive, artifacts are mistaken for facts, and the proponents hold to their theory with great tenacity, which they are able to do by minimizing the significance of counterevidence. A theory can be saved by ignoring the counterevidence and by making the necessary *ad hoc* adjustments. This happened in the case of the reticular versus the neuron theory, in the case of cerebral localization versus cerebral mass action, and also in the case of electrical versus chemical synaptic transmission. Theories in neuroscience are not simply refuted by counterevidence, as Karl Popper's doctrine says they should be (Popper, 1959, 1962). Popper declares that a good theory should specify in advance the terms and conditions for its own refutation. But that is always limited by the knowledge that went into formulation of the theory. In fact new techniques and new data are usually unpredictable. Thus Gaze could not specify in advance that the theory of sliding connections could be refuted by finding that cell death compensates for the observed disparity between retinal and tectal cell production.

A third very important line of research opened by the Gaze school followed the discovery of the binocular visuotectal projection in frogs (Gaze and Jacobson, 1962, 1963). The contralateral retinotectal projection starts developing at embryonic stages and continues developing throughout life, whereas the ipsilateral projection, which is relayed via isthmotectal fibers, develops during metamorphosis (Beazley et al., 1972). The contralateral map forms completely, and the ipsilateral map starts forming, in the absence of visual activity. However, binocular visual activity is required for bringing the contralateral retinotectal and isthmotectal projections into correspondence in *Xenopus* (Keating and Feldman, 1975) but not in Rana (Jacobson and Hirsch, 1973a,b; Skarf and Jacobson, 1974; Beazley, 1979; Kennard and Keating, 1985). In Rana the isthmotectal projection develops in animals reared in the dark (Keating and Kennard, 1987). In Rana, after rotation of one eye no compensatory adjustments occur in either isthmotectal or retinotectal projections (Jacobson and Hirsch, 1973a,b; Skarf and Jacobson, 1974). Contrasting with the rigidity of the system in Rana is its plasticity in

Xenopus: rotation of one eye results within several weeks in reorientation of the isthmotectal map, bringing it into correspondence with the retinotectal map (Keating, 1968, 1974; Gaze et al., 1970). This is called functional adaptation because it requires binocular visual function. However, it does not result in recovery of adaptive visuomotor behavior. Quite rapid progress has recently been made in clarifying some physiological mechanisms of functional interaction between the retinotectal and isthmotectal projections (Udin and Fawcett, 1988). The isthmotectal axons probably release acetylcholine, whereas the retinotecta axons release glutamate. Blockade of glutamate receptors by chronic application of *N*-methyl-D-aspartate (NMDA) antagonists to the tectum shortly after metamorphosis prevents reorientation of the isthmotectal map in *Xenopus* (Scherer and Udin, 1989). This discovery greatly increases the possibility of reducing the phenomenon of functional adaptation to molecular mechanisms. But that remains for the future. If there are lessons we may learn from history, they are that knowledge of the past, however detailed and accurate, is not an infallible guide to the future, and that we should be prepared for the unexpected.

REFERENCES

Anokhin PH (1968): *The Biology and Neurophysiology of the Conditional Reflex* (in Russian). Moscow: Meditsina

Beach DH, Jacobson M (1978a): Patterns of cell proliferation in developing retina of the clawed frog in relation to blood supply and position of choroidal fissure. *J Comp Neurol* 183: 602–614

Beach DH, Jacobson M (1978b): Patterns of cell proliferation in the retina of the clawed frog during development. *J Comp Neurol* 183: 625–632

Beazley LD (1979): Intertectal connections are not modified by visual experience in developing *Hyla moorei*. *Exp Neurol* 63: 411–419

Beazley LD (1981): Retinal ganglion cell death and regeneration of abnormal retinotectal projections after removal of a segment of optic nerve in *Xenopus* tadpoles. *Dev Biol* 85: 164–170

Beazley LD, Darby JE, Perry VH (1986): Cell death in the retinal ganglion cell layer during optic nerve regeneration for the frog *Rana pipiens*. Vision Res 26: 543–556

Beazley LD, Keating MJ, Gaze RM (1972): The appearance, during development, of responses in the optic tectum following stimulation of the ipsilateral eye in *Xenopus laevis*. *Vision Res* 12: 407–410

Bethe A (1903): *Allgemeine Anatomie und Physiologie des Nervensystems*. Leipzig: Verlag von Georg Thieme

Fraser SE, Murray BA, Chuong CM, Edelman GM (1984): Alteration of the retinotectal map in *Xenopus* by antibodies to neural cell adhesion molecules. *Proc Natl Acad Sci USA* 81: 4222–4226

Gaze RM (1959): Regeneration of the optic nerve in *Xenopus laevis*. *J Exp Physiol* 44: 209–308

Gaze RM (1970): *Formation of Nerve Connections*. New York: Academic Press

Gaze RM, Jacobson M (1962): The projection of the binocular visual field on the optic tecta of the frog. *Q J Exp Physiol* 47: 273–280

Gaze RM, Jacobson M (1963): A study of the retino-tectal projection during regeneration of the optic nerve in the frog. *Proc Roy Soc (London) Ser B* 157: 420–448

Gaze RM, Sharma SC (1970): Axial differences in the reinnervation of the goldfish optic tectum by regenerating optic nerve fibres. *Exp Brain Res* 10: 171–181

Gaze RM, Jacobson M, Székely G (1963): The retinotectal projection in *Xenopus* with compound eyes. *J Physiol (London)* 165: 484–499

Gaze RM, Jacobson M, Székely G (1965): On the formation of connections by compound eyes in *Xenopus*. *J Physiol (London)* 176: 409–417

Gaze RM, Keating MJ, Székely G, Beazley L (1970): Binocular interaction in the formation of specific intertectal neuronal connexions. *Proc Roy Soc (London) Ser B* 175: 107–147

Hirsch HVB, Jacobson M (1975) The perfectible brain: Principles of neuronal development. pp 107–137 In: *Handbook of Psychobiology*, Gazzaniga M, Blakemore C, Eds. New York: Academic Press.

Hume D (1739): *A Treatise of Human Nature*. London: John Noon, 2 vol

Jacobson M (1961): The recovery of an electrical activity in the optic tectum of the frog during early regeneration of the optic nerve. *J Physiol (London)* 157: 27–29P

Jacobson M (1962): The representation of the retina on the optic tectum of the frog. Correlation between retinotectal magnification factor and retinal ganglion cell count. *Q J Exp Physiol* 47: 170–178

Jacobson M (1977): Mapping the developing retino-tectal projection in frog tadpoles by a double label autoradiographic technique. *Brain Res* 127: 55–67

Jacobson M (1988): Neural cell adhesion molecule (NCAM) expression in *Xenopus* embryos during formation of central and peripheral neural maps. In: *The Making of the Nervous System*, Parnavelas JG, Stern CD, Stirling RV, Eds. Oxford: Oxford University Press

Jacobson M (1991): *Developmental Neurobiology*, 3rd Edition. New York: Plenum Press

Jacobson M (1993): *Conceptual Foundations of Neuroscience*. New York: Plenum Press

Jacobson M, Gaze RM (1964): Types of visual response from single units in the optic tectum and optic nerve of the goldfish. *Q J Exp Physiol* 49: 199–209

Jacobson M, Hirsch HVB (1973a): Development and maintenance of connectivity in the visual system of the frog. I. The effects of eye rotation and visual deprivation. *Brain Res* 49: 47–65

Jacobson M, Hirsch HVB (1973b): Development and maintenance of connectivity in the visual system of the frog. II. The effects of eye removal. *Brain Res* 49: 67–74

Keating MJ (1968): Functional interaction in the development of specific nerve connexions. *J Physiol (London)* 198: 75P

Keating MJ (1974): The role of visual function in the patterning of binocular visual connections. *Br Med Bull* 30:145–151

Keating MJ, Feldman JD (1975): Visual deprivation and intertectal neuronal connections in *Xenopus laevis*. *Proc Roy Soc (London) Ser B* 191: 467–474

Kennard C, Keating MJ (1985): A species difference between *Rana* and *Xenopus* in the occurrence of intertectal neuronal plasticity. *Neurosci Lett* 58: 365–370

Koelliker RA (1896): *Handbuch der Gewebelehre des Menschen*. Bd. 2, *Nervensystem des Menschen und der Thiere*, 6, Aufl. Leipzig: W. Engelman

Levine R, Jacobson M (1974): Deployment of optic nerve fibers is determined by positional markers in the frog's tectum. *Exp Neurol* 43: 527–538

Locke J (1706): *Essay Concerning Human Understanding*, 5th Edition. London

Maturana HR, Lettvin JY, McCulloch WS, Pitts WH (1960): Anatomy and physiology of vision in the frog (*Rana pipiens*). *J Gen Physiol Suppl* 43: 129–175

Popper KR (1959): *The Logic of Scientific Discovery*. London: Hutchinson

Popper KR (1962): *Conjectures and Refutations: The Growth of Scientific Knowledge*. New York: Basic Books

Rager G, Rager U (1978): Systems matching by degeneration. I. A quantitative electron microscopic study of the generation and degeneration of ganglion cells in the chicken. *Exp Brain Res* 33: 65–78

Scherer WJ, Udin SB (1989): N-Methyl-D-aspartate antagonists prevent interaction of binocular maps in *Xenopus* tectum. *J Neurosci* 9: 3837–3843

Skarf B, Jacobson M (1974): Development of binocularly-driven single units in frogs raised with asymmetrical visual stimulation. *Exp Neurol* 42: 669–686

Sperry RW (1945): The problem of central nervous reorganization after nerve regeneration and muscle transposition. *Q Rev Biol* 20: 311–369

Sperry RW (1951a): Mechanisms of neural maturation. In: *Handbook of Experimental Psychology*, Stevens SS, ed. New York: Wiley

Sperry RW (1951b): Regulative factors in the orderly growth of neural circuits. *Growth Symp* 10: 63–87

Sperry RW (1963): Chemoaffinity in the orderly growth of nerve fiber patterns and connections. *Proc Natl Acad Sci USA* 50: 703–710

Sperry RW (1965): Embryogenesis of behavioral nerve nets. In: *Organogenesis*, DeHaan RL, Ursprung H, eds. New York: Holt, Rinehart and Winston

Straznicky K, Gaze RM (1971): The growth of the retina in *Xenopus laevis*: An autoradiographic study. *J Embryol Exp Morphol* 26: 67–79

Straznicky K, Gaze RM (1972): The development of the tectum in *Xenopus laevis*: An autoradiographic study. *J Embryol Exp Morphol* 28: 87–115

Udin SB, Fawcett JW (1988): Formation of topographic maps. *Ann Rev Neurosci* 11: 289–327

Weiss P (1928a): Eine neue Theorie der Nervenfunktion. Nicht durch gesonderte Bahnen, sondern durch spezifische Formen der Erregung schaltet das Nervensystem mit den Muskeln. *Naturwissenschaften* 16: 626–636

Weiss P (1928b): Erregungspezifität und Erregungsresonanz. *Ergeb Biol* 3: 1–151

Weiss P (1941a): Nerve patterns. The mechanics of nerve growth. *Third Growth Symposium* 5: 163–203

Weiss P (1941b): Self-differentiation of the basic patterns of coordination. *Comp Psychol Monogr* 17: 1–96

Weiss P (1952): Central versus peripheral factors in the development of coordination. *Res Publ Assoc Res Nerv Ment Dis* 30: 3–23

Yoon MG (1976): Progress of topographic regulation of the visual projection in the halved optic tectum of adult goldfish. *J Physiol (London)* 257: 621–643

CHAPTER 2

Development of the Retinotectal Projection in *Xenopus*

PHILIP GRANT

INTRODUCTION

For more than 50 years, beginning with the classic experiments of Sperry, the problem of the formation of ordered neuronal connections in the developing visual system of vertebrates has been an area of much controversy. Sperry's initial experiments (1944) demonstrated the critical role of genetic programming of connectivity ("nature") over the contribution of functional experience ("nurture").

To explain his data, he proposed the chemospecificity hypothesis (1963) which served as a paradigm of the retinotectal field for many years. The matching of position-dependent chemospecific labels acquired by retinal ganglion and tectal target cells during their early differentiation seemed to offer a simple explanation for ordered connectivity. Later experimental work, however, led to observations inconsistent with the hypothesis. The retinotectal recording procedures developed by Gaze and his students made it possible to test the Sperry model directly in a series of size-disparity experiments, among others, where parts of the retina and/or the tectum were eliminated and the resulting regenerated projection could be analyzed (Gaze and Sharma, 1970; Gaze et al., 1963). These studies demonstrated that a rigid chemospecificity, as originally described by Sperry, did not exist. In no case, however, were the data sufficient to eliminate the chemospecificity model.

In fact, new fiber-tracing procedures with horseradish perioxidase (HRP) and fluorescent-labeled vital dyes provided evidence that were consistent with the model (Harris, 1989; Holt, 1984; Bonhoeffer and Huf, 1982). Moreover, gradients of membrane-bound factors that modulate fiber interactions with target cells have been demonstrated in the chick visual system (Bonhoeffer and Huf, 1985; Trisler et al., 1981). Gradients of chemospecific factors do seem to exist on the membranes of tectal cells that promote specific synaptic interactions, negatively as well as positively. For example, factors from posterior tectal membranes in the chick have been shown to induce specific collapse of temporal growth cones without affecting growth cones from nasal retinal fibers (Cox et al., 1990). In the frog, temporal retinal growth cones collapse on contact with nasal retinal axons, suggesting that specific axonal factors also play a negative role in fasciculation (Raper and Grunewald, 1990).

Formation and Regeneration of Nerve Connections
Sansar C. Sharma and James W. Fawcett, Editors
© 1993 Birkhäuser Boston

We now know that nurture (visual function) also plays a key role in creating the precise point-to-point map characteristic of the adult visual projection. Factors inhibiting visual function (pharmacological agents, strobe lights, etc.) have profound effects on the development of the visual map (Schmidt and Edwards, 1983; Schmidt and Eisele, 1985; Reh and Constantine-Paton, 1985).

What was initially thought to be a simple, single mechanism has now become a complicated multistep process in which many different factors influence formation and stabilization of the mature adult map. Although the factors responsible for ordering the visuotectal projection in vertebrates are still unknown, there seems to be general agreement as to the basic developmental steps. At least two developmental projections can be demonstrated in the frog: (1) an early, embryonic "gross" projection, characteristic of the initial phases of optic fiber arrival from the newly differentiating retina onto presumptive thalamic and tectal visual target areas, and (2) a later, "refined" projection seen in the larva and adult, where during growth and differentiation of the visual centers, the fully functional, point-to-point visual map is established. The characteristics of these two projections differ because each is modulated by different developmental factors.

CHARACTERISTICS OF THE EARLY PROJECTION

A common feature of vertebrate development is the early development of the retina. The retina forms shortly after the neural tube closes, long before postsynaptic neurons in the visual centers begin to mature. In *Xenopus*, retinal ganglion cells (RGCs) are the first retinal cells to differentiate (about stage 26/27), whereas other retinal phenotypes mature shortly thereafter, with the photoreceptors usually the last to differentiate. In *Xenopus* RGC axons emerge from the optic cup, enter the optic stalk, and arrive on the immature diencephalic-mesencephalic border of the neural tube about the same time as the retina segregates into nuclear and plexiform layers (Grant et al., 1980).

The early gross projection in *Xenopus* is established between stages 37/38, when the first optic fibers arrive in the diencephalic-mesencephalic region of the neural tube (Holt, 1984), and stages 45–49, when an ordered projection has segregated along both axes on the tectum (i.e., dorsoventral and rostral caudal) (Gaze et al., 1974; Sakaguchi and Murphey, 1985; O'Rourke and Fraser, 1986). Visual function (electroretinogram, ERG) can be detected in the retina at stage 39, and slightly later, the first evidence of an ordered projection is seen as optic fiber arbors from dorsal and ventral retina segregate on the presumptive tectum and pretectal visual neuropils (Holt, 1984; Holt and Harris, 1983; Sakaguchi and Murphey, 1985). The stage at which the temporal and nasal fibers segregate into a projection, on the other hand, is controversial, ranging from stage 39/40 (Holt and Harris, 1983) to stage 49 (Gaze et al., 1974; O'Rourke and Fraser, 1986). This gross projection has the following general features:

1. The early axial segregation of optic fibers is diffuse and defined globally rather than on a point-to-point basis.

2. The diffuse nature of the segregation is in part due to large overlapping arborizations (100–300 μm) of optic fibers in target areas.
3. RGC function is marked by large, diffuse visual fields.
4. The projection is labile, characterized by "shifting connections" (Gaze et al., 1974) during growth of the tectum. Presumably, optic fiber arborizations are actively extended and withdrawn during this process of synaptogenesis.

The first optic fibers to arrive at the presumptive tectum come from the dorsal retina; RGCs differentiate first in this region (Holt, 1984). Fibers from the ventral retina follow shortly thereafter, and the fiber arborizations from these two retinal quadrants seem to segregate in the putative tectum, dorsal fibers projecting laterally while ventral fibers project more medially (Holt, 1984). The initial arbors are large and extend widely over the diencephalic-mesencephalic border region. Nasal and temporal fibers arriving about the same time do not segregate immediately. Their terminals overlap: temporal fibers project to the more rostral regions of the future tectum, while nasal fibers project both rostrally and caudally, their arbors overlapping with temporal fiber arbors in the rostral region. It is only at stages 45–49 that these fibers begin to segregate into their respective domains on the tectum (O'Rourke and Fraser, 1986).

During this embryonic period the optic fiber arbors are large and diffuse (Sakaguchi and Murphey, 1985). They occupy a domain between the rostrally developing diencephalon and the more caudal tectum. It is clear from neuronal birthday studies that the caudal diencephalon is the first to mature (Tay and Straznicky, 1982), with the first cells "born" in stages 22–25, and is the first to show postmitotic cells in presumptive visual centers (i.e., pretectum). The maturation of the tectum occurs later, between stages 35 and 45, with the first postmitotic cells appearing between stages 40 and 45 (Gaze and Grant, 1992b). Thus, the first optic fibers arriving in the diencephalic-mesencephalic boundary at stage 37/38 will initially encounter postmitotic diencephalic cells in maturing visual nuclei, while at the same time other fibers arborize in relatively immature tectal domains, hours before any tectal neurons have become postmitotic (Gaze and Grant, 1992b). If synaptogenesis occurs as the first optic fibers arrive, only the diencephalic cells will be sufficiently mature to serve as target cells. In an ultrastructural study of early synaptogenesis in *Xenopus*, the first synapses did appear in the presumptive pretectal neuropils (Wilson and Gaze, unpublished). These results imply that the evidence of an early retinotectal projection at stage 39/40 (Holt and Harris, 1983) is based on presynaptic recording from optic fibers that may have segregated in the tract and diencephalon but not on the tectum. The earliest postsynaptic recording from the tectum has been reported at stage 49 (Chung et al., 1974), presumably after the early projection is established and after mature, postmitotic tectal cells have developed.

This early projection is labile, expanding across the growing tectum throughout development (Gaze et al., 1974). As the tectum grows rostrocaudally, new optic fibers arrive and seem to displace the preexisting projection; the center of the visual field shifts caudally on the tectum. Presumably this is accomplished by a process of "shifting connections," as optic fiber terminals expand and collapse within the tectal neuropil (Gaze et al., 1974).

FACTORS RESPONSIBLE FOR THE GROSS PROJECTION

The early projection develops from the timing of arrival of the first optic fibers in the diencephalic-mesencephalic region of the brain at stage 37/38 to the ordered segregation of dorsoventral and nasotemporal arbors at stages 45–49. During this period many factors contribute to the map's formation. Those listed below are probably the most important.

Spatiotemporal Ordering of RGC Birthdays

The pattern of growth at the retinal margin results in differential timing of optic fiber arrival in visual centers (Straznicky and Gaze, 1971). This translates into an initial spatial segregation of fibers from dorsal and ventral retina, dorsal fibers rostrolateral and ventral fibers dorsomedial (Holt, 1984). Dorsotemporal fibers, the first RGCs to differentiate, enter the presumptive tectum several hours before those from ventral retina (Holt, 1984). Furthermore, the pattern of marginal growth of the retina means that the youngest retinal cells are always at the periphery, with progressively older cells displaced in gradient fashion toward the center. The central–peripheral age gradient also translates into a spatial distribution of optic fibers within the optic nerve (older fibers in the center, with younger fibers at the periphery) and in the optic tract (older fibers deep within the tract displaced medially toward the midline of the brain; younger, newly arriving fibers superficially displaced in the lateral edge of the optic tract) (Gaze and Grant, 1978).

Fiber Following

The tendency of RGC axons to fasciculate as they converge toward the optic disc results in a radial array of fiber bundles within the optic fiber layer of the growing retina (Grant et al., 1980). Fiber following is also responsible for the ordering of optic fiber bundles within the optic nerve and tract. Nerve cell adhesion molecules (NCAMs) and other matrix molecules are known to play a role in the early outgrowth of optic fibers; antibodies against NCAM interfere with the normal development of retinotectal fibers in the chick (Thanos et al., 1984) and also interfere with the growth of optic fibers over the *Xenopus* tectum (Fraser et al., 1984).

Gradients of Chemoaffinity Labels in the Optic Pathways and Target Areas

The early segregation of fibers from different retinal quadrants could result from the presence in the pathways and in target domains of global gradients (in intercellular spaces and on cell surfaces) of cell adhesion factors, or trophic factors that modulate specific cellular interactions. In the chick visual system, for example, membrane factors differentially affecting the growth of temporal and nasal retinal fibers have been found distributed in a gradient fashion across the tectum (Trisler et al., 1981; Bonhoeffer and Huf, 1982, 1985). The presence of such labels is also suggested by the behavior of optic fibers emerging from abnormally implanted eyes; though arriving on the tectum along ectopic pathways, the fibers project over the tectum in a relatively normal fashion, forming properly oriented maps (Sharma, 1972; Harris, 1989).

Spatiotemporal Patterns of Neuronal Maturation in Visual Centers

Visual centers in diencephalon and mesencephalon, though maturing slightly later than the retina in *Xenopus*, undergo an ordered pattern of cellular differentiation. The birthday studies reveal that diencephalon matures before tectum (Tay and Straznicky, 1982; Gaze and Grant, 1992a) with a caudorostral polarity, while the tectum begins its maturation rostrally and laterally as wedges of neurons at all levels in the tectum differentiate in a caudomedial direction (Straznicky and Gaze, 1972). Some of the first optic fibers arriving at the diencephalic-mesencephalic boundary in the early embryo must turn rostrally toward the maturing diencephalon, while others grow caudally from the boundary to the expanding tectum (Grant and Ma, 1983 and unpublished). Thus, the patterns of target cell maturation seem to be spatially congruent with the order and direction of optic fiber arrival, which should result in ordered patterns of synaptogenesis. These genetically programmed growth processes in visual centers as well as the retina provide a preliminary ordering along two axes that, together with the other factors mentioned above, give rise to the initial gross projection serving the early visual function of the newly hatched tadpole.

CHARACTERISTICS OF THE REFINED PROJECTION

The refined projection develops later in the larva and is completed by the end of metamorphosis. It, too, involves several different stages. The period between stage 49 and stage 54 is characterized by equivalent growth of dorsal and ventral retina, but after stage 54 ventral retina grows much more rapidly, even after metamorphosis (Beach and Jacobson, 1979a). A temporonasal asymmetry in the growth of RGCs is also evident at this stage, extending to the postmetamorphic animal (Tay et al., 1982). The rapid growth of ventral and temporal retina after stage 54 includes development of the first ipsilateral projecting optic fibers, which terminate in several thalamic visual nuclei (Kennard, 1981; Hoskins and Grobstein, 1986). During this same metamorphic period (stages 57–66), the laterally projecting eyes shift dorsally and rostrally simultaneously with the development of binocular vision (Grobstein and Comer, 1977; Keating, 1974). All these processes contribute to the mature, refined retinotectal projection, some of whose properties are:

1. A point-to-point map in both axes;
2. Reduced size of optic fiber arborizations;
3. Well-defined, small visual fields of RGCs;
4. Regional segregation of ipsi- and contralateral fibers in thalamic visual neuropils (no ipsilateral fibers project to the tectum);
5. Stabilization of synapses as both retinal and tectal growth decline, although some plasticity persists as seen in regeneration, etc.

FACTORS RESPONSIBLE FOR THE REFINED PROJECTION

Like the early embryonic projection, development of the mature, fully functional visual system in *Xenopus* requires coordinated expression of many factors as the animal grows through metamorphosis.

Growth of Target Regions Leads to a Reduction in Size of Optic Fiber Arborizations

Optic fibers arrive in future visual neuropil regions at the very early stages of development, in some cases, as in the tectum, before any postsynaptic target cells have begun to mature. It is only later in development that these visual centers undergo their massive increase in cell number and expansive growth. In *Xenopus* the bulk of tectal growth occurs after stage 50 (Straznicky and Gaze, 1972). Newly developed optic fibers continue to arrive on these expanding neuropil areas, while the fibers from older ganglion cells, already arborized within these neuropils, shift their connections to the newly differentiated postsynaptic cells. Shifting connections simply implies a pruning of the large arborizations, resulting, in part, in a reduction in the arbor size and a more precise pattern of connections. A major reduction in arbor size also results from the overall expansion of the visual neuropil areas during this growth period (Sakaguchi and Murphey, 1985). Arbor expansion does not keep pace with the increasing volume of neuropil. As a result, the relative size of arbors is also decreased. The overall effect is to "refine" the anatomical map, to reduce the amount of arbor overlap, and to produce a more precise connectivity pattern.

Chemospecific Labels on Presynaptic Fibers and Postsynaptic Target Cells

This factor of chemospecific cell labeling also plays a role in refining the adult visual projection. It is not clear, however, what these factors are, nor how they function in promoting cell-specific interactions, but the experimental evidence derived from regeneration experiments in both fish and amphibia, although eliminating a rigid specificity of cell matching, is still consistent with the original Sperry hypothesis (Harris, 1989; Willshaw et al., 1983). Pie-slice regions of the embryonic retina, when displaced in different regions of the retina, will usually reconnect with the correct tectal domain, irrespective of the pathway taken. Labels seem to be on tectal cells as well as retinal fibers. They may be the same molecules that help to organize the axes of the early projection, or they may represent a whole new set of chemospecific cell surface cues. To date, none of these chemical factors have been identified.

Synchronous Functional Activity of Neighboring RGCs

This process ensures that synchronously firing terminal arbors are stabilized in visual centers while nonsynchronously firing terminals are unstable and retract, possibly resulting in cell death. Perhaps the most important factor that refines the adult visuotectal map is that of visual function. Contrary to Sperry's original conclusions, it is now clear that in the development of the vertebrate visual system, visual function plays an important, if not critical, role. Interference of visual function during regeneration of the optic nerve in goldfish, either by the use of stroboscopic lights or specific drugs such as tetrodotoxin (TTX), will prevent the reestablishment of the normal visual map (Schmidt and Eisle, 1985; Schmidt and Edwards, 1983). The application of TTX to three-eyed frogs, which develop a pattern of ocular dominance columns in the tectum (fibers from each eye projecting to the tectum are segregated into alternate eye-specific domains), will prevent the formation of such columns (Reh and Constantine-Paton, 1985). More recently, it has been shown that application of specific N-methyl-D-aspartate (NMDA) antagonists which block NMDA receptors will also induce desegregation of the retinotectal strips seen in three-eyed frogs (Cline et al., 1987).

   One feature of such functional processes is the death of those cells which fail to fire synchronously with most terminals in a region; i.e., this map refinement mechanism should lead to RGC death.

Competitive Interactions of Optic Fiber Arborizations for Target Space and/or Trophic Molecules Derived from Target Cells

This process should lead to a segregation and stabilization of connections of those successfully competing cells and the death and elimination of cells that fail to compete successfully. Here, too, the refinement of the visuotectal projection is accompanied by massive cell death in the retina. It is this feature of map refinement that will be discussed in more detail.

CELL DEATH IN THE DEVELOPING *XENOPUS* RETINA: EVIDENCE OF MAP REFINEMENT

Neuronal cell death is a fundamental feature of nervous system development in vertebrates (Cowan et al., 1984; Oppenheim, 1985, 1991; Finlay et al., 1987; Williams and Herrup, 1988). Some of the most dramatic cases of massive neuronal cell death have been reported in the retina of mammals. In the cat 80% of the RGCs are eliminated after birth (Williams et al., 1986), whereas 60–70% are lost in the rat (Crespo et al., 1985), monkey (Rakic and Riley, 1983a), and human (Provis et al., 1985). Most RGCs in mammals develop early in the embryo, usually within a sharply defined developmental window. Optic fibers project to visual centers during embryonic and fetal life, forming, for the most part, the early "gross" projection. For example, before birth (and the beginning of visual function), optic fiber terminals from both eyes overlap within the lateral geniculate (a thalamic visual center), which is still immature and does not show signs of its normal laminar pattern. After birth and the onset of visual function in the monkey, the terminals from the respective eyes segregate into their well-defined laminae within the geniculate. This correlates with the period of massive RGC death, suggesting that most RGCs die from a refinement of the visual projection in this nucleus, presumably as contralateral and ipsilateral fibers displace one another from specific laminae in the geniculate (Rakic and Riley, 1983b).

RGC death in fish and amphibians is controversial; some cell death has been reported in the embryo and larva (Glucksmann, 1940; Jenkins and Straznicky, 1986), but others have reported no evidence of any RGC death (Wilson, 1971; Gaze and Peters, 1961; Jacobson 1976; Easter et al., 1981; Dunlop and Beazley, 1984). Since the retina in these vertebrates continues to grow well into adult life, it has generally been assumed that no cell death occurs. If it does occur, it would be difficult to detect against a background of continuous addition of new RGCs from the growing retinal margin.

CELL DEATH IN THE *XENOPUS* RETINA

To explore the problem of RGC death in the developing retina of *Xenopus*, we used a strategy of identifying a distinct population of postmitotic RGCs in the central retina by labeling the proliferating cells at the outer boundaries of this central population with ^{3}H-thymidine (3HT) or bromodeoxyuridine (BRDU). The *Xenopus* retina grows at the margin, and all postmitotic retinal cells are thereby displaced toward the center. Since the labeled margins persist throughout development into the adult retina, populations of postmitotic central RGCs could be identified at different stages and their fate followed throughout development. In this way we could count the cells at the initial stage of labeling and count them again in identically labeled sibs at frequent intervals during development until after metamorphosis. If no cell death occurred, the number of cells marked in this way should remain constant. Alternatively, if cell death occurred, the number of cells within the labeled boundaries should decrease, since cell migration or transformation (displaced ganglion cells) is relatively unimportant. Using this technique, we could determine which RGCs died and when in development cells were lost.

The results of these analyses are shown in histograms in Figures 2.1–3. RGCs identified in early embryonic stages, as the retina is forming and the first optic fibers begin to grow out (between stages 25/26 and 29/30; Grant et al., 1980), seem to display no reduction in number during the course of development (Figure 2.1). The numbers of RGCs are small (between 100 and 200) and difficult to count

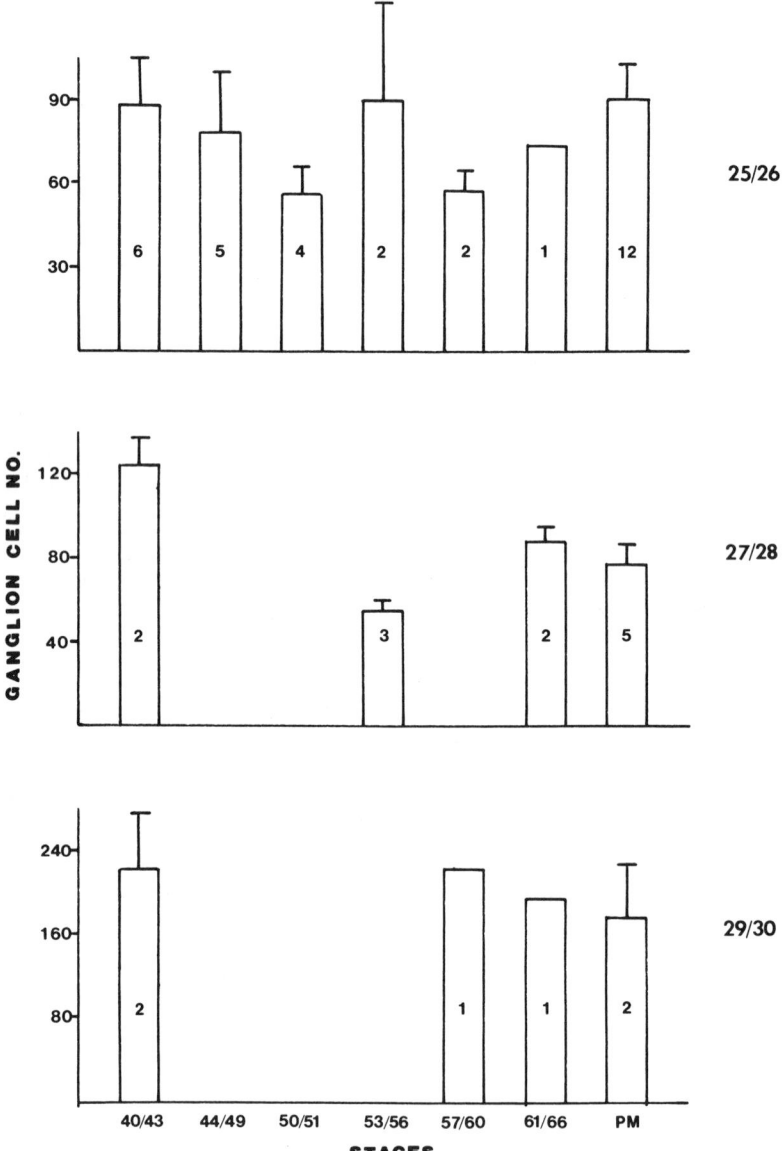

FIGURE 2.1. Histogram showing the changes in numbers of identified postmitotic central RGCs during development. Embryos were injected with BRDU or 3HT at stages 25/26, 27/28, or 29/30 and analyzed at stages shown on the X-axis. Numbers inside boxes are the number of animals analyzed at each stage. Bars represent standard error.

accurately, since labeled and unlabeled cells intermingle at these early stages; there is no distinct labeled boundary at the margins to clearly demarcate the postmitotic central ganglion cells. Since the counts are less reliable than those made in embryos labeled at later stages, it is difficult to conclude that these first cells to differentiate, the so-called pioneers, persist throughout development to 3–6 months postmetamorphosis. If it could be later shown to be true, this would suggest that pioneering optic fibers, the first to arrive at visual neuropils, succeed in surviving throughout development.

The pattern of cell death becomes most evident after stage 31/32, when the optic cup has fully formed (Figure 2.2).RGC number has increased to approximately 500 to 1,500 between stages 31 and 41, when the retina has layered out into distinct plexiform and nuclear layers. At these stages the unlabeled postmitotic central retinal cells are clearly defined by labeled margins, are more easily counted, and can be carefully traced into late larvae and postmitotic animals. The cell counts are much more reliable (counting error 10–15%), and it can be seen that a 50–60% reduction in cell number does occur, beginning at the onset of metamorphic climax (stage 57), persisting unchanged through 3 months postmetamorphosis ($p = 0.001$–0.005, Mannheim Whitney paired analysis). The complete extent of cell loss has occurred by stages 61–66 (late metamorphosis), since cell number does not change in the postmetamorphic animals.

This same pattern of cell loss continues in later larvae, labeled between stages 45 and 49 (Figure 2.3).The most complete analysis is seen at stage 45/46, which begins with approximately 2,500 identified central RGCs, which diminish by 40% through metamorphosis ($p = 0.01$). Although the number of stages sampled in those larvae labeled at stage 49 is few, the pattern is the same; in fact, here we see that of approximately 9,000 identified central postmitotic RGCs, approximately 30% survived through metamorphosis, a loss of 6,000 cells ($p = 0.01$, Mannheim Whitney paired analysis).

The pattern at stage 53/54 is quite different, however. For cells born at this stage the extent of cell loss is only 20%, which is in agreement with the cell loss reported by Jenkins and Straznicky (1986) for the same stage. Approximately 18,000 identified RGCs at stage 53/54 are reduced to 14,000 cells after metamorphosis ($p = 0.01$); 4,000 cells are lost from the central retina at this stage.

DISCUSSION AND CONCLUSIONS

It is clear from this analysis that, as in mammals, there is a loss of RGCs during development in *Xenopus*. Considering the counting error, we can conclude that 50–60% of the RGCs born between stages 29/30 and 49 are lost during metamorphosis, a total of about 5,000 cells from the central retina. Though considerably less than the massive cell death seen in mammalian retinas, it does represent 10% of the adult RGC population. Cell loss begins at the onset of metamorphosis, about stage 57, and is completed before the end of metamorpho-

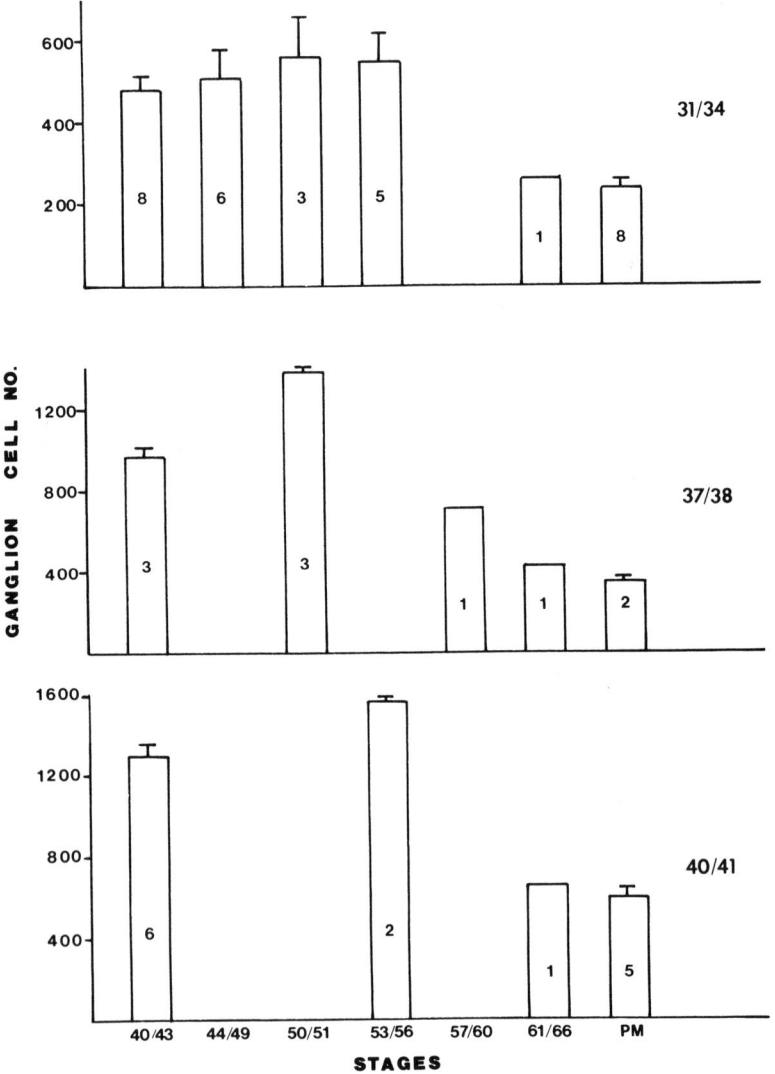

FIGURE 2.2. As in Figure 2.1. Embryos were injected with BRDU or 3HT at stages 31/34, 37/38, or 40/41 and analyzed at stages shown on the X-axis.

sis, since there is no further loss up to 6 months postmetamorphosis, the last stage sampled in this analysis.

It should be pointed out that cells lost after labeling at stage 41 are also included in the population of cells lost after labeling at stage 49 or 53/54. Since the total number of cells lost after labeling at stage 49 is approximately the same as those lost after stage 53/54, it suggests that the same population of cells is involved. In

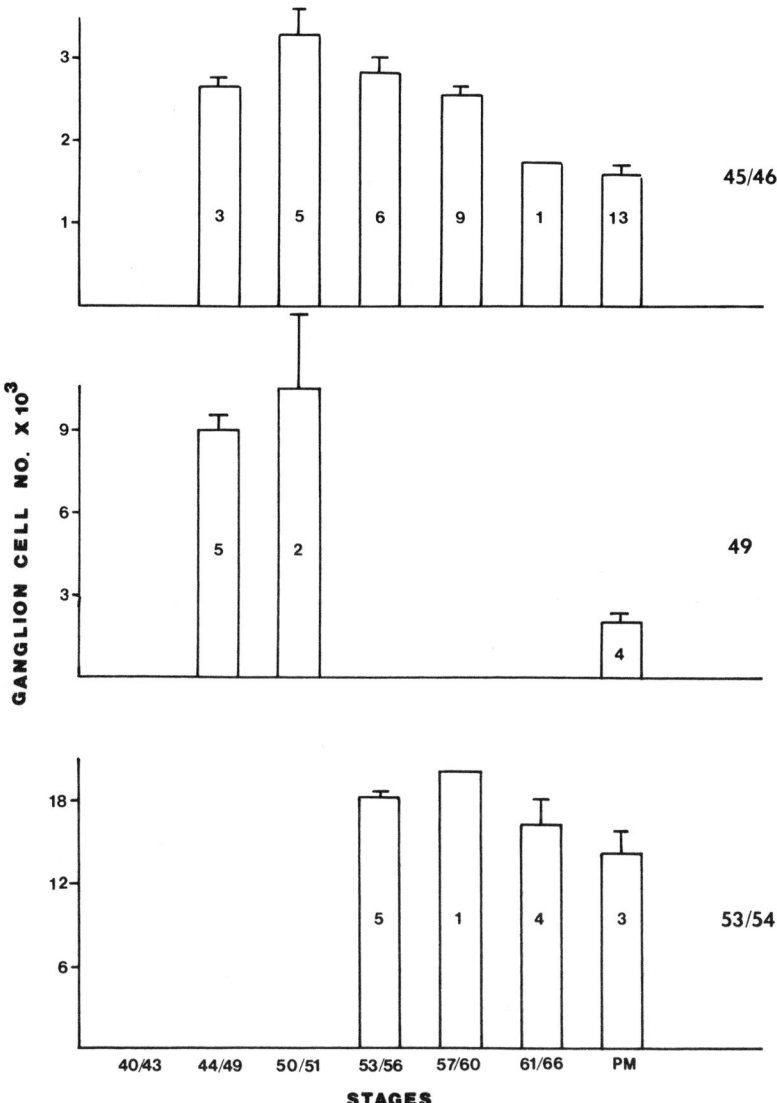

FIGURE 2.3. As in Figure 2.1. Tadpoles were injected with BRDU or 3HT at stages 45/46, 49, or 54 and analyzed at the stages shown on the X-axis.

other words, the 20% of cells that are lost after stage 54, seen by us and by Jenkins and Straznicky (1986) (about 4,000–5,000 cells), are the identical RGCs that are lost after stage 49 (approximately 6,000 cells, well within the counting error of 10–20%). This implies that most, if not all, cells born between stage 49 and stage 53/54, which come to occupy the more peripheral areas of the retina, outside the older central retina, tend to survive. It appears, therefore, that the cells that die are those in the central retina, presumably the oldest cells clustered about the optic disc. Cell loss is not random throughout the retina; it is confined to the center.

Why do 50% of the central cells die during the metamorphic climax while most cells born after stage 49 survive? Many factors have been proposed to account for cell death, ranging from intrinsic, genetically programmed cell death to such extrinsic factors as competition among axon terminals at target sites for trophic factors (Oppenheim, 1985, 1991; Williams and Herrup, 1988). For the visual system in mammals, it has been suggested that a major portion of cell death can be attributed to the competition of ipsilateral and contralateral fibers projecting to the lateral geniculate (Rakic and Riley, 1983a, 1983b). Massive cell death in primates correlates with the segregation of overlapping arborizations from ipsi- and contralateral retinas into their respective laminae within the lateral geniculate. We suggest that a similar explanation accounts for the RGC death seen in the *Xenopus* retina.

Although cell death in *Xenopus* occurs during metamorphic climax, we suggest that high levels of thyroxine are not responsible. In fact, it has been shown that thyroxine stimulates RGC proliferation (Kaltenbach and Hobbs, 1972; Beach and Jacobson, 1979b). Thyroxine stimulation is probably responsible for the asymmetric growth of the retina (ventral retina growing more rapidly than dorsal retina) after stage 54 (Beach and Jacobson, 1979a) and explains the continued high rates of RGC growth seen at these late tadpole stages.

Another factor resulting in cell death may be optic nerve remodeling as the eyes shift from a lateral to a more dorsal position during metamorphosis and as binocular vision begins (Keating, 1974; Grobstein and Comer, 1977). At the same time that binocular vision is acquired during metamorphosis (from stage 54 to stage 66), the fibers from the ipsilateral retina first begin to grow into the thalamic visual centers (corpus geniculatum, nucleus Bellonci, etc.) (Hoskins and Grobstein, 1985; Kennard, 1981) and continue to do so, overlapping with contralateral terminals that had been present from earlier stages of development (Grant and Ma, 1983 and unpublished). Most of these fibers come from the ventral and temporal retina, the principal source of the ipsilateral projection (Hoskins and Grobstein, 1985). Accordingly, we suggest that as in mammals, these projections from ipsilateral and contralateral retinas in the thalamic visual nuclei begin to segregate during metamorphosis to form the adult projection to these centers (Scalia and Fite, 1974).

In view of the changes in eye position during this metamorphic period, ventral retina comes to dominate the visual field and receives most of the visual input. Ipsilateral fibers from this retinal quadrant may, because of enhanced function, compete more successfully with contralateral fibers for targets within thalamic neuropils. This would result in a reduction of central RGCs during metamorphosis.

It has generally been assumed that the continuously growing retinas of fish and amphibians do not exhibit RGC death. For the first time, however, our procedure for distinguishing a central population of postmitotic RGCs has made it possible to follow these cells throughout development and demonstrate that 50% of these are lost during metamorphosis, possibly as a result of mechanisms similar to those found in mammalian retinas.

REFERENCES

Beach DH, Jacobson M (1979a): Patterns of cell proliferation in the retina of the clawed frog during development. *J Comp Neurol* 183: 603–614

Beach DH, Jacobson M (1979b): Influence of thyroxine on cell proliferation in the retina of the clawed frog at different stages. *J Comp Neurol* 183: 615–624

Bonhoeffer F, Huf J (1982): In vitro experiments on axon guidance demonstrating an anterior-posterior gradient on the tectum. *EMBO J* 1: 427–431

Bonhoeffer F, Huf J (1985): Position-dependent properties of retinal axons and their growth cones. *Nature* 315: 409–410

Chung SH, Keating MJ, Bliss TVP (1974): Functional synaptic relations during the development of the retino-tectal projection in amphibians. *Proc Roy Soc Lond B* 187: 449–459

Cline HT, Debski E, Constantine-Paton M (1987) NMDA receptor antagonist desegregates eye specific stripes. Proc Natl Acad Sci USA 84:4342–4345.

Cowan WM, Fawcett JW, O'Leary DDM, Stanfield BB (1984): Regressive events in neurogenesis. *Science* 225: 1258–1265

Cox AC, Muller B, Bonhoeffer F (1990): Axonal guidance in the chick visual system: Posterior tectal membranes induce collapse of growth cones from temporal retina. *Neuron* 2: 31–37

Crespo D, O'Leary DDM, Cowan WM (1985): Changes in the number of optic nerve fibers during late prenatal and postnatal development in the albino rat. *Dev Brain Res* 19: 129–134

Dunlop SA, Beazley LD (1984): A morphometric study of the retinal ganglion cell layer and optic nerve from metamorphosis in *Xenopus laevis. Vision Res* 24: 417–427

Easter SS Jr., Rusoff AC, Kish PE (1981): The growth and organization of the optic nerve and tract in juvenile and adult goldfish. *J Neurosoci* 1: 793–811

Finlay BL, Wikler KC, Sengelaub DR (1987): Regressive events in brain development and scenarios for vertebrate brain evolution. *Brain Behav Evol* 30: 102–117

Fraser SE, Murray BA, Chuong C-M, Edelman GM (1984): Alteration of the retinotectal map of *Xenopus* by antibodies to neural cell adhesion molecules. *Proc Natl Acad Sci USA* 81: 4222–4226

Gaze RM, Grant P (1978): The diencephalic course of regenerating retinotectal fibres in *Xenopus* tadpoles. *J Embryol Exp Morphol* 44: 201–216

Gaze RM, Grant P (1992a) Spatio-temporal patterns of retinal ganglion cell death during *Xenopus* development. J. Comp. Neurol. 315:264–274.

Gaze RM, Grant P (1992b) Development of the tectum and dieucephalon in relation to the time of arrival of the earliest optic fibres in *Xenopus.* Anat. Embryol. 185:599–612.

Gaze RM, Peters A (1961): The development, structure and composition of the optic nerve of *Xenopus laevis* (Daudin). *Q J Exp Physiol* 46: 299–309

Gaze RM, Sharma SC (1970): Axial differences in the reinnervation of the goldfish optic tectum by regenerating optic nerve fibres. *Exp Brain Res* 10: 171–181

Gaze RM, Jacobson M, Szekely G (1963): The retinotectal projection in *Xenopus* with compound eye. *J Physiol* 165: 484–499

Gaze RM, Keating MJ, Chung SH (1974): The evolution of the retinotectal map during development in *Xenopus*. *Proc R Soc Lond B* 185: 301–330

Glucksmann A (1940): Development and differentiation of the tadpole eye. *Br J Ophthalmol* 24: 153–178

Grant P, Ma PM (1983): Development of visual pathways in *Xenopus laevis*: An autoradiographic analysis. *Soc Neurosci Abstr* 9: 759

Grant P, Rubin E, Cima C (1980): Ontogeny of the retina and optic nerve in *Xenopus laevis*: I. Stages in the early development of the retina. *J Comp Neurol* 189: 593–614

Grobstein P, Comer C (1977): Postmetamorphic eye migration in *Rana* and *Xenopus*. *Nature* 269: 54–56

Harris WA (1989): Local positional cues in the neuroepithelium guide retinal axons in embryonic *Xenopus* brain. *Nature* 339: 218–221

Holt CE (1984): Does timing of axon outgrowth influence initial retinotectal topography in *Xenopus*? *J Neurosci* 4: 1130–1152

Holt CE, Harris WA (1983): Order in the initial retinotectal map in *Xenopus*: A new technique for labelling growing nerve fibers. *Nature* 301:150–152

Hoskins SG, Grobstein P (1985): Development of the ipsilateral retinothalamic projection in the frog *Xenopus laevis*. I. Retinal distribution of ipsilaterally projecting cells in normal and experimentally manipulated frogs. *J Neurosci* 5: 911–919

Jacobson M (1976): Histogenesis of retina in the clawed frog with implications for the pattern of development of retinotectal connections. *Brain Res* 103: 541–545

Jenkins S, Straznicky C (1986): Naturally occurring and induced ganglion cell death. A retinal whole-mount autoradiographic study in *Xenopus*. *Anat Embryol* 174: 59–66

Kaltenbach JC, Hobbs AW (1972): Local action of thyroxine on amphibian metamorphosis. V. Cell division in the eye of anuran larvae affected by thyroxine-cholesterol implants. *J Exp Zool* 179: 157–165

Keating MJ (1974): The role of visual function in the patterning of binocular visual connections. Br Med Bull 30:145–151

Kennard C (1981): Factors involved in the development of ipsilateral retinothalamic projections in *Xenopus laevis*. *J Embryol Exp Morphol* 65: 199–217

Oppenheim RW (1985): Naturally occurring cell death during neural development. *Trends Neurosci* 8: 487–493

Oppenheim RW (1991): Cell death during development of the nervous system. *Ann Rev Neurosci* 14: 453–501

O'Rourke NA, Fraser SE (1986): Dynamic aspects of retinotectal map formation revealed by a vital dye fiber-tracing technique. *Devel Biol* 144: 265–276

Provis J, van Driel D, Billson FA, Russel P (1985): Human fetal optic nerve: Overproduction and elimination of retinal axons during development. *J Comp Neurol* 238: 92–101

Rakic P, Riley KP (1983a): Overproduction and elimination of retinal axons in the fetal rhesus monkey. *Science* 219: 1441–1444

Rakic P, Riley KP (1983b): Regulation of axon number in primate optic nerve by prenatal binocular competition. *Nature* 305: 135–137

Raper JA, Grunewald EB (1990): Temporal retinal growth cones collapse on contact with nasal retinal axons. *Exp Neurol* 109: 70–74

Reh TA, Constantine-Paton M (1985): Eye specific segregation requires neural activity in 3 eyed frogs. *J Neurosci* 5: 1132–1143

Sakaguchi D, Murphey R (1985): Map formation in the developing *Xenopus* retinotectal system: An examination of ganglion cell terminal arborizations. *J Neurosci* 5: 3228–3245

Scalia F, Fite K (1974): A retinotopic analysis of the central connections of the optic nerve in the frog. *J Comp Neurol* 158: 455–478

Schmidt JT, Edwards DL (1983): Activity sharpens the map during the regeneration of the retinotectal projection in the goldfish. *Brain Res* 209: 29–35

Schmidt JT, Eisele LE (1985): Stroboscopic illumination and dark rearing block the sharpening of the regenerated retinotectal map in goldfish. *Neuroscience* 14: 535–546

Sharma S (1972): Retinotectal connections of a heterotopic eye. *Nature New Biol* 238: 286–287

Sperry RW (1944): Optic nerve regeneration with return of vision in anurans. *J Neurophysiol* 7: 57–69

Sperry RW (1963) Chemoaffinity in the orderly growth of nerve fiber patterns and connections. Proc Natl Acad Sci USA 50:703–707.

Straznicky C, Gaze RM (1971): The growth of the retina in *Xenopus laevis*: An autoradiographic study. *J Embryol Exp Morphol 26: 67–79*

Straznicky C, Gaze RM (1972): The development of the tectum in *Xenopus laevis*: An autoradiographic study. *J Embryol Exp Morphol* 28: 87–115

Tay D, Hiscock J, Straznicky C (1982): Temporo-nasal asymmetry in the accretion of retinal ganglion cells in late larval and postmetamorphic *Xenopus*. *Anat Embryol* 164: 75–83

Tay D, Straznicky C (1982): The development of the diencephalon in *Xenopus*. *Anat Embryol* 163: 371–388

Thanos S, Bonhoeffer F, Rutishauser U (1984): Fiber-fiber interaction and tectal cues influence the development of the chicken retinotectal projection. *Proc Natl Acad Sci USA* 81: 1906–1910

Trisler GD, Schneider MD, Nirenberg M (1981): A topographic gradient of molecules in retina can be used to identify neuron position. *Proc Natl Acad Sci USA* 78: 2145–2149

Williams RW, Herrup K (1988): The control of neuron number. *Ann Rev Neurosci* 11: 423–453

Williams RW, Bastiani MJ, Lia B, Chalpua LM (1986): Growth cones, dying axons, and developmental fluctuations in the fiber population of the cat's optic nerve. *J Comp Neurol* 246: 32–69

Willshaw DJ, Fawcett JM, Gaze RM (1983): The visuotectal projection made by *Xenopus* pie slice compound eyes. *J Embryol Exp Morphol* 74: 29–45

Wilson MA (1971): Optic nerve fibre counts and retinal ganglion cell counts during development of *Xenopus laevis* (Daudin). *Q J Exp Physiol* 54: 83–91

CHAPTER 3

Axon Pathfinding and the Formation of Maps in the Retinotectal System

JEREMY S.H. TAYLOR

INTRODUCTION

Axons from the ganglion cells of the retina grow from the eye to the central visual relays where they form a very precise series of connections. These connections are arranged so that they form a representation, or map, of the retinal surface. Growing retinal ganglion cell axons have two major feats to achieve: successful navigation to these central visual relays, and connection with the appropriate cells to form the orderly map. This chapter concerns investigations into some of the mechanisms involved in guiding the retinal axons to the optic tectum in the frog *Xenopus* and pathway decisions that may be involved in establishing the retinoptic map.

The axons of retinal ganglion cells grow across the inner surface of the retina and into the optic stalk, leading them to the ventral part of the diencephalon. Once they leave the confines of the optic stalk, the retinal axons are faced with the wide expanse of diencephalic neuroepithelium through which they must navigate. As they grow through the diencephalon, retinal axons always follow a well-defined route: across the midline and then dorsally, growing directly to the contralateral optic tectum (Harris, 1986; Easter and Taylor, 1989; Holt, 1990). To investigate the possible cues responsible for the guidance of retinal axons in the diencephalon, horseradish peroxidase (HRP) has been used to label the projection in a series of embryos, and serial semithin and thin sections have been cut through the brains in order to determine the cellular environment faced by the growing retinal axons (Easter and Taylor, 1989). Rather than forming a discrete entity, "the optic tract," retinal axons join a preexisting pathway, the tract of the postoptic commissure (tPOC). This tract runs from the rostral pole of the brain, where the postoptic commissure lies, through the diencephalon to the midbrain. The fibers of the tPOC are the first structure encountered by the retinal axons as they leave the optic stalk. Retinal axons are always grouped at the rostralmost edge of the tract and are not physically segregated from the nonoptic components of the tract. By following this preformed pathway and maintaining their position at its rostral edge, retinal axons are led to the ventral aspect of the developing tectum.

Formation and Regeneration of Nerve Connections
Sansar C. Sharma and James W. Fawcett, Editors
© 1993 Birkhäuser Boston

The first signs of development in the tPOC are seen at stage 24 as a line of differentiated neurons in the caudolateral part of the diencephalon, extending from the rostral pole of the brain to the midbrain flexure (Figure 3.1a). Axons from these cells interlink, joining one cell with the next, thereby forming a tract. The tract descends to the flexure of the brain where it joins with the major ventrolateral tract of the hindbrain. As development proceeds, the tPOC acts as the main highway for all axon growth in the developing forebrain (Figure 3.1b). Similar observations have been made in the zebrafish (Wilson and Easter, 1991).

To test the dependence of the retinal axons on the tPOC, the caudolateral wall of the diencephalon, containing the early developing tPOC neurons, has been rotated through 90°. This procedure is known to alter the course of the optic tract (Harris, 1989), but it was not known whether this related to abnormal development of the preexisting tPOC. After a rotation of 90° to the caudolateral diencephalon, the tPOC developed with a horizontal component inside the graft. Axons from the cells grew at their usual time and in their normal descending fashion until they reached the discontinuity at the graft borders. At the graft borders, the axons commonly restored the continuity of the pathway, resulting in the formation of dramatically sinuous tracts. In some cases the grafted axons failed to rejoin the tract, while the remaining tPOC axons succeeded in forming a nearly normal tract by bypassing the graft. The patterns of retinal axon outgrowth that developed after such operations corresponded largely with these patterns of altered tPOC development (Figure 3.2). Serial 2-μm sections taken through operated brains confirmed that the retinal axons still fasciculated with the rostral edge of the tPOC, which they followed either to the tectum or to a blind ending in the rostral diencephalon. Three-dimensional reconstruction of these operated brains revealed the basic scaffold of axon tracts and the relationship of the optic tract to the tPOC. These observations suggest that the tPOC is the main guidance cue used by the retinal axons as they grow through the diencephalon.

The ultimate test of the reliance of retinal axons upon the tPOC would be to selectively delete the cells giving rise to the tPOC. Unfortunately, selective deletion of these cells is not feasible, but even if possible it might not yield a definitive answer. As described, the tract forms by interlinkage of neurons across undifferentiated neuroepithelium. The outgrowth of the initial tPOC axons is stereotyped and suggests that some local specialization of the neuroepithelium may exist favoring axon outgrowth. Deletion of the tPOC cells, or of all the other early components of the tPOC, may still leave the permissiveness of this region of neuroepithelium for axon growth intact and would therefore still allow retinal axons to follow their normal course through the diencephalon.

Once in the vicinity of the developing tectum, retinal axons leave the tPOC, arborize, and terminate. Some cue, probably emanating from the tectum, lures the axons away from the tract. The search for such a cue has been fraught with tantalizing suggestions and no really conclusive answers (Harris et al, 1985, 1987; Harris, 1986; Taylor, 1990; Hankin and Lund, 1991). Hide-and-seek can be played with the retinal axons and the tectum. The eye can be grafted to an abnormal

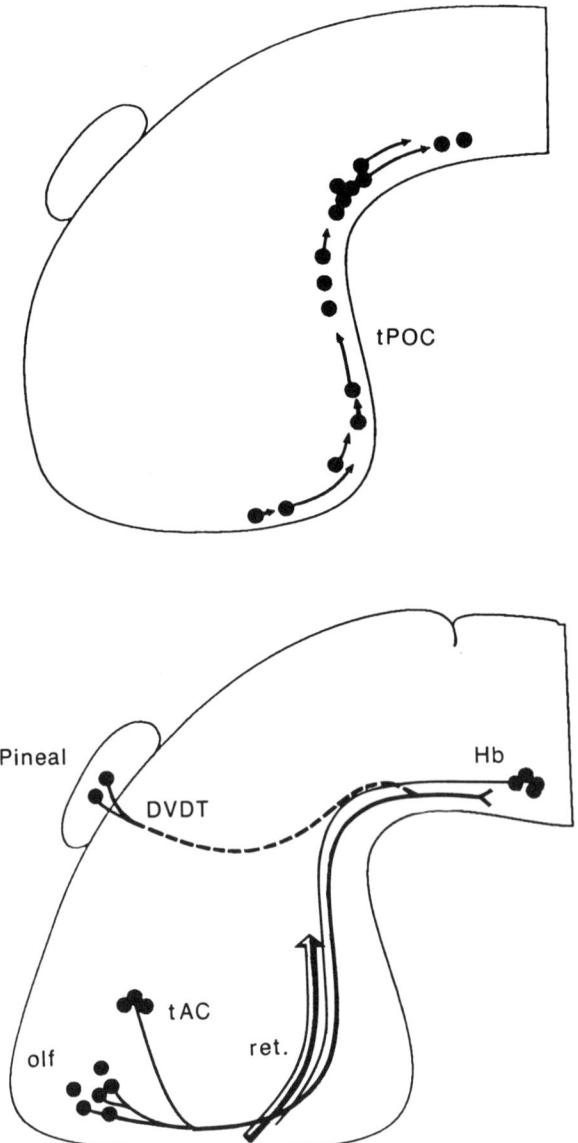

FIGURE 3.1 The first neurons to develop in the forebrain are a chain of cells which lie in the caudolateral part of the diencephalon. Axons from these neurons interlink with other neurons and axons in the chain, forming a descending pathway, the tPOC. As development proceeds, this tract forms the major highway for axon growth in the diencephalon. Olfactory (olf) axons, ascending hindbrain axons (Hb), retinal axons (ret), and axons from the pineal, which enter the tPOC from the dorsolateral diencephalic tract (DVDT), all follow specific trajectories which include fasciculation with the tPOC.

position relative to the tectum (Harris, 1986), or the tectum moved within the developing brain (Taylor, 1990). In each case the retinal axons usually locate their target.

During normal development the retinal axons locate the tectum over a very short distance–the few tens of microns from the tPOC to the tectum. However, in experimental cases they must navigate over a distance of 100 μm or more. One explanation of this longer-distance navigation is that the growth cones may be responding to a diffusible cue emanating from the tectum (Harris, 1986; Taylor, 1990). As yet there is no evidence for a diffusible cue. Coculture experiments have failed to show any directed outgrowth of retinal axons (Jack and Taylor, unpublished; Harris et al, 1985; but see Smalheiser et al, 1981).

In experimentally manipulated brains the routes followed by axons to displaced tecta are very well defined. This observation led to an alternative suggestion that the retinal axons fasciculate, as normal, with a preexisting axon pathway, but that this was altered by the operation (Taylor, 1990). By labeling the early scaffold of pathways in such operated brains, we now know that this is unlikely as there is little effect upon the normal pattern axon tracts in the developing brain.

In normal development it is quite conceivable that rather than a diffusible cue, filopodial contact could mediate the process of target recognition. Growth cones which have been observed, using a variety of techniques, within the developing frog central nervous system (CNS) are within the size range 20–50 μm (HNK-1— Nordlander, 1989; Taylor, 1991; HRP—Roberts and Clarke, 1982; Nordlander, 1987; Easter and Taylor, 1989; Taylor, 1990; intracellular injection—Holt, 1990; intracellular labeling following blastomere injection—Jacobson and Huang, 1985; Huang and Jacobson, 1986; DiI labeling—Harris et al, 1987; Taylor, 1990). The size of retinal ganglion cell axon growth cones is consistent with local sensing of a cell surface cue.

Once the retinal axons have reached the tectum, they are faced with the task of forming a retinotopically ordered series of terminations. One suggestion as to how the map is established is that retinal axons grow directly into the appropriate region of the tectum and terminate in a generally retinotopic fashion. This establishes a rough map, which is then refined through a process of activity-mediated synaptic reorganization.

In late larval and adult *Xenopus* we know that the axons within the optic tract are arranged in an order that predicts their termination pattern (Steedman, 1981; Fawcett and Gaze, 1982; Fawcett et al, 1984; Taylor et al, 1985). This organization is established within the region of the chiasm from a nonretinotopic ordering of axons within the optic nerves (Taylor, 1987). Two questions arise: (1) whether the retinoptic order of axons in the optic tracts is responsible for the formation of an orderly pattern of termination, and (2) what governs the process of segregation of axons from different retinal origins in the optic tract.

Initially, the retinotectal projection map is only crudely ordered (Gaze et al, 1974; Holt and Harris, 1983; O'Rourke and Fraser, 1990). For ventral and dorsal retinal axons there is a generally appropriate retinotopicity, with their terminals occupying the medial and lateral regions of the tectum. Nasal and temporal axons

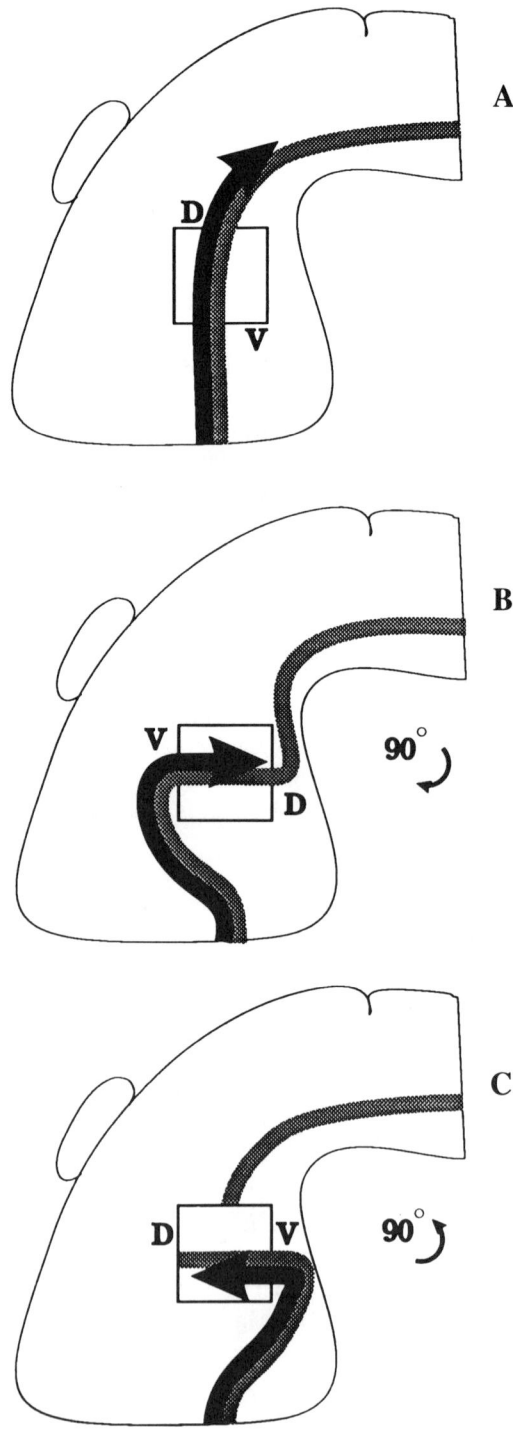

are not retinotopically organized; their terminals are completely overlapping (O'Rourke and Fraser, 1990). It was unclear how the pattern of axons within the optic nerve and tracts relates to this initial termination pattern. HRP and fluorescent dyes have been used to label selected groups of axons as they are growing to the tectum. Axons from each retinal quadrant leave the eye and enter the optic stalk in retinotopic position (Figure 3.3A). As they pass down the stalk there is some decay in the order, but as they approach the chiasm there is a retinotopic bias to the projection (Figure 3.3B). As the axons join with the tPOC, the order changes such that dorsal and ventral retinal axons occupy the caudal and rostral aspects of the tract, respectively (see also Holt and Harris, 1983). Nasal and temporal axons grow with temporal axons tending to aggregate in the central part of the tract, as is found in older animals (Figure 3.3c). This suggests that the very first retinal axons are resorting in the region of the chiasm, according to some information contained within their pathway.

As already described, the ingrowing retinal ganglion cell axons fasciculate with the preexisting tract, the tPOC, accumulating at the rostral edge. Within this region, retinal axons reorganize to become retinotopically ordered. To see whether other component axons of nonretinal systems are also differentially partitioned within the tPOC, HRP has been used to label some of the component projections over a range of developmental stages (Taylor and Dar, 1990). Retinal axons, as already described, are strongly partitioned in the rostral aspect of the tract and are relatively late arrivals. Axons of the olfactopeduncular tract, which are a very early addition to the tract, lie in the caudal region of the tract, and are displaced centrally and caudally as other components are added to the tract. An ascending hindbrain projection, which develops at an intermediate stage between the olfactory and retinal axons, occupies the midregion of the tract. Axons from the pineal organ join the tract at the flexure and descend into the hindbrain. These results suggest that each type of axon entering the tPOC selects a specific trajectory and grows in a specific subregion of the tract. The segregation of axons may partially reflect their sequence of addition, but it is also likely to result from an active selection of a subregion of the tract in which to grow. This type of fiber–fiber recognition has been well established in the invertebrate CNS (Raper et al, 1983; Bastiani et al, 1984; Harrelson and Goodman, 1988). In the vertebrate

---

FIGURE 3.2. A: The normal relationship of retinal axons (black) to those of the tPOC (gray). Retinal axons lie at the rostral edge of the tPOC and follow the tract to the position of the developing tectum. The lateral wall of the diencephalon can be rotated, before the development of the tPOC, through 90° either clockwise (B) or anticlockwise (C). The tPOC now develops as a sinuous pathway through the diencephalon, which in some cases is a continuous tract, but in others, especially after anticlockwise rotations, is blind-ending. Retinal axons which subsequent grow through the diencephalon fasciculate with the rostral edge of the tPOC and follow its aberrant course which may lead either to the tectum or to an inappropriate part of the brain.

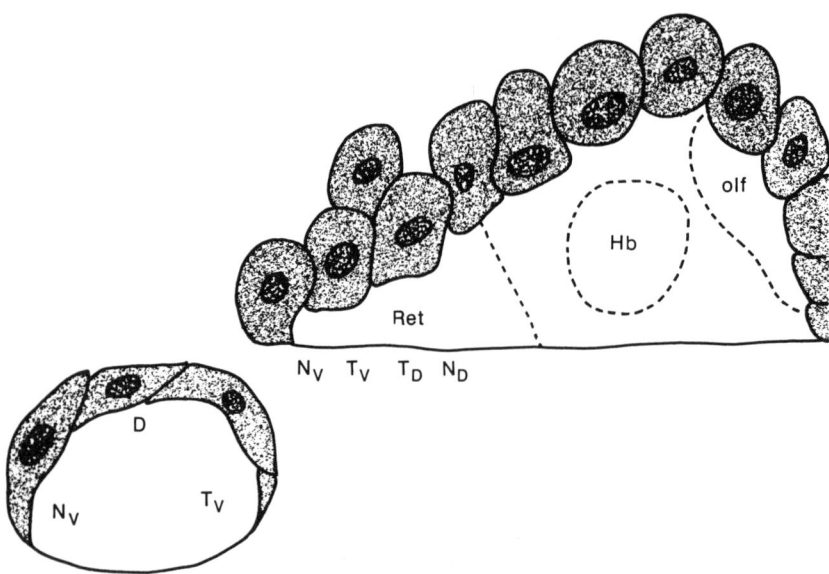

FIGURE 3.3. Within the optic stalk, retinal axons are retinotopically ordered, reflecting their direct passage from the retina into the stalk. As they pass into the ventral diencephalon and join with the tPOC, a re-sorting of the axons occurs, producing a retinotopic order which predicts the pattern of termination in the tectum. This segregation of different components of the retinal projection parallels the segregation of all components of the tPOC into different subregions of the tract. Retinal axons (ret) lie rostrally, with ventral axons (N_VT_V) rostral of dorsal axons (T_DN_D. Nasal (N) and temporal (T) axons have an overlapping distribution, with temporal axons aggregating in the centre of the tract. Hindbrain ascending axons (Hb) lie in the midregion of the tract, while the early-developing olfactopeduncular (O4) projection lies at the caudalmost edge.

CNS similar axon guidance decisions have been suggested to be based not only upon positive cues (Dodd and Jessell, 1988; Dodd et al, 1988) but also upon inhibitory signals (Kapfhammer and Raper, 1987; Raper and Kapfhammer, 1990). We are currently trying to look at the interactions between these difference components of the tPOC using *in vitro* confrontation assays.

CONCLUSIONS

The central role of the tPOC in axon guidance through the developing diencephalon has been described. This tract, which originates as a chain of neurons with descending axons, forms the major pathway for longitudinal axon growth in the developing diencephalon. The reliance of retinal axons upon the tPOC has been examined by experimental deviation of the tract. Retinal axons faithfully follow the abnormal route of the deviated tPOC, maintaining their normal spatial arrangement within the tract. The retinal axon and axons of other component systems that use this pathway are segregated into specific subregions of the tract, with retinal axons also exhibiting an internal segregation across the dorsoventral axis of the projection. This process of segregation is thought likely to be based upon differential interactions of growing retinal axons with other fiber components of the tPOC. The segregation of retinal axons across the dorsoventral axis of the projections is responsible for the retinotopic arrangement in which axons invade the optic tectum, which is a major contributory factor in establishing a retinotoptic projection.

Acknowledgments

I would like to express my immense gratitude to Mike Gaze for his guidance and inspiration in my work on the retinotectal system. I also thank the Wellcome Trust for financial support and Terry Richards for drawing the figures. J.S.H.T. is in receipt of a Wellcome Vision Research Fellowship.

REFERENCES

Bastiani MJ, Raper JA, Goodman CS (1984): Pathfinding by neuronal growth cones in grasshopper embryo: III. Selective affinity of the G growth cone for the P cells within the A/P fascicle. *J Neurosci* 4: 2311–2328

Dodd J, Jessell TM (1988): Axon guidance and the patterning of neuronal projections in vertebrates. *Science* 242: 700–708

Dodd J, Morton SB, Karagogeos D, Yamamoto M, Jessell TM (1988): Spatial segregation of axonal glycoprotein expression on subsets of embryonic spinal neurons. *Neuron* 1: 105–116

Easter SS Jr., Taylor JSH (1989): The development of the *Xenopus* retinofugal pathway: Optic fibres join a pre-existing tract. *Development 107: 553–573*

Fawcett JW, Gaze RM (1982): The retinotectal fibre pathways from normal and compound eyes in *Xenopus*. *J Embryol Exp Morphol* 72: 19–37

Fawcett JW, Taylor JSH, Gaze RM, Grant P, Hirst E (1984): Fibre order in the normal *Xenopus* optic tract, near the chiasma. *J Embryo Exp Morphol* 83: 1–14

Gaze RM, Keating MJ, Chung SH (1974): The evolution of the retinotectal map during development in *Xenopus*. *Proc R Soc Lond B* 185: 301–330

Hankin M, Lund RD (1991): How do retinal axons find their targets in the developing brain? *Trends Neurosci* 6:224–228

Harrelson AL, Goodman CS (1988): Growth cone guidance in insects: Fasciclin II is a member of the immunoglobulin superfamily. *Science* 242: 700–708

Harris WA (1986): Homing behaviour of axons in the embryonic vertebrate brain. *Nature* 320: 266–269

Harris WA (1989): Local positional cues in the neuroepithelium guide retinal axons in *Xenopus* brain. *Nature* 339: 218–221

Harris WA, Holt CE, Bonhoeffer F (1987): Retinal axons with and without their somata, growing to and arborising in the tectum of *Xenopus* embryos: A time-lapse video study of single fibre *in vivo*. *Development* 101: 123–133

Harris WA, Holt CE, Smith T, Galenson N (1985): Growth cones of developing retinal cells *in vivo*, on collagen surfaces and in collagen matrices. *J Neurosci Res* 13: 101–122

Holt CE (1990): A single cell analysis of early retinal ganglion cell differentiation in *Xenopus*: From soma to axon tip. *J Neurosci* 9: 3123–3145

Holt CE, Harris WA (1983): Order in the initial retinotectal map in *Xenopus*: A new technique for labelling growing nerve fibres. *Nature* 301: 150–152

Huang S, Jacobson M (1986): Neurites show pathway specificity but lack directional specificity or predetermined lengths in *Xenopus* embryos. *J Neurobiol* 17: 593–604

Jacobson M, Huang S (1985): Neurite outgrowth traced by means of horseradish peroxidase inherited from neuronal ancestral cells in frog embryos. *Dev Biol* 110: 102–113

Kapfhammer JP, Raper JA (1987): Collapse of growth cone structure on contact with specific neurites in culture. *J Neurosci* 7: 201–212

Nordlander RH (1987): Axonal growth cones in the developing amphibian spinal cord. *J Comp Neurol* 263: 485–496

Nordlander RH (1989): HNK-1 marks earliest axonal outgrowth in *Xenopus*. *Dev Brain Res* 50: 147–153

O'Rourke NA, Fraser SE (1990): Dynamic changes in optic fiber terminal arbors lead to retinotopic map formation: An in vivo confocal microscopic study. *Neuron* 5: 159–171

Raper JA, Bastiani MJ, Goodman CS (1983): Pathfinding by neuronal growth cones in grasshopper embryos. II Selective fasciculation onto specific axonal pathways. *J Neurosci* 3: 31–41

Raper JA, Kapfhammer JP (1990): The enrichment of a neuronal growth cone collapsing activity from embryonic chick brain. *Neuron* 2: 21–29

Roberts A, Clarke JDW (1982): The neuroanatomy of an amphibian embryo spinal cord. *Phil Trans Roy Soc Ser B* 296: 195–212

Smalheiser NR, Peterson ER, Crain SM (1981): Neurites from mouse retina and dorsal root ganglion explants show specific behavior within co-cultured tectum or spinal cord. *Brain Res* 208: 499–505

Steedman JG (1981): Pattern formation in the visual pathways of *Xenopus laevis*. Ph.D. thesis, University of London

Taylor JSH (1987): Fibre organisation and reorganisation in the retinotectal projection of *Xenopus*. *Development* 99: 393–410

Taylor JSH (1990): The directed growth of retinal axons towards surgically transposed tecta in *Xenopus*; An examination of homing behavior by retinal ganglion cell axons. *Development* 108: 147–158

Taylor JSH (1991): The early development of the frog retinotectal projection. *Development* S2:95–104

Taylor JSH, Dar S (1990): Spatial restriction of component axons in the developing post-optic commissure of *Xenopus laevis*. *Eur J Neurosci Suppl* 3: 207

Taylor JSH, Willshaw DJ, Gaze RM (1985): The distribution of fibres in the optic tract after contralateral translocation of an eye in *Xenopus*. *J Embryol Exp Morphol* 85: 225–238

Wilson SW, Easter SS Jr. (1991): Stereotyped pathway selection by growth cones of early epiphysial neurons in the embryonic zebrafish. *Development* 112: 723–746

CHAPTER 4

In Vitro Investigations of Fiber–Fiber and Fiber–Glial Cell Interactions in the Development of the Amphibian Visual System

DOUGLAS GOODAY

The correct functioning of the visual system depends on the development of an accurate neural map of the retina on the optic tectum. During the development of these highly ordered neural connections between eye and brain in amphibia, the nerve fibers from different areas of the retina grow along stereotyped routes in the visual pathway to terminate in retinotopic order on the optic tectum. This precision implies that the retinal fibers are directed in their growth by finely tuned guidance mechanisms, probably mediated through a sequence of interactions with the substrates they encounter as they grow to the optic tectum.

Anatomical evidence based on tracing axonal trajectories with horseradish peroxidase (HRP) in *Xenopus* reveals segregation of fibers from dorsal, ventral, nasal, and temporal retina within the optic tract when the fibers have left the optic chiasma (Fawcett and Gaze, 1982). Similar segregation of fibers has also been observed in other species of amphibia: *Rana nigromaculata* (Fujisawa et al., 1981a), *Rana pipiens* (Reh et al., 1983) and *Cynops* pyrrhogaster (Fujisawa et al., 1981b). In *Xenopus* fibers are distributed such that nasal fibers occupy the entire width of the optic tract, temporal fibers form a coherent, fasciculated group in the center of the tract, and dorsal and ventral fibers occupy lateral and medial aspects of the tract. Using similar methods, little quadrant-specific grouping of fibers has been found in the *Xenopus* optic nerve (Taylor, 1987). These neuroanatomical tracing studies reveal that, in *Xenopus* at least, the beginning of fiber ordering is in the optic tract, and a degree of fiber segregation has already taken place in the retinal projection even before fibers have reached the optic tectum.

The environment of the optic tract must fulfill two main functions regarding the retinal projection: it must guide retinal fibers in the direction of the tectum, particularly pioneering fibers, and it must provide some kind of molecular information that permits or instructs fibers from the different retinal quadrants to segregate and take particular routes to the tectum.

The disparity between the fasciculation patterns of nasal and temporal fibers in the optic tract led Gaze and Fawcett (1983) to suggest that temporal fibers have properties which make them fasciculate more than fibers from other parts of the retina. This tendency for temporal fibers to fasciculate is seen in normal animals and in fibers regenerating from both normal and compound eyes (Fawcett and

Formation and Regeneration of Nerve Connections
Sansar C. Sharma and James W. Fawcett, Editors
© 1993 Birkhäuser Boston

Gaze, 1982; Gaze and Fawcett, 1983). An *in vitro* experiment performed by
Bonhoeffer and Huf (1985) using chick retinal tissue supports this view and shows
that the tendency to fasciculate is an intrinsic property of temporal fibers and is
manifest when they are forced to grow on a substrate of other retinal fibers, but not
when they grow on a uniform culture substrate coated with laminin.

This experiment has been performed in our laboratory using *Xenopus* retina, and
the same result has been obtained (Figure 4.1).Explants of nasal and temporal

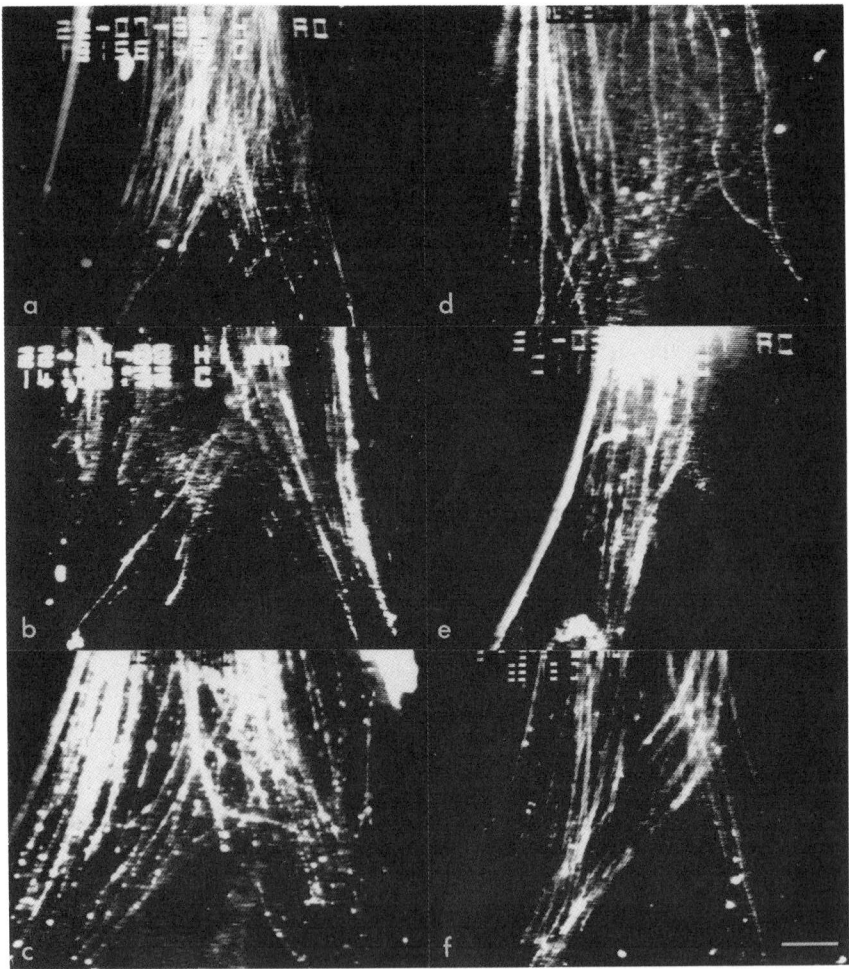

FIGURE 4.1. Photographs of fluorescently labeled fibers from nasal and temporal third
explants in the selective fasciculation experiment. The images were enhanced using a SIT
camera and photographs were taken from the video screen. In **a, b,** and **c,** fibers from the
nasal explant grow down both arms of the Y. In **d, e,** and **f,** the temporal fibers exhibit a
preference for growing down the temporal track. Scale bar represents 100 μm.

retina are cultured, and the outgrowing neurites are restricted to growing down the arms of a Y-shaped track. When the neurites have reached the junction point of the Y, a third, fluorescently labeled explant of either nasal or temporal retina is added and the neurites grow to meet those from the first two explants. At this juncture a solution of anti-laminin antiserum is added to the culture. This has the effect of denying the fibers further access to the laminin substrate so they are forced to grow along other fibers. Observation of the culture under the fluorescence microscope reveals whether the labeled fibers from the third explant have any preference for growth on like or unlike fibers. Out of 22 cultures in which the third explant was from temporal retina, five showed no preference for growth on either nasal or temporal fibers. In the remaining 17 cultures, all or most of the temporal fibers grew down the temporal track (Table 4.1).

The results of this experiment tell us that the fasciculation behavior is intrinsic to temporal fiber and is not due to an extrinsic factor, such as attraction to the center of the tract by some diffusible agent. However, in order to fasciculate the substrate in the optic tract must presumably be favorable, temporal fibers are not seen as a fasciculated group in the optic nerve.

Selective fasciculation could be due either to preferential adhesion between temporal fibers or to the avoidance of unlike fibers by temporal fibers. The experiments of Raper and Grunewald (1990) suggest that the latter is the case. In mixed cultures of nasal and temporal chick retina, time-lapse recording of growth cone–neurite encounters showed that temporal neurites retracted from, and were inhibited from growing along, nasal fibers but did not retract from other temporal fibers. Nasal growth cones showed no retraction behavior.

The environment of the diencephalic optic tract must differ from that of the optic nerve for the temporal fibers to reveal their fasciculative tendencies, and for the dorsal and ventral fibers to diverge. An indication of a possible underlying cause of this change in substrate comes from the work of Maggs and Scholes (1986), who demonstrated that there was a change in the intermediate filament expression of glial cells between the optic nerve and optic tract in cichlid fish. Optic nerve glial cells contained a vimentin-like intermediate filament and lacked glial fibrillary acidic protein (GFAP), whereas in the tract the glia were positive for GFAP and lacked the vimentin-like protein; this change in glial cell type coincides with a rearrangement of fiber order in the visual projection. Similar changes in glial cell

TABLE 4.1. Summary of results of the selective fasciculation experiment

	Total	Third Explant	
		Nasal	Temporal
Number of results	30	8	22
Number without difference	13	8	5
Number with difference	17	0	17
Preferred track (N = nasal, T = temporal)		N	T
		0	17

phenotype have been found in the visual system of the ferret (Guillery and Walsh, 1987). Bearing in mind the already well-documented role of radial glia as mediators of axonal guidance in the mammalian cortex and cerebellum (for a review see Hatten, 1990), diencephalic glia would be good candidates for mediators of fiber sorting in the optic tract. The results of culture experiments in which explants of retina were grown on monolayers of glia from the *Xenopus* diencephalon do not support this idea, however; nasal and temporal fibers have the same growth patterns (Figure 2) (Gooday, 1990).

The next consideration in the development of the visual system is the topographic mapping of the retina on the tectum so that the order of origin of the fibers in the retina is reflected in the order of the axon terminals on the tectum. Most current theories relating to the formation of the ordered visual map rely on the existence of graded positional markers on the tectum. Much of the research in this field was inspired by the work of Sperry and colleagues who discovered the phenomenon of specific fiber–target recognition after experimentally altering the spatial relationships of eye and tectum (Sperry, 1943, 1963; Attardi and Sperry, 1963).

More recent *in vivo* experimental evidence for the existence of directional tectal cues comes from tectal rotation experiments in goldfish (Yoon, 1973) and *Xenopus* (Taylor and Stirling, 1989). In both these experiments rotation of the tectum or tectal precursor led to the subsequent development of a rotated visual map.

The *in vitro* experiments of Bonhoeffer and Huf (1980, 1982) showed that axons from explants of nasal and temporal retina responded differently when presented with a choice of growing along cells or plasma membrane extracts isolated from the front and back of the optic tectum. Nasal fibers grew equally well on both substrates, but temporal fibers preferred to grow on cells or membranes from the front tectum only. Division of the tectum into five slices showed that the property of the tectal cells was graded from front to back of the tectum: temporal fibers always preferred to grow on the more anterior membranes. More recently this group of workers have modified the assay to present retinal fibers with rostral and caudal tectal membrane extracts deposited by suction on Nucleopore filters in alternating stripes (Walter et al., 1987a), and similar results have been obtained to those above. However, when given no choice of membrane, temporal fibers will grow as well as nasal fibers on carpets of posterior tectal membranes. Similar experiments have been performed with goldfish tissue (Vielmetter and Stuermer, 1989). This alternating-stripe assay has been more amenable to biochemical investigations, which are discussed below.

These experiments lend a lot of weight to the theory of graded positional markers, but the use of plasma membrane extracts made from whole tissues does not give an indication of the source of any positional marker in terms of cell type.

When segments of *Xenopus* retina were cultured on monolayers of GFAP-positive glial cells from the optic tectum, comparable results to those of Bonhoeffer and Huf (1980, 1982) and Walter et al. (1987a) were seen. Nasal and temporal retinal fibers had identical growth patterns on monolayers of diencephalic glia, and grew as numerous long fine fascicles and individual fibers. On glia isolated from the optic tectum, nasal fibers showed the above pattern of growth. Temporal fibers, however, were restricted to growing in a few, very short, highly fasciculated bundles (Figure 4.2) (Gooday, 1990). When nasotemporal strips of *Xenopus* retina, attached to Millipore filters, were cultured on tectal glial monolayers, a prolific outgrowth of fibers was only seen in the nasal half of the strip (Figure 4.2) (Jack et al., 1991). This property of tectal glial, which restricts

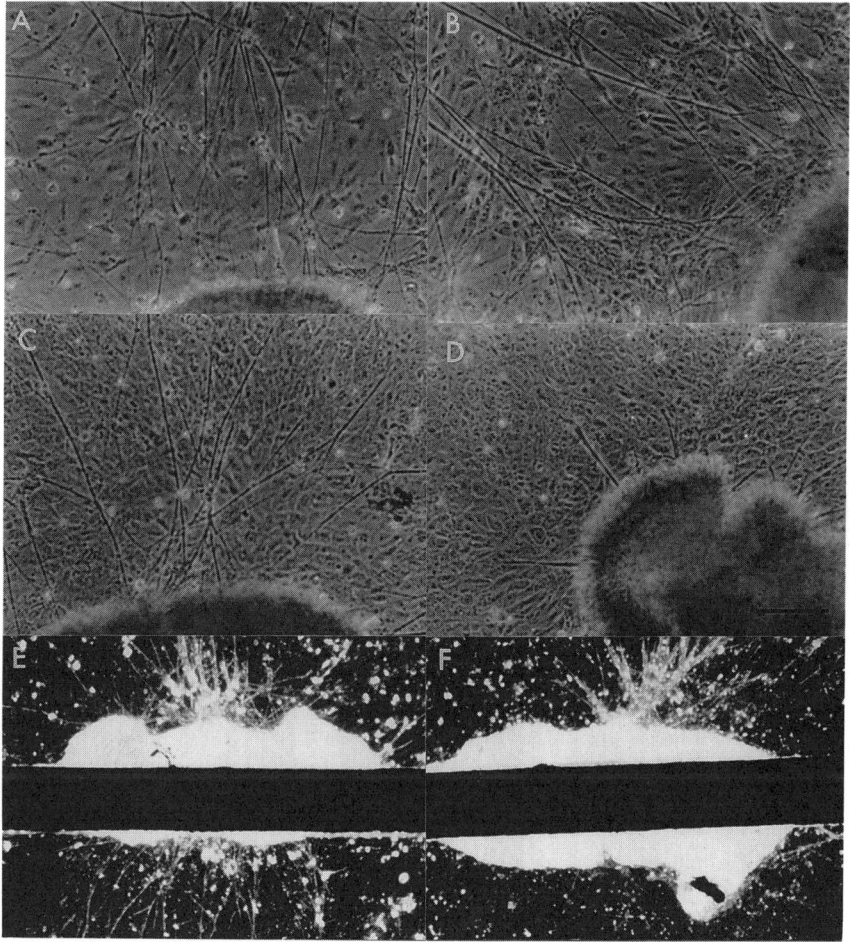

FIGURE 4.2. A-D: Photographs of nasal and temporal retinal explants growing on monolayers of diencephalic glial cells **A** and **B**, and on monolayers of tectal glial cells **C** and **D**. All the photographs were taken 4 days after explanation; **E** and **F**: Growth of fibers from strips of retina, cut nasotemporally, on (**E**) rostral tectal glia and (**F**) caudal tectal glia. On the caudal glial monolayer, fibers grow only from the nasal part of the strip. The width of the filter strip is 230 μm.

the growth and induces the fasciculation of temporal retinal fibres, was more concentrated in glia from the caudal end of the tectum than in rostral tectal glia (Jack et al., 1991). In these experiments it was not possible to section the *Xenopus* tectum into more than three parts, due to its small size, and to ensure no overlap of the tissues, the middle part was discarded. It is not possible, therefore, to say whether this temporal retinal fiber-inhibiting effect we see on tectal glia is graded across the tectum, as has been demonstrated for chick and goldfish. We can only say that there is a difference between the rostral and caudal ends of the tectum, but nonetheless our results are in agreement with those mentioned above.

This differential response of fibers from nasal and temporal sides of the eye to glial cells cultured from the rostral and caudal sections of the optic tectum could be caused by a number of mechanisms. The results of the above experiments are reminiscent of a sorting-out phenomenon, and could therefore be explained on the basis of differential adhesion caused by the relative strengths of fiber–fiber and fiber–substrate interactions for nasal and temporal fibers. The experimental masking of the adhesion molecule N-CAM by the implantation of agarose inserts containing antisera to N-CAM into the *Xenopus* optic tectum lends some support to this idea (Fraser et al., 1988). This experiment resulted in anatomical and physiological disruption of the retinotectal map. Although the specific masking of N-CAM leads to alteration of the map, no graded distribution of N-CAM or of any of the other acknowledged adhesion molecules has been detected in the optic tectum or in the retinal projection. N-CAM may therefore serve a stabilizing role rather than forming a gradient of adhesion along which the retinal projection could sort out, and we must look for an alternative model to explain the sorting out of the retinal projection.

A simple mechanism that would sort out the projection in the nasotemporal orientation would be to retard or inhibit the growth of temporal fibers with respect to nasal fibers. The experiments of Walter et al. (1987b) show that heat treatment of the posterior membranes in the stripe assay results in a loss of discrimination by the temporal fibers. The activity can also be removed by treatment of posterior membranes with phosphatidylinositol-specific phospholipase C (Walter et al., 1990); treatment of rostral tectal membrane preparations had no effect. This suggests that a factor present in the posterior tectum acts to inhibit the growth of temporal fibers. The results of our experiments with glial cell monolayers in vitro suggest that the glial cells at the caudal end of the tectum are an inhibitory substrate for temporal fibers but not nasal fibers.

The phenomenon of neurite inhibition was examined by Bray et al. (1980) and Kapfhammer et al. (1986) who noted that in coculture retinal and sympathetic neurites tended to occupy distinct territories and did not mix or cross. Later studies by Kapfhammer and Raper (1987a, 1987b) investigated the cell biology of this phenomenon and showed that this avoidance was caused by the collapse and retraction of growth cones on contact with heterotypic neurites (as already mentioned above as a mechanism for selective fasciculation). This work suggests that growth cone inhibition could serve as a guidance mechanism for neurites during development (for a review see Patterson, 1988). It also indicates that growth cone collapse is an integral part of neurite inhibition *in vitro*, and may thus serve as a useful indicator for analyzing whether inhibition is operating during a particular developmental event.

We have analyzed encounters between retinal growth cones and isolated tectal glial cells *in vitro* to see whether the results outlined above could be explained by the inhibitory theory (Johnston and Gooday, 1991).

Twenty encounters of each of six types were analyzed between nasal and temporal neurites and glial cells from the diencephalon and rostral and caudal ends of the optic tectum. The results are presented in Table 4.2. In all encounters between temporal neurites and glial cells from the caudal part of the optic tectum, the growth cones collapsed and retracted on contact with the glial cell (Figure 4.3). In all the other types of encounter, only a few retractions were observed. This suggests that in the *Xenopus* tectum, temporal retinal neurites are prevented from growing to the caudal tectum, an inappropriate target for them, by inhibitory interactions with glial cells. Tectal glial cells, therefore, appear to aid the establishment of the ordered visual projection in the nasotemporal axis during normal development by providing positional information for the ingrowing retinal neurites which enables nasal and temporal fibers to segregate and locate their appropriate positions on the tectum.

Neuronal inhibition is involved in the formation of the retinotectal projection in chicks and in the development of other neural systems. Davies et al. (1990) found a glycoprotein fraction in chick somites that causes the growth cones of cultured dorsal root ganglion neurons to collapse. They proposed that this molecule restricts the growth of sensory and motor nerves to the anterior half of each sclerotome and so brings about the vertebrate pattern of segmental spinal nerves. Raper and Kapfhammer (1990) found collapsing activity in extracts of whole chick brain of putative molecular weight 50 kilodaltons, and Cox et al. (1990) demonstrated that cell membrane extracts from posterior tectum, and to a lesser extent anterior tectum, caused the collapse of temporal retinal growth cones, whereas nasal growth cones remained unaffected by either preparation. A 33-kilodalton glycoprotein has been identified in these extracts which appears to be responsible for the collapsing activity (Stahl et al., 1990).

All these extracts have so far been prepared from whole tissues, and the location of the inhibitory molecules in terms of cell type is at present unknown. Our results

TABLE 4.2. Number of growth cones to collapse and retract (r) out of a total (t) for each class of encounter

| | Origin of retinal explant | | | |
| | Nasal | | Temporal | |
Origin of glia	t	r	t	r
Diencephalon	19	2	21	2
Rostral tectum	19	0	23	3
Caudal tectum	21	2	20	20

from experiments using pure cultures of glial cells indicate that such molecules are present in glial cell membranes. This is perhaps surprising, bearing in mind the role of glial cells in providing positive guidance for neurons during development of other central nervous system (CNS) structures such as the cerebellum. Recently much attention has been focused on the role of oligodendrocytes in preventing the

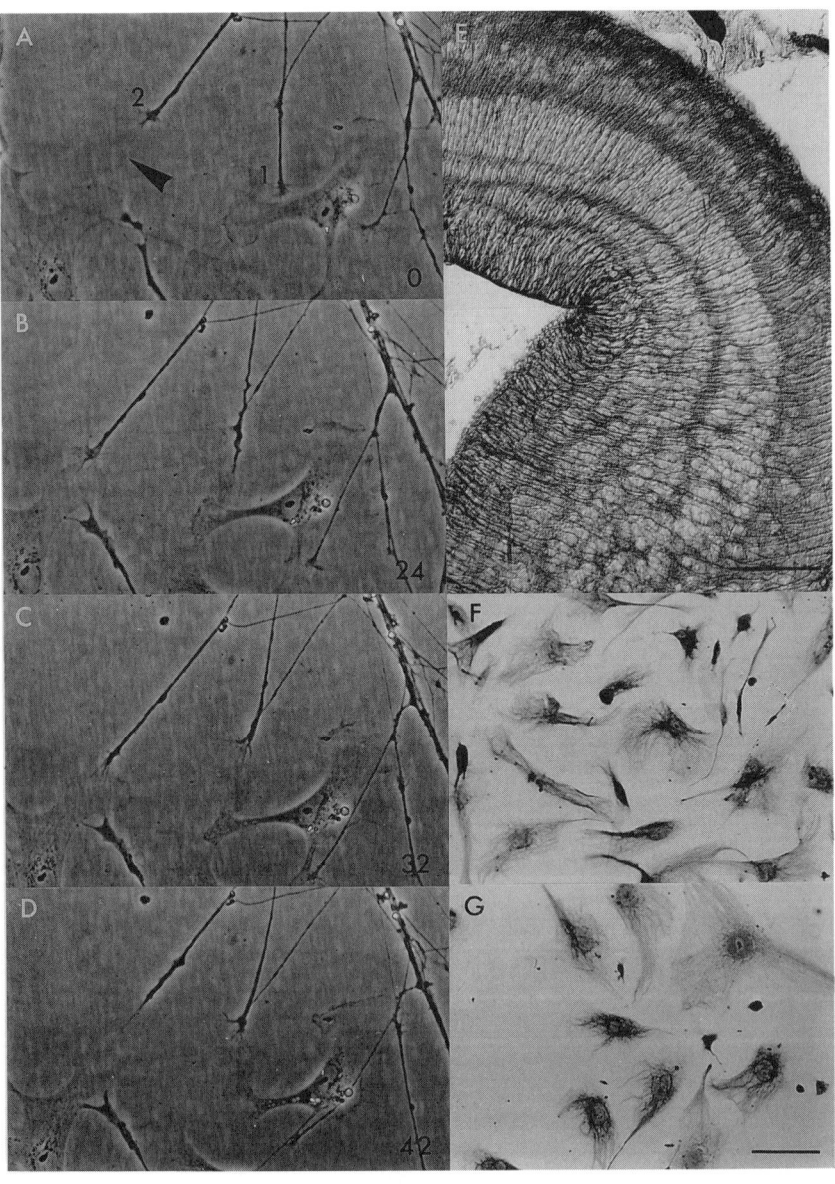

regeneration of CNS neurons in mammals. This is thought to be due to two components of CNS myelin of 35 and 250 kilodaltons (Schwab and Caroni, 1988). Mammalian oligodendrocytes have been shown to cause the collapse of peripheral neurite growth cones *in vitro* (Fawcett et al., 1989; Bandtlow et al., 1990). The oligodendrocytes of goldfish, however, support the growth of fish and chick retinal neurites *in vitro* (Bastmeyer et al., 1991). This suggests that a fundamental difference exists between the oligodendrocytes of amniotes and anamniotes in their ability to support neurite growth, and this may be part of the reason why the central nervous tissue of fish and amphibia can regenerate and restore functional connections after injury. The cells in our cultures are either fibroblastlike or elongated and bipolar in shape and are unlike mammalian oligodendrocytes morphologically. They resemble more the mammalian type 1 astrocyte and stain with a commercially available anti-GFAP antiserum (clone G-A-5) (Figure 4.3). The only cells in the *Xenopus* tectum to label with this antibody are radial glial cells (Figure 3). The developmental lineages of the glial cells of fish and amphibia are not known, and unlike the situation in the rat optic nerve, little antigenic characterization has been done, as many of the antisera used to identify mammalian glial cell types do not cross-react with amphibian glia (unpublished observations). Therefore we are reluctant to categorize the cell type of our cultured glia. They may be equivalent to radial glia, and it is highly probable that growing retinal axons encounter the end feet of these glia as they grow through the most superficial part of the tectum.

In conclusion, the establishment of neuronal ordering during development of the visual projection in *Xenopus* would seem to be a multistaged process, with different fiber groupings becoming segregated at different places in the optic pathway, and precision of ordering increasing at each stage. Glial cells appear to be important at several stages in the development of the projection and provide positional information to the retinal fibers once they reach the tectum.

FIGURE 4.3. A-D: Sequence of four photographs of encounters between growth cones of temporal retinal neurites and glial cells isolated from the caudal end of the optic tectum. The numbers at the bottom right of each photograph refer to the time in minutes which had elapsed since the sequence was started. **A**: Growth cone **1** is in contact with a glial cell, and growth cone **2** is advancing toward another cell. A fine, stretched process can be seen extending between growth cone **2** and the cell (arrowed), indicating that the growth cone has already contacted and retracted from this cell. **B**: Growth cone **1** has collapsed and retracted from the glial cell, and the axon has a wavy, concertina-like appearance. Growth cone **2** has contacted its glial cell. **C**: Growth cone **1** has been reorganized and growth cone **2** has collapsed. **D**: Growth cone **1** has advanced toward the glial cell for a second time, and growth cone **2** has retracted from its glial cell again. A stretched process can again be seen between the growth cone and the glial cell. **E**: Cryostat section of *Xenopus* optic tectum labeled with anti-GFAP to show the radial glial organization. Scale bar represents 50 μm. **F** and **G**: Cultured diencephalic and tectal glial cells stained with anti-GFAP. Scale bar represents 100 μm.

REFERENCES

Attardi DG, Sperry RW (1963): Preferential selection of central pathways by regenerating optic fibers. *Exp Neurol* 7: 46–64

Bandtlow C, Zachleder T, Schwab M (1990): Oligodendrocytes arrest neurite growth by contact inhibition. *J Neurosci* 10: 3837–3848

Bastmeyer M, Beckmann M, Schwab ME, Steurmer CAO (1991): Growth of regenerating goldfish axons is inhibited by rat oligodendrocytes and CNS myelin but not by goldfish optic nerve tract oligodendrocyte like cells and fish CNS myelin. *J Neurosci* 11: 626–640

Bonhoeffer F, Huf J (1980): Recognition of cell types by axonal growth cones *in vitro*. *Nature* 288: 162–164

Bonhoeffer F, Huff J (1982): In vitro experiments on axon guidance demonstrating an anterior-posterior gradient on the tectum. *EMBO J* 1: 427–431

Bonhoeffer F, Huf J (1985): Position-dependent properties of retinal axons and their growth cones. *Nature* 315: 409–410

Bray D, Wood P, Bunge RP (1980): Selective fasciculation of nerve fibers in culture. *Exp Cell Res* 130: 241–250

Cox EC, Muller B, Bonhoeffer F (1990): Axonal guidance in the chick visual system: Posterior membranes induce collapse of growth cones from the temporal retina. *Neuron* 4: 31–37

Davies JA, Cook GMW, Stern CD, Keynes RJ (1990): Isolation from chick somites of a glycoprotein fraction that causes collapse of dorsal root ganglion growth cones. *Neuron* 4: 11–20

Fawcett JW, Gaze RM (1982): The retinotectal fibre pathways from normal and compound eyes in *Xenopus*. *J Embryol Exp Morphol* 72: 19–37

Fawcett JW, Rokos J, Bakst L (1989): Oligodendrocytes repel axons and cause axonal growth cone collapse. *J Cell Sci* 92: 93–100

Fraser SE, Carhart MS, Murray BA, Chuong C-M, Edelman GM (1988): Alterations in the *Xenopus* retinotectal projection by antibodies to *Xenopus* N-CAM. *Dev Biol* 129: 217–230

Fujisawa H, Watanabe K, Tani N, Ibata Y (1981a): Retinotopic analysis of fiber pathways in amphibians. I. The adult newt *Cynops pyrrhogaster*. *Brain Res* 206:9–20

Fujisawa H, Watanabe K, Tani N, Ibata Y (1981b): Retinotopic analysis of fiber pathways in amphibians. II. The frog *Rana nigromaculata*. *Brain Res* 206: 21–26

Gaze RM, Fawcett JW (1983): Pathways of *Xenopus* optic fibres regenerating from normal and compound eyes under various conditions. *J Embryol Exp Morphol* 73: 7–38

Gooday DJ (1990): Retinal axons in *Xenopus laevis* recognise differences between tectal and diencephalic glial cells *in vitro*. *Cell Tissue Res* 259: 595–598

Guillery RW, Walsh C (1987): Changing glial organisation relates to changing fibre organisation in the developing optic nerve of ferrets. *J Comp Neurol* 265: 203–217

Hatten ME (1990): Riding the glial monorail: A common mechanism for glial-guided neuronal migration in different regions of the developing mammalian brain. *Trends Neurosci* 13: 179–184

Jack JL, Gooday DJ, Wilson MA, Gaze RM (1991): Retinal axons in *Xenopus* show different behavior patterns on various glial substrates *in vitro*. *Anat Embryol* 183: 193–203

Johnston AR, Gooday DJ (1991): *Xenopus* temporal retinal neurites collapse on contact with glial cells from caudal tectum in vitro. *Development* 113:409–417

Kapfhammer JP, Grunewald BE, Raper JA (1986): The selective inhibition of growth cone extension by specific neurites in culture. *J Neurosci* 6: 2527–2534

Kapfhammer JP, Raper JA (1987a): Collapse of growth cone structure on contact with specific neurites in culture. *J Neurosci* 7: 201–212

Kapfhammer JP, Raper JA (1987b): Interactions between growth cones and neurites growing from different tissues in culture. *J Neurosci* 7: 1595–1600

Maggs A, Scholes J (1986): Glial domains and nerve fibre patterns in the fish retinotectal pathway. *J Neurosci* 6: 424–438

Patterson P (1988): On the importance of being inhibited, or saying no to growth cones. *Neuron* 1: 263–267

Raper JA, Grunewald EB (1990): Temporal retinal growth cones collapse on contact with nasal retinal axons. *Exp Neurol* 109: 70–74

Raper JA, Kapfhammer JP (1990): The enrichment of a neuronal growth cone collapsing activity from embryonic chick brain. *Neuron* 4: 21–29

Reh TA, Pitts E, Constantine-Paton M (1983): The organisation of the fibres in the optic nerve of normal and tectum-less *Rana pipines*. *J Comp Neurol* 218: 282–296

Schwab ME, Caroni P (1988): Oligodendrocytes and CNS myelin are non-permissive for neurite growth and fibroblast spreading in vitro. *J Neurosci* 8: 2381–2393

Sperry RW (1943): Effect of 180 degree rotation of the retinal field on visuomotor coordination. *J Exp Zool* 92: 263–279

Sperry RW (1963): Chemoaffinity in the orderly growth of nerve fibre patterns and connections. *Proc Natl Acad Sci USA* 50: 703–709

Stahl B, Muller B, von Boxberg Y, Cox EC, Bonhoeffer F (1990): Biochemical characterisation of a putative axonal guidance molecule of the chick visual system. *Neuron* 5: 735–747

Taylor JSH (1987): Fibre organization and reorganization in the retinotectal projection of *Xenopus*. *Development* 99: 393–410

Taylor JSH, Stirling RV (1989): Investigations of the formation of ordered retinotectal projections to rotated tecta. In: *Systems Approaches to Developmental Neurobiology, NATO ASI Series A, vol 192*, Raymond PA, Easter SS, Innocenti GM, eds. New York: Plenum Press

Vielmetter J, Stuermer CAO (1989): Goldfish retinal axons respond to position-specific properties of tectal cell membranes in vitro. *Neuron* 2: 1331–1339

Walter J, Kern-Veits B, Huf J, Stolze B, Bonhoeffer F (1987a): Recognition of position-specific properties of tectal cell membranes by retinal axons *in vitro*. *Development* 101: 685–696

Walter J, Henke-Fahle S, Bonhoeffer F (1987b): Avoidance of posterior tectal membranes by temporal retinal axons. *Development* 101: 909–913

Walter J, Muller B, Bonhoeffer F (1990): Axonal guidance by an avoidance mechanism. *J Physiol (Paris)* 84: 104–110

Yoon MG (1973): Retention of the original topographic polarity by the 180 degree rotated tectal reimplant in young adult goldfish. *J. Physiol. (London)* 233: 575

CHAPTER 5

NMDA Receptors and Intertectal Plasticity in *Xenopus*

Susan B. Udin

INTRODUCTION

Many factors contribute to the development of the orderly connections character-istic of the normal brain. Gaze and his colleagues have helped to deepen our understanding of the interplay of temporal factors, chemoaffinity, and axon–axon interactions in the establishment of connections in one of the canonical models, the retinotectal system. Gaze's laboratory also played a major role in the development of our understanding of the role of sensory experience by their demonstration of the dramatic plasticity of the ipsilateral eye's input to the tectum. I will briefly review some of the work on binocular connections initiated by Gaze, extended by Keating and his co-workers, and continued in my own laboratory. These studies show that visual input is the dominant factor in bringing the two eyes' representa-tions into register and have led to the hypothesis that ipsilateral terminals become stabilized at locations where they find retinotectal terminals with correlated firing patterns. The *N*-methyl-D-aspartate (NMDA) receptor, located on tectal cell dendrites, mediates the interaction between the two sets of inputs.

ESTABLISHMENT OF BINOCULAR MAPS

The existence of an input to the tectum from the ipsilateral eye was first demonstrated with electrophysiological recording which showed that maps from both eyes were in register in the superficial layers of the tectum (Gaze and Jacobson, 1962). In the normal adult most of the tectum receives binocular input, reflecting the fact that placement of the two eyes on the top of the head produces a great deal of binocular overlap. In contrast, tadpoles have laterally placed eyes with negligible binocular overlap. Eye position changes gradually over a period of months. This developmental change raised some questions. First, are the maps in register throughout the period when the eyes are changing position? Recordings from normal developing frogs show that the tectum receives matching maps that increase in extent concomitantly with changes in eye position (Grant and Keating, 1986). The ipsilateral maps recorded at stage 60 occupy only a sliver of tectum

Formation and Regeneration of Nerve Connections
Sansar C. Sharma and James W. Fawcett, Editors
© 1993 Birkhäuser Boston

near the rostral margin, while those recorded at the end of metamorphosis cover the rostral 40% and those recorded in adults occupy about 80% of the tectum. Given that there is a dynamically changing congruence of the maps, what factors keep the ipsilateral map in register with the contralateral map as the eyes change position? Several lines of evidence indicate that multiple factors are involved. Chemoaffinity cues and temporal factors play a role, but visual input is the overriding factor.

THE ROLE OF VISUAL ACTIVITY

Comparisons of developing frogs reared normally or visually deprived indicate the possible roles of these factors. The ipsilateral eye's map can first be detected electrophysiologically at about stage 60, when the eyes begin to shift dorsofrontally in the head. Visual deprivation does not interfere with the development of a normal contralateral map, nor does it prevent the ipsilateral map from beginning to form an orderly, matching map. These observations suggest that chemoaffinity cues and/or temporal factors (see below) establish the earliest ordering of the ipsilateral projection.

Later, the role of visual input becomes manifest. Under normal circumstances, the ipsilateral map expands caudally, always maintaining registration with the contralateral map, as the eyes shift to their final locations. This orderly progression fails to occur if the frogs are reared in the dark (Feldman et al., 1971). If the animals are maintained in the dark, their ipsilateral maps show increasing disorder, particularly along the rostrocaudal tectal axis. Still later, ipsilateral units become impossible to isolate from background noise, indicating either atrophy or dispersal of terminal branches. Thus, dark-rearing studies offer strong support for a crucial role of visual input during development.

Visual input is not merely permissive; it is actually instructive. The pattern of visual input to the two eyes determines where ipsilateral units will terminate in the tectum, and abnormal input can induce correspondingly altered connections. This phenomenon is demonstrated by rearing frogs with a rotated eye (Gaze et al., 1970). This maneuver allows a normal amount of visual input but disrupts the relative orientations of the inputs to the two eyes (see Figure 5 1A-C). Maps in *Xenopus* reared with a rotated eye show compensation for the rotation, such that each eye's ipsilateral map is in register with the map from the other eye. Consider an animal reared with a 180° clockwise rotation of the right eye. The left tectal lobe receives a 180° rotated contralateral map as a simple result of the mechanical rotation of the right eye. The left lobe also receives a map from the normally oriented left eye, but that map is not normally oriented; instead it is rotated 180° to match the map from the rotated eye. The right tectal lobe receives a normal contralateral map from the normally oriented left eye, while the ipsilateral map from the rotated eye reorganizes to match this normal orientation. These results indicate that abnormal visual input overrides the other cues that help to produce normal ipsilateral maps. The mechanism underlying this behavior

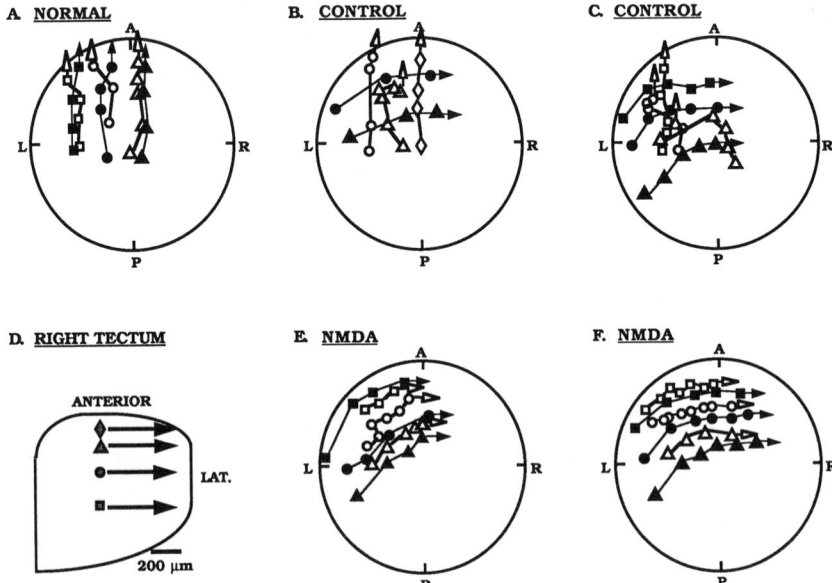

FIGURE 5.1. Visual field maps recorded at 11–12 months after metamorphosis. **A**: Normal. **B** and **C**: Left eye rotated at 8 months post-metamorphosis. **D**: Schematic dorsal view of tectum showing locations of the electrode penetrations. **E** and **F**: Left eye rotated and NMDA (0.4 mm) applied at 8 months after metamorphosis. Closed symbols, contralateral eye; open symbols, ipsilateral eye; A, anterior; P, posterior; L. left; R, right; LAT., lateral. Reprinted with permission of the AAAS from Udin SB, Scherer WJ (1990): Restoration of the plasticity of binocular maps by NMDA after the critical period in *Xenopus*. *Science* 249:669–672. Copyright 1990 by the AAAS.

probably involves correlation of activity patterns, such that ipsilateral units make and break connections until they encounter loci with retinotectal axons with matching receptive fields and activity patterns. Synapses made at locations where the activity patterns do not match would destabilize the ipsilateral units, while synapses formed at locations where the firing patterns are correlated would be stabilized.

THE NUCLEUS ISTHMI AS A LINK FOR BINOCULAR VISION

These fascinating results all are based upon electrophysiological studies. Anatomical studies began much later because the route by which ipsilateral input reached the tectum was known only in part. Gaze and Keating had shown that the pathway from one eye to its ipsilateral tectum involves a relay through the opposite tectal lobe and a recrossing at the base of the diencephalon (Keating and Gaze, 1970). These clues implied that there should be a projection from the tectum to a relay nucleus and thence to the other tectum. The diencephalon seemed like the logical place for such a relay nucleus, but no such nucleus was found.

The "missing link" was not found until the late 1970s, when the horseradish peroxidase (HRP) and proline autoradiography tracing techniques appeared on the scene (Gruberg and Udin, 1978). The link is the nucleus isthmi, a midbrain structure located beneath the caudal end of the tectum, considerably more caudal than anyone had suspected for a relay nucleus. The pathway for ipsilateral visual input to the tectum is illustrated in Figure 5.2. Further evidence that the nucleus isthmi is the source of the ipsilateral map comes from the topography of its connections with the two tecta: by comparing the topography of the electrophysiologically recorded retinotectal map with the connectivity revealed by localized injections of HRP, we determined that the tectoisthmotectal projection links positions with corresponding receptive fields in the two tecta. In addition, the isthmotectal projection terminates in two bands straddling the upper and lower limits of the retinotectal laminae, and these two bands coincide with the layers in which ipsilateral visual input can be recorded.

DEVELOPMENT OF THE NUCLEUS ISTHMI AND ITS CONNECTIONS WITH THE TECTUM

The discovery of the isthmotectal link allowed us to begin to investigate the anatomical patterns underlying the electrophysiological results described above. One set of questions relates to the initial development of the isthmotectal

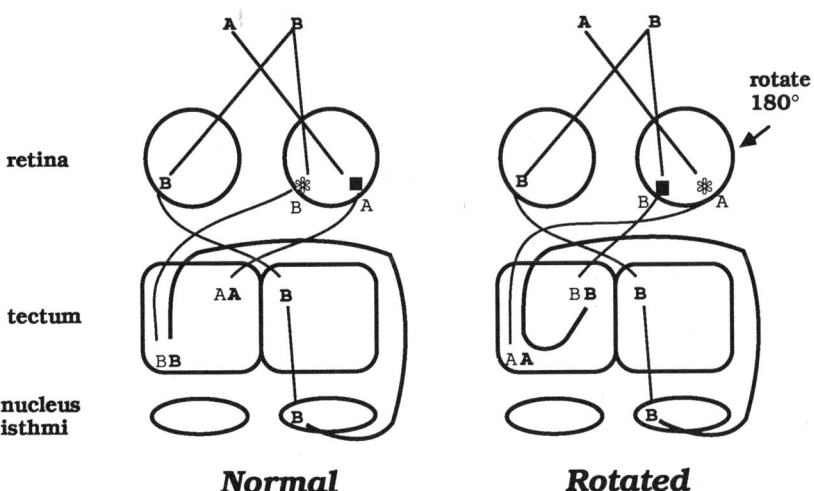

FIGURE 5.2. Pathway for binocular input to the left tectal lobe. Left: normal projection for an axon relaying input about visual field location. Right: altered projection from isthmotectal axon relaying visual field **B** after rotation of the right eye.

projection. Is the gradual expansion of the ipsilateral map during normal development a reflection of the gradual genesis of the nucleus isthmi? We addressed this question by using tritiated thymidine to determine cell birthdays. We and others found that the nucleus isthmi has a very protracted development, beginning at about stage 29 and continuing to about stage 62, with the greatest number of cells born in the mid-50s stages (Udin and Fisher, 1985; Dann and Beazley, 1982; Tay and Straznicky, 1980). Thus, the great majority of isthmotectal cells are born before the ipsilateral map begins to expand, and the number of cells in the nucleus isthmi is not the factor that limits the extent of the ipsilateral map.

We also investigated the time at which the axons of those cells reach the tectum. Injection of HRP into the tectum disclosed that isthmic cells project to both tecta as early as stage 52. Moreover, anterograde filling of isthmotectal axons by injection of HRP into the nucleus isthmi reveals those axons are not restricted to the small rostral region of the tectum where ipsilateral units are found electrophysiologically. Instead, many isthmotectal axons extend well caudal to the "binocular" zone. They do not show up in electrophysiological maps because they are too sparsely branched to generate sufficient localized current to be detected over the background noise inherent in the extracellular recording process. As development proceeds, more and more of these axons form terminal arbors and take on the appearance characteristic of normal adult isthmotectal arbors. This pattern is consistent with the hypothesis that isthmotectal axons make and break connections during development, with branches tending to be retained in locations where the retinotectal activity matches the activity of the isthmotectal axon.

Nonvisual cues also have a role in promoting this pattern of growth. Even before binocular cues become available, the isthmotectal projection has a crude mediolateral organization. This aspect of order is achieved when the axons distribute themselves along the rostral border of the tectum (Udin, 1990). From that point each axon grows caudally into the tectum, with relatively little deviation medially or laterally. Under normal rearing conditions, this pattern sets each axon up in the correct "corridor" within which it normally will terminate in the adult. This mediolateral order may be the result of chemoaffinity cues and/or temporal effects. Isthmotectal axons may recognize the same chemoaffinity cues in tectal tissue that are postulated to help establish the retinotectal map. Another possibility is that the mediolateral order arises from the temporospatial gradients of tectal and isthmic development. The firstborn isthmotectal axons enter the tectum near its firstborn part, the rostrolateral zone, and later-born isthmotectal axons enter the tectum at progressively younger (more medial) locations along the rostral tectal margin (Udin and Fisher, 1985; Straznicky and Gaze, 1972). This relationship promotes the development of normal topography. Thus, either temporal or chemical factors could establish mediolateral order and could contribute to the order observed in maps of dark-reared frogs, in which the ipsilateral maps are relatively orderly along the mediolateral axis, reflecting a maintenance of the order with which they entered the tectum.

ISTHMOTECTAL CONNECTIONS AFTER EYE ROTATION

We also have examined whether the topography of the isthmotectal projection is different from normal in frogs that are reared with one eye rotated and that therefore have rotated ipsilateral maps. Our evidence indicates that the axons first grow toward their normal terminal locations and then become rerouted to new sites. This pattern was first inferred from the results of studies in which we used small tectal injections of HRP to reveal the the locations of the isthmic cell bodies projecting to a given tectal location in response to different degrees of eye rotation. The results confirmed that eye rotation does change the topographic relationship between the nucleus isthmi and the tectum. The pattern of labeling showed that the tectoisthmic projection was normal but that the isthmotectal projection was changed in a manner consistent with the new topography so that cells which normally would be labeled after a tectal injection in a particular location would now be labeled after a tectal injection in a different location (Udin and Keating, 1981). However, an unexpected result was that the "new" labeling pattern did not replace the normal pattern but instead was superimposed upon it, so that a single tectal injection in a frog with an eye rotation labeled two groups of isthmic cells rather than just one: one group corresponded to the normal projection while the other group corresponded to the reorganized projection. We infer that these two patterns indicate that isthmotectal axons first grow to approximately normal locations and later continue their growth to the new locations. The dual labeling pattern results from uptake of HRP by axons traversing their normal terminal sites en route to new locations and from uptake of HRP by terminals of other cells which have replaced the original population. This interpretation is consistent with subsequent electrophysiological studies of maps in the first few weeks after eye rotation; the maps indicate that an essentially normal pattern of connections is formed before the ipsilateral map reorients to match the input from the contralateral eye.

A second anatomical study employed anterograde filling of isthmotectal axons in normal and eye-rotated frogs and demonstrated that normal isthmotectal axons travel in essentially parallel rostrocaudal trajectories from their point of entry into the tectum to their termination zones, whereas axons in eye-rotated frogs show significant crisscrossing and even looping as they travel toward their new terminal zones (Udin, 1983). Together these results strengthened the conclusion that the ability of the ipsilateral map to rearrange to match virtually any retinotectal input is a consequence of a physical rewiring of the isthmotectal projection.

ISTHMOTECTAL SYNAPSES

As mentioned above, the underlying hypothesis applied to explain these results is that ipsilateral and contralateral activity stabilizes ipsilateral axon branches where the activity patterns are correlated and destabilizes the branches in other locations.

If ipsilateral axons are to be influenced by contralateral axons, what is the physical means by which the influence is transmitted? To determine whether isthmotectal axons are influenced directly by retinotectal axons via axoaxonal connections or indirectly by connections onto tectal cell dendrites, we used electron microscopy to investigate the connections of isthmotectal axons labeled with HRP (Udin et al., 1990). We found that isthmotectal axons make conventional asymmetric contacts onto tectal cell dendrites and that there are no morphologically identifiable axoaxonal connections between retinotectal and isthmotectal axons. Therefore, the interaction is probably by means of convergent connections of retinotectal and isthmotectal axons onto dendrites.

THE ROLE OF NMDA RECEPTORS DURING DEVELOPMENT

How are firing patterns translated into signals that stabilize or destabilize isthmotectal axons? The NMDA receptor may be a key element (for a review of this subject, see Debski et al., 1990). The NMDA receptor is one type of glutamate receptor, and it has the unusual characteristic of being a calcium channel that is gated by a combination of ligand binding and voltage change. (see Figure 5.3.) NMDA receptors are opened as a consequence of retinotectal activity. When retinotectal axons fire, they release glutamate (Van Deusen and Meyer, 1990; Debski and Constantine-Paton, 1990), which depolarizes the postsynaptic dendrite by activating non-NMDA-type glutamate receptors. NMDA receptors probably do not contribute to the initial depolarization because they are quickly blocked by magnesium ions. When the dendrite is sufficiently depolarized, however, the magnesium ions are expelled from the channel and the NMDA receptor channels admit sodium and calcium. The calcium is hypothesized to be especially important because it triggers a wide range of responses. One such response may be the release of a substance such as arachidonic acid (Williams et al., 1989) or nitric oxide (East and Garthwaite, 1991), which could affect nearby terminals such as those from the nucleus isthmi. If an isthmotectal axon were active at the proper time, it also would have a relatively high internal free calcium concentration, which could interact synergistically with the postulated substance from the dendrite to stabilize that terminal.

This model can be tested by blocking the action of the NMDA receptor during development. We investigated whether such blockade would prevent reorientation of the ipsilateral map in animals reared with a rotated eye (Scherer and Udin, 1989). To block NMDA receptors we applied thin slabs of slow-release polymer impregnated with the blocker DL-2-amino-5-phosphonovaleric acid (APV) to the surface of the tectum for the first 3 months after metamorphosis. This treatment did prevent the ipsilateral map from rotating to match the contralateral map. The APV did not, however, prevent the development of any ipsilateral projections. Instead, an ipsilateral map developed with a roughly normal orientation and with a degree of disorder comparable to that seen in dark-reared animals.

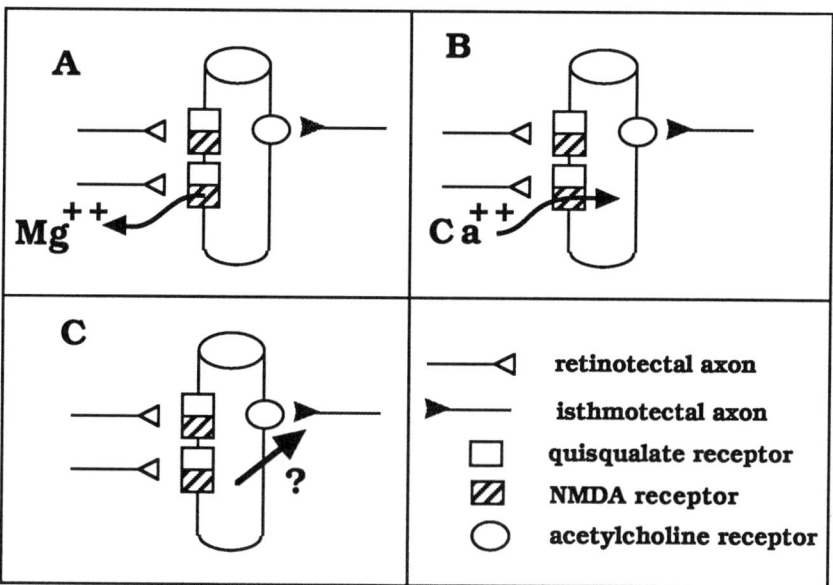

FIGURE 5.3. Mechanisms possibly underlying activity-dependent stabilization of isthmotectal synapses. Retinotectal axons produce depolarized tectal cell dendrites via non-NMDA [quisqualate or α-amino-3-hydroxy-5-methyl-isoxadole-4-propionic acid (AMPA)] receptors. The depolarization expels Mg^{2+} ions from NMDA receptors, which then respond to binding of glutamate by allowing influx of cations such as Na^+ and Ca^{2+}. Calcium may induce release of a substance that stabilizes a recently activated isthmotectal terminal. Reprinted with permission of the AAAS from Udin SB, Scherer WJ (1990): Restoration of the plasticity of binocular maps by NMDA after the critical period in *Xenopus*. *Science* 249: 669–672. Copyright 1990 by the AAAS.

To determine whether APV blocked plasticity by specifically blocking the NMDA receptor rather than by reducing overall tectal activity, we assessed the degree to which APV altered tectal output in response to flashes of light (Scherer and Udin, 1991a). The right lobe was exposed to APV for 3 to 4 weeks. The activity in the treated lobe was determined indirectly, by taking advantage of the relay from the right lobe via the right nucleus isthmi to the left lobe. That pathway, of course, is the route by which ipsilateral eye input reaches the left lobe, so we could monitor changes in the right lobe by recording ipsilateral eye activity in the left lobe. Activity levels were assessed by recording the total number of action potentials per flash from isthmotectal terminals. There was no significant difference between normal and APV-treated frogs. This result supports the interpretation that APV interferes with reorganization of the ipsilateral map by selectively blocking the NMDA receptor.

RESTORATION OF PLASTICITY BY NMDA TREATMENT

We extended these results by investigating the possible role of NMDA receptors in the changes associated with the critical period of isthmotectal plasticity (Udin and Scherer, 1990). The critical period in this system is defined as the period during which a change in eye position elicits an orderly rearrangement of the ipsilateral map. Plasticity diminishes gradually, and the end of the critical period is defined as the time when the ipsilateral map fails to reorganize in response to eye rotations of 30°, the smallest change that is within the resolution of mapping techniques. Depending upon rearing conditions, the end of the critical period is about 3 months postmetamorphosis.

Is the loss of plasticity related to diminished function of the NMDA receptor system? We reasoned that NMDA receptors might diminish in number, kinetics, or some other characteristic and that chronic treatment with NMDA might boost the activity of the system enough to restore plasticity. To test the effects of NMDA, we tested for plasticity in frogs 8 months postmetamorphosis. After rotating the left eye by 90° clockwise, we implanted Elvax impregnated with 0.1 or 0.4 mM NMDA on the right tectum. Three months later we found that ipsilateral maps in the NMDA-treated tectum had rotated to come into register with the rotated contralateral maps. Control frogs were treated with drug-free implants, and their ipsilateral maps showed no reorientation (see Figure 5.1). Thus, chronic treatment with NMDA restored plasticity to a level indistinguishable from that seen after comparable rotations during the critical period.

How does NMDA restore plasticity? Our working hypothesis is that NMDA prolongs the opening of NMDA channels when appropriate visual input depolarizes tectal cell dendrites. Calcium influx would thereby be boosted to reach a critical threshold level required for plasticity. For such a mechanism to promote orderly reorganization of the ipsilateral map, the NMDA concentration would have to be neither too low nor too high. Too low a concentration would fail to trigger stabilization processes, and too high a concentration might stabilize inappropriate connections by activating dendrites even when the visual input was weak. Our mapping experiments suggest that we managed to find the correct concentrations of NMDA for our purposes. This interpretation is further supported by our studies of the effects of this chronic treatment on tectal output, using the same approach described above for studies of APV. We found that NMDA does boost tectal responsiveness, as indicated by a 35% increase in tectal output to the isthmotectal relay in response to flashes of light (Scherer and Udin, 1991b). Also, the NMDA did not increase spontaneous firing and therefore was not likely to reinforce inappropriate connections.

TEMPORAL FEATURES AND BINOCULAR ACTIVITY

Another issue raised by the studies of NMDA receptors in this system is whether their kinetics are appropriate for the timing of the events in the tectum. In particular, we are confronted by the potential problem of the synaptic delay between the onset of contralateral and ipsilateral activity. Ipsilateral input is delayed relative to contralateral input by transmission time of at least two extra synapses plus axonal conduction time, so we cannot build a model that requires simultaneous onset of activity in the two populations of afferents. We have measured the delay of onset of firing in response to flashes of light, and we find that ipsilateral activity lags behind contralateral activity by about 10 msec (Scherer and Udin, 1991b). However, this delay does not mean that there is no period of simultaneous firing of both sets of axons. Flashes of light elicit bursts of spikes, and there typically is a period of 40–60 msec during which both ipsilateral and contralateral units are firing. These parameters fit comfortably with what we know about NMDA receptor kinetics. NMDA receptors generally open after a delay of a millisecond or more and can stay open for tens of milliseconds (Dale and Roberts, 1985; but see also D'Angelo et al., 1990). Therefore, we envision a sequence in which a visual stimulus activates retinotectal input, leading to an influx of calcium via NMDA receptors a few milliseconds later. A few milliseconds after that, some retrograde messenger is released and reaches the isthmotectal terminal soon after it begins to fire. The events that occur during these critical milliseconds are still unknown and remain a challenge for the future.

SUMMARY

Binocular maps in the tectum of *Xenopus* frogs display remarkable plasticity during a critical period of early life. The formation of orderly matching maps requires binocular visual input, and abnormal relationships between the orientations of the two eyes lead to systematic reorganization of the ipsilateral map. These developmental changes reflect the changing connectivity of the isthmotectal projection. Correlations of firing patterns of contralateral and ipsilateral units are hypothesized to be the means by which visual input selectively stabilizes appropriate isthmotectal terminals while destabilizing others. The NMDA receptor, located on tectal cell dendrites, plays a pivotal role in this correlation. It is activated by contralateral input and triggers processes that affect ipsilateral axons. Evidence supporting this view comes from experiments which show that blocking the NMDA receptor during development prevents the ipsilateral map from coming into register with the contralateral map. In addition, the normal loss of plasticity in late juvenile frogs is reversed by infusion of NMDA, which may reactivate processes normally functional only during the critical period.

Acknowledgments

I thank Memet Cirpili and Dr. Warren Scherer for their comments on this manuscript. This work was supported by US PHS Grant EY-03470 and Basic Research Grant 1-1192 from the March of Dimes Birth Defects Foundation.

REFERENCES

Dale N, Roberts A (1985): Dual-component amino-acid-mediated synaptic potentials: Excitatory drive for swimming in *Xenopus* embryos. *J Physiol (Lond)* 363: 35–59

D'Angelo E, Rossi P, Garthwaite J (1990): Dual-component NMDA receptor currents at a single central synapse. *Nature* 346: 467–470

Dann JF, Beazley LD (1982): The development of connections between the isthmic nucleus and the tectum in *Xenopus* and *Limnodynastes* tadpoles. *Neurosci Lett* 33: 107–113

Debski EA, Cline HT, Constantine-Paton M (1990): Activity-dependent tuning and the NMDA receptor. *J Neurobiol* 21: 18–32

Debski EA, Constantine-Paton M (1990): Evoked pre- and post-synaptic activity in the optic tectum of the cannulated tadpole. *J Comp Physiol A* 167: 377–390

East SJ, Garthwaite J (1991): NMDA receptor activation in rat hippocampus induces cyclic GMP formation through the L-arginine-nitric oxide pathway. *Neurosci Lett* 123: 17–19

Feldman JD, Gaze RM, Keating MJ (1971): The effect on intertectal neuronal connections of rearing *Xenopus* in total darkness *J Physiol (Lond)* 212: 44–45P

Gaze RM, Jacobson M (1962): The projection of the binocular visual field on the optic tecta of the frog. *Q J Exp Physiol* 47: 273–280

Gaze RM, Keating MJ, Székely G, Beazley L (1970): Binocular interaction in the formation of specific intertectal neuronal connexions. *Proc Roy Soc Lond B* 175: 107–147

Grant S, Keating MJ (1986): Normal maturation involves systematic changes in binocular visual connections in *Xenopus laevis*. *Nature* 322: 258–261

Gruberg ER, Udin SB (1978): Topographic projections between the nucleus isthmi and the tectum of the frog *Rana pipiens*. *J Comp Neurol* 179: 487–500

Keating MJ, Gaze RM (1970): The ipsilateral retinotectal pathway in the frog. *Q J Exp Physiol* 55: 284–292

Scherer WJ, Udin SB (1989): N-Methyl-D-aspartate antagonists prevent interaction of binocular maps in *Xenopus* tectum. *J Neurosci* 9: 3837–3843

Scherer WJ, Udin SB (1991a): Chronic effects of NMDA and APV on tectal output in *Xenopus laevis*. *Vis Neurosci* 6: 185–192

Scherer WJ, Udin SB (1991b): Latency and temporal overlap of visually-elicited contralateral and ipsilateral firing in *Xenopus* tectum during and after the critical period. *Devel Brain Res* 58: 129–132

Straznicky C, Gaze RM (1972): The development of the tectum in *Xenopus laevis*: An autoradiographic study. *J Embryol Exp Morphol* 28: 87–115

Tay D, Straznicky C (1980): The development of the nucleus isthmi in *Xenopus*: An autoradiographic study. *Neurosci Lett* 16:313–318

Udin SB (1983): Abnormal visual input leads to development of abnormal axon trajectories in frogs. *Nature* 301: 336–338

Udin SB (1990): Development of orderly connections in the retinotectal system. In: *Science of Vision*, Leibovic KN, ed. New York: Springer-Verlag

Udin SB, Fisher MD (1985): The development of the nucleus isthmi in *Xenopus laevis*: I. Cell genesis and formation of connections with the tecta. *J Comp Neurol* 232: 25–35

Udin SB, Keating MJ (1981): Plasticity in a central nervous pathway in *Xenopus*: Anatomical changes in the isthmotectal projection after larval eye rotation. *J Comp Neurol* 203: 575–594

Udin SB, Scherer WJ (1990): Restoration of the plasticity of binocular maps by NMDA after the critical period in *Xenopus*. *Science* 249: 669–672

Udin SB, Fisher MD, Norden JJ (1990): Ultrastructure of the crossed isthmotectal projection in *Xenopus* frogs. *J Comp Neurol* 292: 246–254

Van Deusen EB, Meyer RL (1990): Pharmacologic evidence for NMDA, APB and kainate/quisqualate retinotectal transmission in the isolated whole tectum of goldfish. *Brain Res* 536: 86–96

Williams JH, Errington ML, Lynch M, Bliss TVP (1989): Arachidonic acid induces a long-term activity-dependent enhancement of synaptic transmission in the hippocampus. *Nature* 341: 739–742

CHAPTER 6

Plasticity of Binocular Visual Connections in the Frog: From R.M. Gaze to NMDA

SIMON GRANT, STEPHEN G. BRICKLEY, AND MICHAEL J. KEATING

The adult brain contains numerous topographic representations of the receptor epithelia, underpinned by highly ordered sets of neuronal connections that link each receptor surface to various target structures within the central nervous system (CNS). The development of receptotopic maps has been extensively studied in respect of the connections from the retina to the midbrain optic tectum in both lower vertebrates and mammals. In general accord with the seminal ideas of Sperry (1963) and their subsequent elaboration by Gaze and colleagues, the formation of the retinotectal map seems to be governed entirely by intrinsic developmental processes. These processes, involving mainly cell surface interactions, guide outgrowing retinal ganglion cell axons in an orderly fashion down selected pathways to the tectum, and cause them to distribute their terminations so as to more-or-less faithfully replicate the topology of their parent cell bodies. Temporally correlated patterns of neural activity between neighboring retinal ganglion cells and between adjacent tectal neurons may impart additional topological information to these connections, serving to refine the spatial order of the initial map (Cline and Constantine-Paton, 1989; Chapter 14, this volume). Activity generated by the neuronal elements themselves (i.e., intrinsically), however, appears sufficient for the map refinement, since the ordering of retinotectal connections is unperturbed by the absence of extrinsic activity arising through visual stimulation (Keating et al., 1986; Olson and Meyer, 1991).

By contrast, the formation of integrative topographic maps from two related receptor surfaces seems to utilize a different class of activity-dependent process that requires sensory experience for its completion. In the mammalian visual cortex, for example, the development of connections responsible for integrating inputs from the two eyes is well known to be influenced by early experience. This raises the question as to what special features of developing binocular maps require experience-dependent activity. In the context of binocular cortical connections, one speculation (Pettigrew, 1974) is that the early experience-dependent phase satisfies a vital maturational requirement that cannot be accommodated by intrinsic processes alone— specifically, to match the substrates underlying binocular vision to pronounced and unpredictable changes in relative eye positions that occur during growth of the head. In this chapter we review the evidence that

Formation and Regeneration of Nerve Connections
Sansar C. Sharma and James W. Fawcett, Editors
© 1993 Birkhäuser Boston

similar experience-dependent processes are generalized to a system of binocular connections in the frog, *Xenopus laevis*, and that these processes have been secured to counter this very developmental problem.

BINOCULAR VISION AND THE "INTERTECTAL" SYSTEM IN THE FROG

The evolution of frontally directed eyes with overlapping fields of view has occurred several times during vertebrate phylogeny, suggesting that the arrangement confers important advantages. The best understood of these is the computation of visual depth from stereoscopic cures generated by the binocular parallax. For this advantage to be realized, it seems essential that visual centers within the brain receive binocular input from points of retinal correspondence. This arrangement contributes to establishing which components in the two retinal images correspond in object space and, ultimately, to the "singleness" of binocular vision.

In frogs such as *Xenopus*, the paired lobes of the optic tectum constitute the major visual center. Using the current electrophysiological mapping techniques, which involved presenting localized visual stimuli to each eye and recording the impulses generated with microelectrodes inserted into these optic lobes, Gaze (1958b) reported that these structures in adult frogs contain this necessary substrate for binocular vision. He had previously shown that the full extent of each eye in the frog maps, in a point-to-point fashion, onto its contralateral tectum, reflecting a high degree of topographic precision in the direct retinotectal projection (Gaze, 1958a). His new finding indicated that "each eye projects [topographically] to both optic lobes; and any point in space within the overlapping parts of the visual field is represented, binocularly, at one point on one optic lobe and at another point on the other" (Gaze, 1958b).

At the time, the source of the spatially aligned input from each eye to its ipsilateral tectum was unknown. It is now apparent that the arrangement described by Gaze obtains as follows: visual input from any given point in the binocular field is delivered by an eye first via the direct projection to its opposite tectum and then back across the midline by a commissural system of "intertectal" connections, which relays the information to the point on the ipsilateral tectum itself receiving direct visual input from the corresponding point in space (for further details, see Chapter 5, this volume).

COMPENSATION FOR NORMAL DEVELOPMENTAL CHANGES IN RETINAL CORRESPONDENCE

The precise pattern of intertectal connections necessary to achieve the spatial alignment of binocular maps on each tectum depends upon the precise pattern of retinal correspondence in the two eyes. This, in turn, is influenced by their relative positions. Figure 6.1 indicates that the *Xenopus* intertectal system is thus

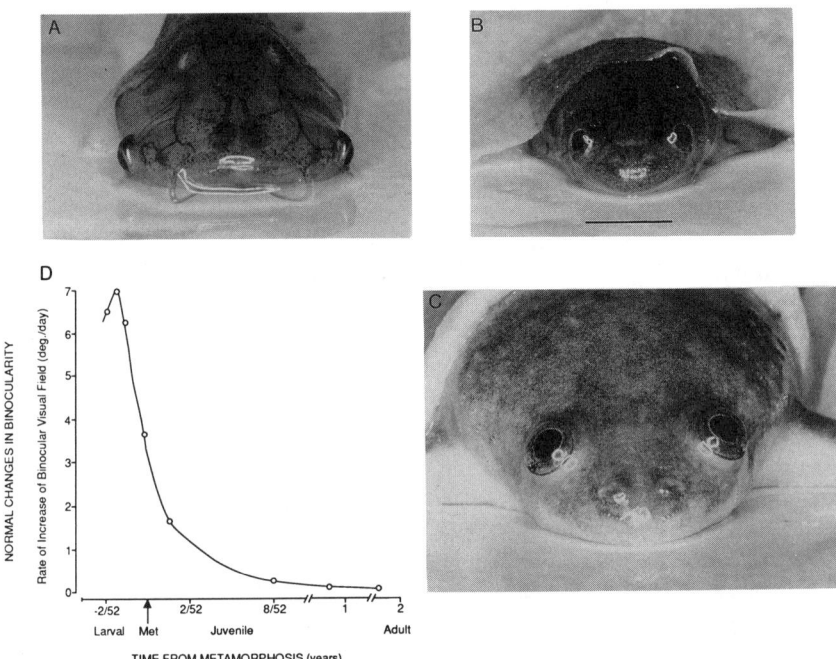

FIGURE 6.1. Growth-related changes in eye position in developing *Xenopus laevis*. The photographs show frontal views of the head and eyes at three developmental stages: (A) stage 58 of late larval life; (B) about 2 weeks later at metamorphic climax (stage 66); (C) 1 year after metamorphic climax. Scale bar in (B) = 5 mm, and also applies to the other photographs. During this developmental period, the eyes migrate from lateral to frontodorsal positions in the head. As a consequence the binocular visual field enlarges. (D) shows the rate of this enlargement, expressed in degrees/day, and plotted as a function of age. Reprinted with permission of Oxford University Press from Keating MJ, Grant S (1992): The critical period for experience-dependent plasticity in a system of binocular visual connections in *Xenopus laevis*: Its temporal profile and relation to normal developmental requirements. *Eur J Neurosci 4*, 27–36.

confronted with a considerable developmental problem. At larval stages (Figure 6.1A) the eyes face laterally, affording minimal binocular overlap, but beginning about 2 weeks prior to metamorphic climax, growth-related alterations in skull shape and size cause them to migrate frontally and dorsally in the head (Figure 6.1B). The ocular migration continues after metamorphic climax (Figure 6.1C) and persists into adulthood some 2 years later. These changes produce a progressive expansion of the region of binocular overlap throughout the developmental period spanning late larval to adult life (Figure 6.1D).

We have used electrophysiological mapping techniques to examine how the *Xenopus* intertectal system responds to this developmental problem (Grant and Keating, 1986, 1989a). Our major finding was that at all developmental ages, from the onset of metamorphic climax (stage 60) until adulthood, binocular maps on the tectum are in appropriate spatial alignment. That is, the alignment is maintained despite a progressive expansion in the tectal representation of the binocular field and a radical alteration in the tectal projections of corresponding points in the two retinae. Analysis of developmental changes in the form of the ipsilateral visuotectal map indicated that the plasticity was effected by a continuous expansion and adjustment in the detailed order of connections in the intertectal system that exactly compensated for the shifts in retinal correspondence.

We have also examined the contribution of visual experience to this normal developmental plasticity (Grant and Keating, 1989b). Animals were deprived of vision, by rearing them in total darkness, to different metamorphic and postmetamorphic ages, and then subjected to a mapping experiment. Our major finding was that the ipsilateral visuotectal maps in these animals always showed significant signs of disorder, and that the degree of disorder increased in severity with the period of visual deprivation. As a result, the binocular maps on the tectum never achieved appropriate spatial alignment. Analysis of the changing form of the ipsilateral visuotectal map indicated that intrinsic developmental processes generate an intertectal system that is initially quite well organized and that they program much of its normal expansion with age. But visual experience is required by the system if it is to match the precise pattern of its connections to the shifts in retinal correspondence that result from changing eye position.

COMPENSATION FOR EXPERIMENTALLY INDUCED CHANGES IN RETINAL CORRESPONDENCE

In fact, historically, the first indication that the development of the *Xenopus* intertectal system might be governed by processes other than intrinsic ones arose from the observations of Gaze et al. (1965) on the form of the ipsilateral visuotectal map in animals reared from embryonic life with one "compound" eye. They found that the projection from this eye to its ipsilateral tectal lobe was normally organized rather than compound—as would have occurred if the system were a fixed set of connections faithfully conveying information from one tectal lobe to the other—and, even more strikingly, that the ipsilateral visuotectal projection from the normal eye was "compound." As a result, the binocular maps on the two tectal lobes were in spatial alignment, despite the quite anomalous set of retinal correspondences engendered by the embryonic surgery. Keating (1968) later pointed out that this arrangement implied the formation of novel intertectal linkages between precisely those points on the two tectal lobes receiving correlated patterns of neural activity from the two retinae, and he suggested that the arrangement might thus be a product of binocular visual experience. This suggestion was later substantiated by Keating, Gaze, and others, who demonstrated that the intertectal system was also able to compensate for other experimental alterations in retinal correspondence, effected through larval eye rotation or translocation (Gaze et al., 1970; Beazley, 1975), but that the compensation was blocked in the absence of vision (Keating and Feldman, 1975).

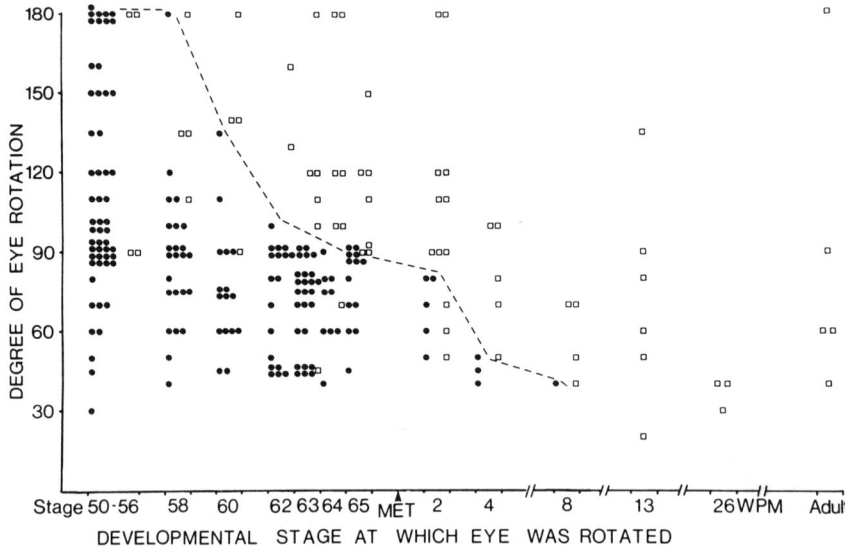

FIGURE 6.2. Data from 238 experimental animals showing the age-dependent response of the intertectal system to eye rotations of different sizes. The abscissa shows the age at eye rotation: MET = metamorphic climax (stage 66); WPM = weeks postmetamorphosis. The ordinate shows the size of the eye rotation present at terminal mapping. Each symbol depicts the result obtained in one animal; filled circles = animals showing spatially aligned binocular maps, indicative of intertectal plasticity; open squares = animals showing misaligned binocular maps, indicating no change in the intertectal system. The dotted line denotes the transition point between the two classes of result obtained from groups of age-matched animals. Reprinted with permission of Oxford University Press from Keating MJ, Grant S (1992): The critical period for experience-dependent plasticity in a system of binocular visual connections in *Xenopus laevis*: Its temporal profile and relation to normal developmental requirements. *Eur J Neurosci 4*, 27–36.

The Critical Period

A requirement for experience in the normal maturation of binocular connections in the mammalian visual cortex has been inferred largely through the observation that abnormal vision during early "critical" phases of development may result in long-lasting defects in their functional organization. We have proposed that the ability of the *Xenopus* intertectal system to compensate for experimentally induced alterations in eye position reflects an exaggeration of capacities originally secured to compensate for the ocular migration that normally occurs with growth. If so, then the temporal profile of any such critical period for intertectal plasticity should resemble the profile of these changes (Figure 6.1D).

To test this prediction we have challenged the system at different developmental stages, from midlarval to adult life, to adjust to surgically induced eye rotations of varying degree (Keating and Grant, 1992). Obviously, the larger the degree of eye rotation, the greater the plasticity required to achieve binocular alignment of the tectal maps. Figure 6.2 summarizes the results obtained; the abscissa shows 13 developmental stages at which eye rotations were performed and the ordinate shows the degree of rotation produced. In the vast majority of animals receiving a rotation up to stage 58 of larval life (approximately 2 weeks before metamorphic climax), the intertectal system compensated for the rotation, even if it was of maximal size (i.e., 180°). By contrast, none of the animals operated upon at 3 months postmetamorphosis or later showed any plasticity, even if the degree of rotation was comparatively small. At intervening ages the outcome was more mixed, though with a very strong tendency for those animals of a given age showing intertectal plasticity to be the ones with the smaller degrees of eye rotation.

The maximal sensitivity of the system existing at a particular age can be described by determining the largest degree of eye rotation that can be compensated for at that age (Figure 6.2; broken line). Sensitivity to eye rotation clearly begins to decline between stages 58 and 60, at about the time that the intertectal connections first become functional, and thereafter it continues to fall with age. The rate of this decline is characterized by a precipitous drop in the 2-week period before metamorphic climax and a more gradual fall in the first months of juvenile life, similar to the profile of the normal developmental enlargement of the binocular field (Figure 6.1D). It seems, therefore, that the ability of the *Xenopus* intertectal system to compensate for the anomalous retinal correspondences produced by eye rotation is not arbitrarily set, but is related to its normal developmental requirements. Some support for this conclusion can be inferred from studies of this same system in other species of frog in which normal growth-related changes in eye position are relatively minor. In both *Rana pipiens* (Grobstein and Comer, 1977) and *Hyla moorei* (Beazley, 1979), there is a little or no change in eye position between metamorphosis and adulthood, and the intertectal system of these anuran species displays no ability to compensate for eye rotation (Jacobson, 1971; Beazley, 1979; Kennard and Keating, 1985).

Although under normal circumstances there is clearly a developmental decline in the plasticity of the *Xenopus* intertectal system, we have found that the decline is not rigidly dependent on the animal's age, but is influenced by vision itself (Grant et al., 1992). In animals dark-reared for the duration of the normal "critical" period, between stage 58 of larval life until 3 months after metamorphosis (cf. Figure 6.2), or even through this period until 1 or 2 years postmetamorphosis, rotations instituted at these ages are followed by intertectal reorganization if exposure to a normal visual environment is permitted after the operation. Moreover, in animals subjected to prolonged deprivation into late juvenile or adult life, the intertectal system appears to be capable of reestablishing alignment of binocular tectal maps during the later period of normal vision, even when the eye rotation present is a maximal one of 180° or close to it. This suggests that the mechanisms underlying intertectal plasticity in dark-reared *Xenopus* remain suspended in a maximal or near-maximal state, akin to that normally observed only in larval or early metamorphic animals, while they remain visually deprived.

VISION AND THE MECHANISMS OF INTERTECTAL PLASTICITY

It is known that intertectal plasticity following eye rotation in *Xenopus* is effected by terminal arbors of axons in the crossed isthmotectal portion of the pathway leaving their normal sites of tectal termination and selecting novel ones appropriate to the rotationally disparate visual input (Udin and Keating, 1981; Udin, 1983). This process obviously requires the identification, by these arbors, of the termination sites that are newly appropriate. Keating (1968) first suggested that mechanisms involving the detection of coincident neural activity resulting from binocular visual stimulation are used to make this identification. Crossed isthmotectal arbors carry neural activity evoked by stimulation of a particular region of the binocular field of the ipsilateral eye. We currently suppose that the initial impetus for these arbors to disconnect from their normal termination sites arises when they are in contact with tectal cells that are not coincidentally activated from the same region of the binocular field by retinotectal arbors derived from the opposite eye. We also suppose that they subsequently reestablish connections at new sites by seeking the location at which coactivation of tectal cells by the same visual stimulus in the two eyes will occur. As we have discussed elsewhere (Grant and Keating, 1989a, 1989b, 1992), the "coincidence detection" hypothesis adequately accounts for the normal developmental plasticity of this pathway, for the disruptions that occur in it when vision is denied, and for the novel patterns on connectivity induced by eye rotation.

What factors contribute to the normal developmental decline in intertectal plasticity and to the prevention of this decline by dark-rearing? One important element is the *N*-methyl-D-aspartate (NMDA)-type of glutamate receptor. Scherer and Udin (1989) have shown that application to the tectum of the selective NMDA receptor antagonist, DL-2-amino-5-phosphonovaleric acid (APV), for most of the normal critical period blocks intertectal compensation to larval eye rotation in *Xenopus*. Further, the ipsilateral visuotectal maps in these APV-treated animals were found to exhibit a degree of topographic disorder. In both these respects, therefore, APV treatment produced effects analogous to those observed in dark-reared animals, suggesting that a common mechanism is involved: specifically, that intertectal plasticity following eye rotation requires a vision-dependent activation of NMDA receptors present in the optic tectum, these being obligatory intermediaries of the coincidence detection processes mentioned above. Nonactivation of NMDA receptors, whether by APV-induced blockade or by dark-rearing, would then attenuate the ability of crossed isthmotectal arbors to recognize either their normal tectal termination sites or any new ones appropriate to an eye rotation.

Udin and Scherer (1990) have also been able to restore eye-rotation-induced intertectal plasticity in *Xenopus* that have already passed through the normal critical period, by chronically infusing the tectum with NMDA. Taken together, these results imply that vision-dependent activation of NMDA receptors is followed by a gradual change in their properties which, in turn, may underlie the normal decline in the plasticity of the system. This change could, perhaps, involve

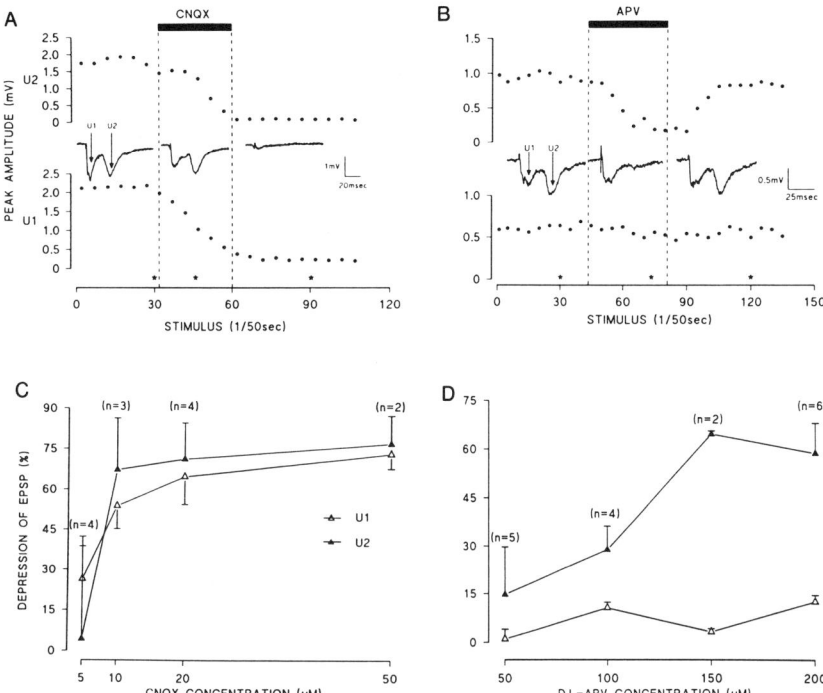

FIGURE 6.3. Effects of CNQX and APV on the U1 and U2 postsynaptic field potentials recorded from the adult tectum *in vitro*. (A,B): Plots of the peak amplitude of the U1 and U2 components of the evoked response are shown from experiments in which these AMPA and NMDA receptor antagonists were applied to the bathing medium during the period indicated by the solid bar. Typical postsynaptic potentials, recorded at the time points marked by an asterisk, are shown. In all traces negativity is downward, but for clarity, the peak amplitudes are plotted as positive values. (C,D): Dose-dependency of effects (mean ± SD) on the U1 and U2 waves. Brains were dissected out into a Perspex bath and superfused with standard frog Ringer's solution at 21–22°C, aspirated with 95%O_2/5%CO_2. Drugs were applied through individual superfusion lines with a common entry point into the bath. Current (100 mA) was passed (square wave pulses; 20–50 μsec duration; repetition rate, 1/50 sec) using bipolar concentric stimulating electrodes inserted into the optic tract. Conventional recordings were made with glass microelectrodes (2% Pontamine Sky Blue in 0.5 M sodium acetate), at a depth within the tectal neuropil (approximately 150 μm) that optimized the U2 component of the response.

a down-regulation in their number and/or sensitivity. Exogenous supplementation of receptor agonist in post-critical-period juvenile animals may then compensate for the loss of NMDA receptor function, causing the system to revert to an earlier phase of higher sensitivity to eye rotation. Dark-rearing may thus prevent the normal decline in intertectal plasticity simply by delaying the advent of the trigger (i.e., vision) that engages the changes in NMDA receptor mechanisms.

Future Directions

Operationally, the studies discussed above have been spearheaded by mapping experiments that have hardly changed in character since Gaze began utilizing this approach over 30 years ago. This approach has, to date, been the optimal one for defining much of the phenomenology of the experience-dependent changes that can occur in the intertectal system during normal development and following experimental perturbation. The comparatively recent demonstration that intertectal plasticity involves, at some level, NMDA-type glutamate receptor mechanisms affords a particular impetus to future studies. In particular, it facilitates formulation of specific mechanistic questions about how the process of "coincidence detection" might operate, within what temporal constraints, and on which neural elements the processes might especially impinge (Chapter 5, this volume).

To these ends, we have begun to investigate the nature of these receptor mechanisms using an *in vitro* preparation of the adult *Xenopus* tectum, in which most of the visual pathways remain intact. Postsynaptic field potentials recorded from the tectal neuropil following optic tract stimulation in these preparations display characteristics (number of potentials, latency, depth- and intensity-dependency) similar to those previously described *in vivo* by Chung et al. (1974). Initial studies (Figure 6.3) have focused on the U1 and U2 components of the evoked response, which have mean poststimulus latencies to peak amplitude of about 10 and 30 msec, respectively. These were thought to be generated by two distinct classes of unmyelinated retinal fibers (Chung et al., 1974), though our preliminary evidence indicates that the very long-latency U2 wave is more likely a polysynaptic potential.

We have found that bath application of 6-cyano-2,3-dihydroxy-7-nitro-quinoxaline (CNQX), a selective antagonist of the α-amino-3-hydroxy-5-methyl-isoxazole-4-propionic acid (AMPA)-subtype of glutamate receptor, produces a dose-dependent decrease in the amplitude of both components of the evoked response (Figure 6.3A,C). This observation is consistent with recent evidence that glutamate may be responsible for retinotectal transmission in lower vertebrates (Langdon and Freeman, 1987). An unexpected finding was that at higher concentrations of CNQX, the onset of the depression of the U1 wave preceded that of the U2 component. Bath application of the NMDA-receptor antagonist, APV, produces a reversible, dose-dependent decrease in the amplitude of the U2 wave with negligible effect on U1 (Figure 6.3B,D). We interpret these findings as suggesting that NMDA receptors contribute little to the U1 potential, but that the longer-latency U2 component is primarily NMDA receptor-mediated. The effects of CNQX upon the U2 component are reconcilable with this interpretation. In particular, it seems that U2 declines only after U1 has dropped to approximately 50% of its control amplitude, and then this potential is rapidly abolished, as if its expression were contingent upon some critical level of prior depolarization, a well-known requirement of NMDA receptor-mediated events.

It is interesting to note that the onset of visually evoked activity in the tectum following stimulation of the ipsilateral eye is delayed, on average, by some 10–30 msec compared to that evoked directly from the contralateral eye (Scherer and Udin, 1991; Brickley and Grant, unpublished observations), this delay being a reflection of the circuitous nature of the intertectal pathway. The appearance, following retinal stimulation, of the long-latency, APV-sensitive U2 potential would thus coincide, *in vivo*, with the arrival of inputs via this commissural pathway, and it could, therefore, provide a substrate for the vision- and NMDA-dependent coincidence detection processes thought to underlie the plasticity of this pathway.

CONCLUDING REMARKS

The ordering of neuronal connections in both the retinotectal and the intertectal pathways of the frog involves activity-dependent processes mediated by NMDA receptor mechanisms. In the monocular projection from retina to tectum, spontaneously generated activity is sufficient to effect the ordering of the underlying connections. The refinement of detailed order in the connections of the intertectal pathway, mediating the binocular projection to this midbrain structure, utilizes a different class of activity-dependent mechanism that requires sensory experience. It should be emphasized that the operations thought to underlie both forms of developmental plasticity are similar to the classical postulates of Hebb (1949): that synaptic stability increases when there is temporal correlation of firing between pre- and postsynaptic elements and decreases when there is not. These postulates evolved in the search that synaptic stability increases when there is temporal correlation of firing between pre- and postsynaptic elements and decreases when there is not. These postulates evolved in the search for a neural explanation of behavioral changes during classical and operant conditioning. Indeed, current theorists (e.g., Rescorla, 1988) continue to emphasize that learning is critically dependent on the assimilation of correlated patterns of sensory presentation. One might suggest, therefore, that the neural plasticity required for learning and adaptive behavior is but a special expression of mechanisms initially acquired to solve developmental problems. Developmental plasticity would then differ from these forms of adult plasticity principally because it can be demonstrated at rather peripheral levels of sensory processing and because it is usually confined to early periods of ontogeny.

Acknowledgments

This work was supported by the Medical Research Council. We thank Dr. John H. Williams for his comments on the manuscript.

REFERENCES

Beazley LD (1975): Development of intertectal neuronal connections in *Xenopus*. The effects of contralateral transposition of the eye and eye removal. *Exp Brain Res* 23: 505–518

Beazley LD (1979): Intertectal connections are not modified by visual experience in developing *Hyla moorei*. *Exp Neurol* 63: 411–419

Chung SH, Bliss TVP, Keating MJ (1974): The synaptic organization of optic afferents in the amphibian tectum. *Proc Roy Soc Lond [Biol]* 187: 421–447

Cline HT, Constantine-Paton M (1989): NMDA receptor antagonists disrupt the retinotectal topographic map. *Neuron* 3: 413–426

Gaze RM (1958a): The representation of the retina on the optic lobe of the frog. *J Exp Physiol* 43: 209–214

Gaze RM (1958b): Binocular vision in frogs. *J Physiol (Lond)* 143: 20P

Gaze RM, Jacobson M, Szekely G (1965): On the formation of connexions by compound eyes in *Xenopus J Physiol (Lond)* 176: 409–417

Gaze RM, Keating MJ, Szekely G, Beazley LD (1970): Binocular interaction in the formation of specific intertectal neuronal connections. *Proc Roy Soc Lond [Biol]* 175: 107–147

Grant S, Dawes EA, Keating MJ (1992): The critical period for experience-dependent plasticity in a system of binocular visual connections in the frog, *Xenopus laevis*: Its extension by dark-rearing. *Eur J Neurosci* 4: 37–45

Grant S, Keating MJ (1986): Normal maturation involves systematic changes in binocular visual connections in *Xenopus laevis*. *Nature* 332: 258–261

Grant S, Keating MJ (1989a): Changing patterns of binocular visual connections in the intertectal system during development of the frog, *Xenopus laevis*. I. Normal maturational changes in response to changing binocular geometry. *Exp Brain Res* 75: 99–116

Grant S, Keating MJ (1989b): Changing patterns of binocular visual connections in the intertectal system during development of the frog, *Xenopus laevis*. II. Abnormalities following early visual deprivation. *Exp Brain Res* 75: 117–132

Grant S, Keating MJ (1992): Changing patterns of binocular visual connections in the intertectal system during development of the frog, *Xenopus laevis*. III. Modifications following early eye rotation. *Exp Brain Res 89*: 383–396

Grobstein P, Comer C (1977): Post-metamorphic eye migration in *Rana* and *Xenopus*. *Nature* 269: 54–56

Hebb DO (1949): *The Organization of Behaviour*. New York: Wiley

Jacobson M (1971): Absence of adaptive modification in developing retinotectal connections in frogs after visual deprivation or disparate stimulation of the eyes. *Proc Natl Acad Sci USA* 68: 528–532

Keating MJ (1968): Functional interaction in the development of specific nerve connexions. *J Physiol (Lond)* 198: 75–77P

Keating MJ, Feldman JD (1975): Visual deprivation and intertectal neuronal connections in *Xenopus laevis*. *Proc Roy Soc Lond [Biol]* 191: 467–474

Keating MJ, Grant S (1992): The critical period for experience-dependent plasticity in a system of binocular visual connections in *Xenopus laevis*: Its temporal profile and relation to normal developmental requirements. *Eur. J. Neurosci. 4*: 27–36

Keating MJ, Grant S, Dawes EA, Nanchahal K (1986): Visual deprivation and the maturation of the retinotectal projection in *Xenopus laevis J Embryol Exp Morphol* 91: 101–115

Kennard C, Keating MJ (1985): A species difference between *Rana* and *Xenopus* in the occurrence of intertectal neuronal plasticity. *Neurosci Letts* 58: 365–370

Langdon RB, Freeman JA (1987): Pharmacology of retinotectal transmission in the goldfish: Effects of nicotinic ligands, strychnine, and kynurenic acid. *J Neurosci* 7: 760–772

Olson MD, Meyer RL (1991): The effect of TTX-activity blockade and total darkness on the formation of retinotopy in the goldfish retinotectal projection. *J Comp Neurol* 303: 412–423

Pettigrew JD (1974): The effect of visual experience on the development of stimulus specificity by kitten cortical neurones. *J Physiol (Lond)* 237: 49–74

Rescorla RA (1988): Behavioural studies of Pavlovian conditioning. *Ann Rev Neurosci* 11: 329–352

Scherer WJ, Udin SB (1989): N-methyl-D-aspartate anatagonists prevent interaction of binocular maps in *Xenopus* tectum. *J Neurosci* 9: 3837–3843

Scherer WJ, Udin SB (1991): Latency and temporal overlap of visually-elicited contralateral and ipsilateral firing in *Xenopus* tectum during and after the critical period. *Develop Brain Res* 58: 129–132

Sperry RW (1963): Chemoaffinity in the orderly growth of nerve fiber patterns and connections. *Proc Natl Acad Sci USA* 50: 703–710

Udin SB (1983): Abnormal visual input leads to development of abnormal axon trajectories in frogs. *Nature* 301: 336–338

Udin SB, Keating MJ (1981): Plasticity in a central nervous pathway in *Xenopus*: Anatomical changes in the isthmo-tectal projection after larval eye rotation. *J Comp Neurol* 203: 575–594

Udin SB, Scherer WJ (1990): Restoration of the plasticity of binocular maps by NMDA after the critical period in *Xenopus*. *Science* 249: 669–672

CHAPTER 7

Experimental Manipulation of the Developing Rodent Visual System

RAYMOND D. LUND, JEFFREY D. RADEL, AND KATHLEEN T. YEE

An important review by Gaze in 1960 drew attention to the neurospecificity experiments of Sperry (1943a, 1943b, 1944, 1948, 1951) done some years earlier and, most importantly, introduced new studies that were fundamental for the large body of work that has used the retinotectal system for examining the origins of neural specificity and plasticity. These experiments took advantage of the accessibility of the amphibian visual system to experimental manipulation at early stages of development, as well as of the opportunity to examine the repercussions of manipulations in mature animals on patterns of regeneration (see Gaze, 1959).

This work stimulated us to approach similar issues in the mammalian visual system, with special emphasis on rodents. A number of questions arose from the frog studies:

1. Was it possible to address similar questions in mammals? Surprisingly little was known of the organization of the visual pathways in mammals at that time. Furthermore, experimental manipulations during development had not been explored, and it was generally believed that regeneration after injury was not a normal property of mature mammalian optic pathways.
2. Would the events defined in animals like frogs and fish, in which visual behaviors were relatively inflexible, generalize to mammals, in which there was clear evidence of adaptive visual responses?
3. Would the different organization of the mammalian visual system, with hierarchical and parallel processing, and not only segregation, but also convergence of channels at various levels, make it amenable to posing different sorts of specificity questions from those associated with topographic specificity?

Our early studies focused on the basic anatomy of the rodent visual system, including localization of optic axon termination in the superior colliculus (Lund, 1969) and the lateral geniculate nucleus (Lund and Cunningham, 1972), two regions that receive parallel topographical encoded inputs from the eye, as well as topographic matching between cortex and superior colliculus (Lund, 1964). This work also provided the foundation for two studies showing how the primary optic pathway could be modified. In the first it was found that the uncrossed optic

Formation and Regeneration of Nerve Connections
Sansar C. Sharma and James W. Fawcett, Editors
© 1993 Birkhäuser Boston

pathway was diminished in albino rats compared with normally pigmented rats (Lund, 1965). Subsequently this was identified as an anomaly associated with hypopigmentation of the eye (Guillery et al., 1975) and specifically of the pigment epithelium, due to the albino gene or to hypopigmentation color dilution genes (Wise and Lund, 1976). The observation was generalized to a wide variety of animals including humans (Guillery et al., 1975). It was possibly the first system in which a subtle change in axon distribution was shown to have a genetic correlate, although how the gene product leads to alterations in the pattern of optic projections is still unclear.

The second set of studies again focused on the question of laterality of optic projections and showed that while the uncrossed retinotectal projection originates only from the lower temporal retina in mature rats, this projection arises from the entire retina at perinatal times and distributes over the entire area of the superior colliculus (Land and Lund, 1979). There is a progressive loss of this exuberant projection during the first postnatal week. If, however, one eye is removed (Lund and Lund, 1971, 1973) or rendered nonfunctional during this period (Simons and O'Leary, 1990), the exuberant projection is retained and remains broadly distributed over the ipsilateral superior colliculus in a "mirror" map (Thompson, 1979).

These observations suggest that the laterality of distribution of primary optic connections is not rigidly determined, and that for a small group of axons, this depends on an interaction with the axons from the opposite eye. Neonatal enucleation and other manipulations of the eye, besides having an impact on the organization of the projection from the remaining eye, also influence higher-order connections, such as the callosal pathway between the visual cortices (Lund et al., 1978; Cusick and Lund, 1982). More recent work has shown a similar diffuse projection that is subsequently refined in the topographic representation of the retina on the tectum (Simons and O'Leary, 1991).

These studies show that as the optic pathways become established, the detailed patterning of connections may be modified, depending on interactions occurring between axons and terminals carrying information from each eye.

One problem with studies on intact animals is that it is difficult to examine the effect on pathway organization of factors such as spatial disposition of axons as they approach target areas and relative developmental timing between interacting elements. Clearly these issues can be approached *in vitro*; but while invaluable for addressing certain questions, culture studies are not without limitations. It is not always possible to create the conditions confronted by growing axons *in vivo*. Cells can express different antigens *in vivo* and *in vitro*, and one of the important tests of specificity, namely that of eliciting appropriate functional responses, is hard to achieve in culture. An important tool in nonmammalian neurobiology has been the transplantation of tissue to the central nervous system of a region of the brain taken from the same animal or from a different animal. Such transplantation was inadvertently achieved in mammals in a study involving tectal lesions made in early fetal rats (Miller and Lund, 1975), and this prompted us to examine systematically the use of transplantation as a way to study the constraints that operate in the formation of specific connections.

Several areas of investigation have emerged from the initial studies (Lund et al., 1991). We have been able to study patterns of optic axon outgrowth when the axons are confronted with abnormal conditions. We have been able to approach the question of specificity of connections, with the hope of identifying the substrates and interactions of various aspects of specificity. We have examined the functional consequences of introducing ectopic tissue and how an input relayed through a transplant interacts with that introduced through host optic pathways. Finally we have begun to examine how far the developmental process can progress in the mature brain. This work is summarized in the present review.

OUTGROWTH PATTERNS

In the course of normal development, optic axons initially leave the retina on E13 (mice) or E14 (rats). They reach the optic chiasm on E14 (mice) or E15 (rats) and arrive at the superior colliculus on E15 (mice) or E16 (rats) (Godement et al., 1984; Bunt et al., 1983). The first axons grow on the surface of the colliculus where they form en passant synapses (Lund and Bunt, 1975), but by birth the stratum griseum superficiale has formed superficial to the optic fiber layer, and it is in this synaptic layer that the optic axons form their terminal arbors (Sachs and Schneider, 1984).

Retinae taken from mice (E12) and rats (E13) can be dissected from their investing layers and implanted in various locations of neonatal rat brains (Figure 7.1), where their early outgrowth can be studied using a variety of labels, including antibodies, fiber stains, and fluorescent dyes. The first outgrowth from the retinae can be recognized within a day of transplantation, and by 3 days substantial outgrowth can be seen (Hankin and Lund, 1990). When placed on the surface of the brain, either the brainstem or the cortex, fascicles of axons can be seen growing for considerable distances in the region just below the surface (Sefton et al., 1991). Unless an appropriate target is available, these axons disappear within a week: axons growing on the surface of the cortex will only survive when a region such as tectum or lateral geniculate nucleus is cotransplanted with the retina (Sefton et al., 1987; Sefton and Lund, 1988). Outgrowth from transplants placed deep in the brain has been seen only in three specific circumstances. First, there is a transient outgrowth around the graft and into the

---

FIGURE 7.1 A: Transplanted rat retina (Tp) located on the dorsal surface of the cerebellum (cer) of a host rat brain, shown in coronal section and stained with Cresyl Violet. An enlargement of the boxed region of the transplant is shown in **B**, with arrows indicating retinal cells similar in size to the type I class of retinal ganglion cells found in normal rat retinae. GCL, ganglion outer nuclear layer; os, photoreceptor outer segments; IPL, inner plexiform layer; INL, inner nuclear layer; ONL, outer nuclear layer. Although not labeled, the outer plexiform layer exists as a thin, acellular region located between the outer and inner nuclear layers. *see figure next page*

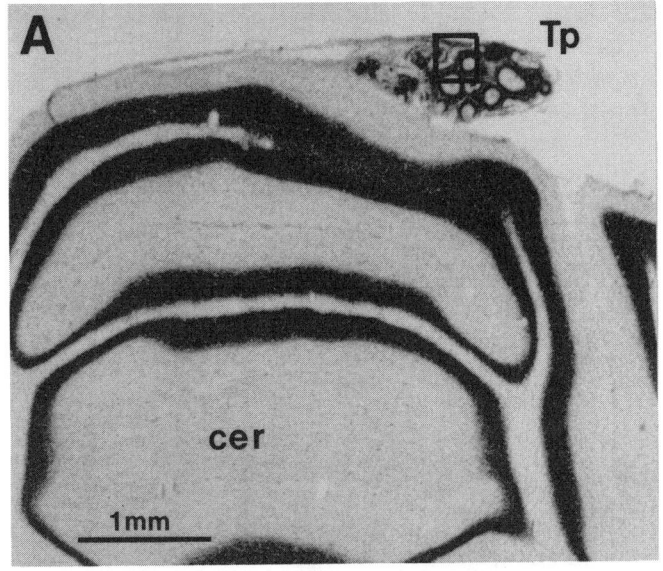

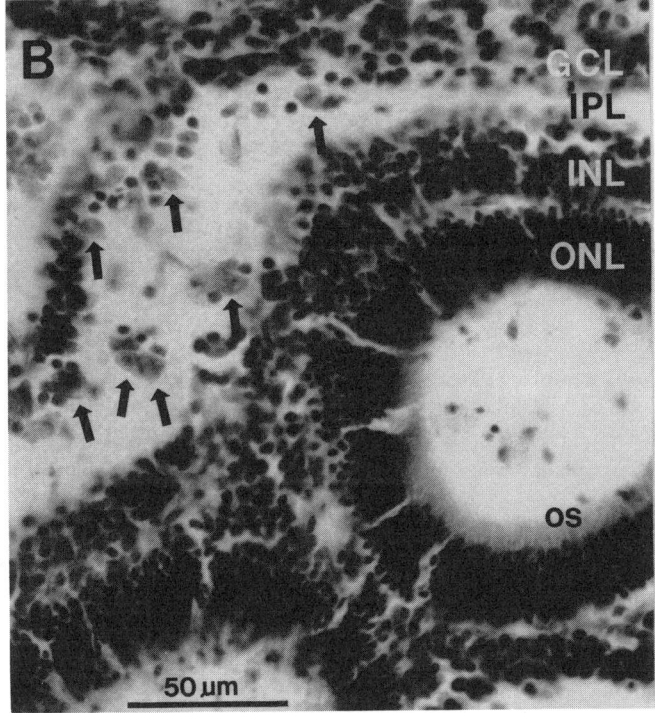

FIGURE 7.1.

implantation site that persists for a few days. Second, if the graft is placed close to a target region, as for example in the midbrain parenchyma within 1 mm of the tectum, a targeted outgrowth toward the tectum is seen that does not appear to follow any prescribed substrate (Hankin and Lund, 1987). Third, outgrowth has been seen along deep fiber tracts. For example, transplants placed in the white matter underlying the cerebral cortex do show transient outgrowth, which can be sustained if they are located close to a normal optic axon target, as when they are sufficiently rostral to be close to the lateral geniculate nucleus.

These observations suggest several points. First, there is no obligatory pathway for optic axon outgrowth. The axons appear to grow in the region deep to the pial surface, as well as in white matter tracts, even in locations in which optic axons would not normally be found. Second, access to a target region within a limited time period is necessary to sustain the projection. Third, close proximity to a target may support outgrowth that appears to be target-directed, possibly in response to neurotropic (or neurotrophic) cues, rather than substrate-dependent cues. Fourth, the surface of the wound site associated with introduction of the graft appears to support outgrowth transiently, but this is clearly not sustained. Fifth, areas of gray matter do not alone support optic axon outgrowth.

These observations have led to the suggestion that there are two major components to the formation of the primary optic pathway. One involves a substrate-dependent growth that is relatively nonselective and is not obviously target-directed, and the other is a target-dependent effect that may serve to focus axons on a target once the axons are in the target's immediate vicinity, promote local axonal sprouting, and change the program of the parent cell from one of axonal extension to one of target dependency.

SPECIFICITY OF CONNECTIONS

The normal optic pathway shows a highly predictable pattern of central connections. Axons within the optic nerve have their origin in subclasses of ganglion cells that respond to particular components of the visual signal and are often morphologically distinct. The optic axons distribute to as many as a dozen brainstem regions (Sefton and Dreher, 1985). It is clear that individually or in subsets, each region is associated with a different functional response to visual information, and there is evidence, especially in larger animals, that not all ganglion cell subclasses project to each visual center in the brainstem. Besides this regional and functional parcellation, optic axons also show a number of other defined patterns of organization. Whether they project ipsilaterally or contralaterally in the brain relates to the position of the parent ganglion cells on the retina. The axons derived from each eye project in a topographically ordered manner on the brain, either as a map of the whole retina or of the contralateral field. To achieve correlation of the input from each eye, they often lie in register with respect to visual field representation, as in the lateral geniculate nucleus and the intermediate layers of the primary visual cortex, where they are segregated one from another in laminae or stripes. The reason for segregation is unclear at present. However, it is apparent that the precise map emerges from a more diffuse representation as a result of an interactive process across the region. Perhaps such an interaction would be harder to achieve if inputs from both eyes were also completely integrated. Within the target regions, the axon groups show highly stereotypic patterns of termination with characteristic arbor patterns and terminal synaptic configurations.

One question to which we have devoted a considerable amount of attention is the degree to which an ectopic transplanted retina is capable of expressing the same level of specificity as a normal optic projection, irrespective of age differences between retina and host brain, and of abnormal graft location. While we have not provided critical evidence for the parallel channels that exist in the normal optic nerve, a study of ganglion cell morphology revealed several of the subclasses that can be seen in the normal retina: ganglion cells were observed in transplants having dendritic arbors and soma sizes that corresponded to the types I, II, and III cells and the displaced ganglion cells seen in normal rat retina. Furthermore, dendrites were often distributed close to the ganglion cell layer and to the inner nuclear layer, a feature that has been correlated with physiological response characteristics (Perry et al., 1985).

The transplant projections also show a selectivity of innervation that closely compares with normal. Transplanted mouse retinae placed on the surface of the brainstem of neonatal rats show highly specific projections identified using several monoclonal antibodies, anti-M4 and anti M-6, specific for mouse neurons to a number of visual centers (Figure 7.2). There is always a projection to the superior colliculus and frequently inputs to the pretectum (olivary pretectal nucleus and nucleus of the optic tract), accessory optic system, and lateral geniculate nucleus (outer shell of the dorsal division). We were unable to show input to the ventral lateral geniculate nucleus, intergeniculate leaflet, and suprachiasmatic nucleus, even when transplants were placed adjacent to these nuclei. The projections are always more evident if the host optic input is removed at the time of transplantation (Radel et al., 1990a).

More recent studies have shown that target nuclei such as the ventral lateral geniculate nucleus, the intergeniculate leaflet, and the deeper part of the dorsal lateral geniculate nucleus can be more reliably or heavily innervated by altering the transplant paradigm. Such changes include placing retinae into much younger hosts (E18; Figure 7.3; Yee et al., 1991), placing more than one retina into neonatal rats (Yee et al., 1991), placing retinae in the internal capsule rather than on the surface of the brainstem, and using mice as hosts. In none of these cases was the suprachiasmatic nucleus innervated. We have also failed to demonstrate a persistent innervation to regions to which optic axons would not normally project, even in cases (e.g., cerebellum, inferior colliculus) in which the transplant was embedded in them, or when the implanted retina replaced the natural afferents (as when implanted in the dorsal columns). This suggests that connectivity is not simply the end point in a process of axon guidance in which specificity is inherent in the guidance process rather than in axon–target relations (McLoon et al., 1985).

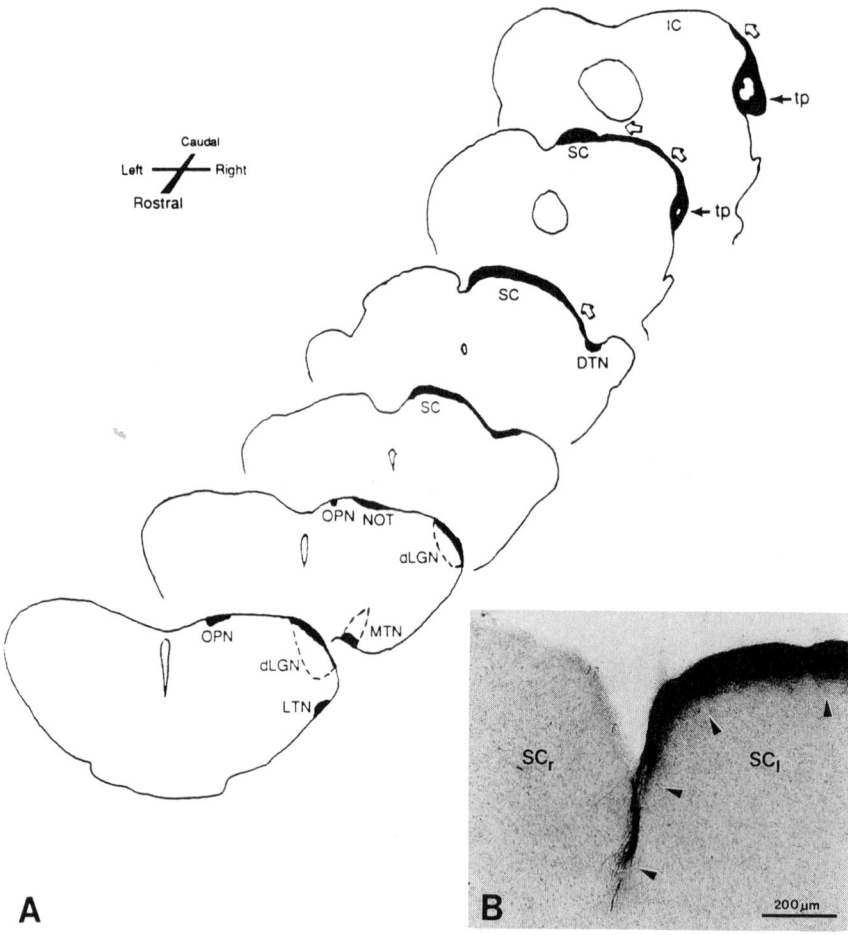

FIGURE 7.2. A: Camera lucida drawings of selected coronal sections through the midbrain of a host rat, showing the pattern of fiber projections for a transplanted mouse retina (tp) located on the lateral surface of the left inferior colliculus (IC). The right eye of the host was removed on the first postnatal day, at the time of transplantation. At maturity, monoclonal antibodies specific for mouse neural tissue (αM4 and αM6) were used to identify the transplant and its projections. The route taken by transplant fibers is indicated by open arrows; labeling was not observed in nonvisual nuclei. OPN, olivary pretectal nucleus; SC, superior colliculus; DTN, dorsal terminal nucleus of the accessory optic tract; IC, inferior colliculus; dLGN, dorsal division of the lateral geniculate nucleus; MTN, medial terminal nucleus of the accessory optic tract; LTN, lateral terminal nucleus of the accessory optic tract; NOT, nucleus of the optic tract. Reprinted with permission of Wiley-Liss, a division of John Wiley and Sons, Inc. from Radel JD et al. (1990a); Proximity as a factor in the innervation of host brain regions by retinal transplants. *J Comp Neurol* 300: 211–229. Copyright © 1990 J.D. Radel. **B**: Photomicrograph showing the medial portions of the right superior colliculus (SC_r) and left superior colliculus (SC_l) of another, mature host rat, which also had the right eye removed at the time of transplantation. Projections from the transplanted mouse retina, indicated by arrowheads, were identified using specific antibodies for mouse neurons.

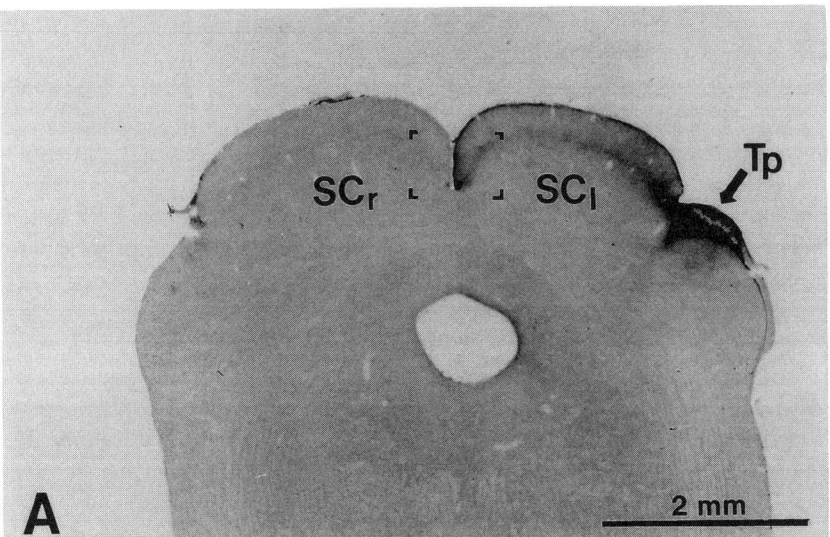

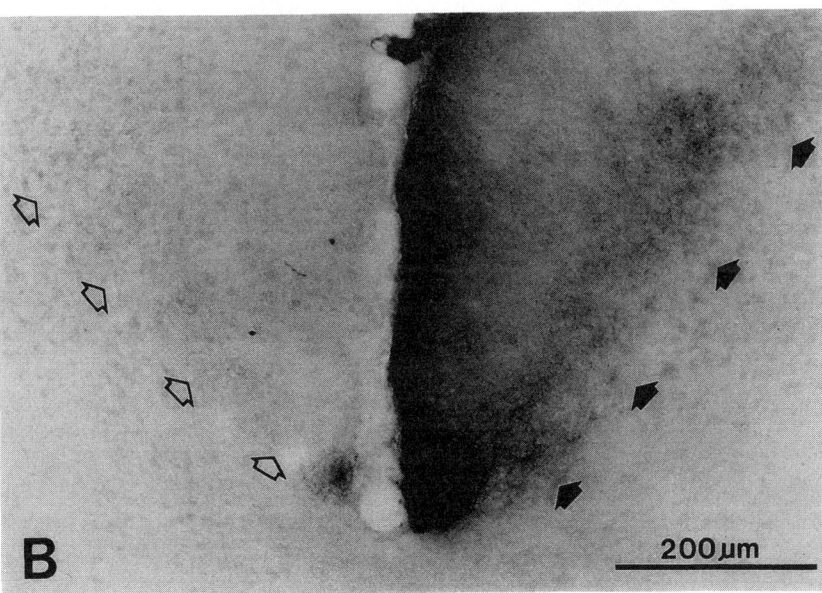

FIGURE 7.3 A: Coronal section through the midbrain of a mature host rat stained with antibodies specific for mouse neurons, showing the transplanted mouse retina (Tp) and its projections. The transplantation surgery took place *in utero*, with the host on the 18th gestational day and the transplant on the superior colliculus. SC_r, right superior colliculus; SC_l, left superior colliculus. **B:** Enlargement of the region in **A** enclosed by brackets. Arrows indicate the location of the stratum opticum layer in the right (open arrows) and left (filled arrows) superior colliculi.

In contrast to regional specificity, expression of which appears to be relatively independent of context, the side of the brain to which axons from a transplanted retina project does not appear to be intrinsically specified. Axons innervate the side of the brain to which they have access, and there are frequent cases of bilateral innervation, especially from grafts located at the midline (Radel et al., 1990).

Another aspect of specificity, and that to which most attention has been given in comparable studies in nonmammalian vertebrates, is that of topographic order. This has been examined for the retinal transplant input to the superior colliculus, by studying labeling patterns in transplants from small injections of two retrogradely transportable dyes at rostral and caudal poles of the colliculus (Galli et al., 1989). If topographic order had been maintained, it would be expected that there would be two clusters of ganglion cells on opposite poles of the transplant, labeled with each dye, respectively. If there was random order, one would see many double-labeled cells. Instead, we found single-labeled cells with either dye scattered throughout the retina: there was an equally likely chance for two neighboring cells to be marked with the same or different dyes. Double-labeled cells were seen only when the injection sites overlapped. These observations suggest that individual ganglion cells make relatively restricted terminal arbors, but that near-neighbor ganglion cells are equally likely to project to widely separated or adjacent points on the superior colliculus. This result was somewhat surprising, given the many transplant studies in amphibians showing the restoration of a topographically ordered projection. There are several possible explanations for the failure to establish normal topography in the rodent studies. Matching "labels" may no longer be expressed at the time when transplant axons reach the colliculus, and evidence from tissue culture studies suggests this to be the case (Godement and Bonhoeffer, 1989). Intrinsic retinal polarity cues may be lost by the transplantation process. Studies with one antibody specific for dorsal retina, however, show that this is not the case. Dolce, a marker for developing normal dorsal retina (McCaffery et al., 1990) was used to identify projections from transplanted dorsal or ventral hemiretinae. While labeling remained restricted to dorsal hemiretinae, projections from hemiretinae extend across the entire host superior colliculus and are not restricted to topographically appropriate regions (Yee and Lund, 1991). There may be too few axons to produce a coherent map: one estimate showed about 10,000 axons emanating from a graft, compared with 120,000 in a normal optic nerve (Lund and Simons, 1985). Since transplant axons frequently run in small bundles, rather than one nerve, they may attempt to map independently, much like the two halves of a compound eye. These axons enter the tectum by an anomalous route, running in a predominantly coronal rather than sagittal plane, and this may influence their ability to form a coherent map. Finally, in light of tetrodotoxin (TTX) studies, activity may be necessary to establish a refined map (Meyer, 1983; Simons and O'Leary, 1990). While we do not know whether normal spontaneous activity patterns develop in the retinae after transplantation, it is apparent that the morphological features develop over a normal time course and that an induced response—pupilloconstriction of the host eye to transplant stimulation—was first elicited at approximately the same maturational stage as normal (Lund et al., 1990; Radel et al., 1992).

One further aspect of normal organization, the segregation of inputs from the two eyes, is not evident between transplant and host optic inputs, in contrast to amphibians where a transplanted retinal projection is segregated from the host input as stripes across the tectum (Constantine-Paton and Law, 1978). In rats, if both host eyes remain intact, the transplant projection is confined mainly to the surface of the colliculus, with some input entering by way of the stratum opticum. If one host eye is removed, the transplant projection spreads evenly through the upper half of the superior colliculus (McLoon and Lund, 1980; Radel et al., 1990a, 1991a). The lack of segregation in stripe formation may be due to several factors. It is possible that transplants are electrically silent. Although this has yet to be tested directly, as indicated above, the lack of spontaneous activity during transplant development would be somewhat surprising, given that normal fetal retinae are spontaneously active (Galli and Maffei, 1988). The absence of a topographic map and hence of the "recognition" of spatial barriers might be an important factor, but this will be hard to investigate unless a way is found to achieve a topographic representation. A further possibility exists relating to the relative timing of ingrowth of transplant axons and host input. At the time when transplant axons invade the superior colliculus, the host input has already been present for 10 days (Lund and Bunt, 1975; Bunt et al., 1983); it is possible that the host projection is at a sufficient competitive advantage to overwhelm the transplant afferents. This issue has been approached in two ways: (1) by transplanting two retinae at birth and determining whether their inputs interact one with another, and (2) by transplanting retinae at fetal ages, reducing developmental disparity with the host optic input. Presently there is no evidence that inputs from the two transplanted retinae or from host and transplant segregate (Yee and Lund, 1991).

Overall these observations indicate that while transplant axons go a long way toward creating a normal central optic pathway, it is far from perfect. However, the anomalies encountered provide opportunities to examine substrates of normal development and to generate further experiments to test the hypotheses raised.

FUNCTION

The studies summarized above demonstrate that embryonic retinae transplanted to intracranial locations express many of the properties of normal retinae, including the stereotypic patterns of connectivity. Given the similarities to normal, we were interested in examining whether transplanted retinae were capable of transducing photic input and relaying this information to the host brain. In turn, does the host brain have the capacity to receive the transplant input and use this information to elicit an appropriate response? These questions have been addressed in several different sets of experiments in which embryonic retinae are placed on the dorsal midbrain surface of newborn rat hosts prior to the overgrowth of the colliculi by the developing cortex. As the cortex grows, typically about half of the transplants are displaced caudally, coming to rest on the anterior edge of the cerebellum. Such transplants can then be exposed at maturity via a small craniotomy. This preparation was employed by Simons and Lund (1985) to demonstrate that intensity-dependent gross potential responses could be recorded from transplanted retinae when illuminated, and that unit responses specific to light-on, light-off and to ambient light levels could also be recorded from the superior colliculus when transplants were illuminated. Examination of cortical responsiveness has subsequently defined a subdivision of area 18a that is activated by transplant illumination, most likely mediated through the superior colliculus and lateral posterior nucleus of the thalamus (Craner et al., 1989). A further study (Craner et al., 1990) has shown that photic stimulation of retinal transplants also induces c-fos activity in the superior colliculus, indicating that besides mediating these transient physiological responses, the transplant connections are capable of causing longer-lasting metabolic changes in host neurons through the induction of the c-fos gene to express c-fos protein.

Taken together, these findings indicate that transplanted retinae can, in fact, respond appropriately to changes in light level, and that information regarding the photic environment can be relayed from a transplanted retina through several synapses to the host brain. The question of whether a host rat has the capacity to interpret transplant-derived activity in a useful manner has been addressed in a series of other experiments, using a simple reflex (pupilloconstriction), a conditioned response (conditioned suppression of lever pressing), and unconditioned responses (light avoidance in an open field and a startle response) to assess the impact of light-induced transplant activity on host behavior. Pupilloconstriction in normal rats serves to regulate pupil diameter in response to light presented to either eye. Although pupil diameter can be influenced by a number of factors other than changes in light levels, the base diameter and constriction to light are highly consistent for individual animals if testing conditions are carefully controlled. The pupillary light reflex is subserved by a relatively simple neural circuit in rodents, involving a primary optic input to the olivary pretectal nucleus (OPN), a relay to the Edinger-Westphal nucleus, preganglionic motor output along the oculomotor nerve to the ciliary ganglion, and postganglionic output to the iris musculature (Scalia, 1972; Campbell and Lieberman, 1985; Loewy, 1990). Because of both the circumscribed nature of the neural circuit mediating this response and the predictability of the stimulus–response relationship, we examined whether this system could be used to examine the interaction between transplant connectivity and function. Damaging the OPN eliminates pupilloconstriction to light (Carpenter and Pierson, 1973; Young et al., 1991), while direct electrical stimulation of the nucleus results in pupilloconstriction (Trejo and Cicerone, 1984). Our anatomical studies demonstrated that the OPN usually received a substantial projection from transplanted retinae and that transplants located on the anterior margin of the cerebellum were accessible for direct illumination. By visualizing and illuminating transplanted retinae directly in host rats, we were able to demonstrate that transplant illumination resulted in pupilloconstriction of the host eye. Figure 7.4 illustrates the neural circuit subserving the pupillary light reflex and how a transplanted retina is incorporated into this pathway. It also shows transplant-mediated responses elicited by a series of stimuli graded in intensity. Pupilloconstriction occurred even when the host optic nerve was cut, rendering the eye unresponsive to illumination (Klassen and Lund, 1987, 1988). Subsequent work has shown that the response is highly predictable for individual animals at a fixed stimulus intensity, that it is not altered by repetitive presentations of stimuli, and that it does not change when tested over periods of 1 to 2 months (Radel et al., 1990b; Lund et al., 1990). Individual animals vary with respect to the effectiveness with which transplant stimulation affects host pupilloconstriction. There are several reasons for this. In one study, we found that the amplitude and velocity of constriction of the transplant-mediated pupillary light response related directly to the density of innervation of the OPN by transplant axons (Klassen and Lund, 1990). We have also shown that the presence of a host optic input to the OPN (even when the host eye is maintained in darkness) limits transplant effectiveness (Radel et al., 1991b).

When the host input is removed entirely, there is rapid enhancement of the transplant response (Radel et al., 1991b). Even though the host eye was maintained in darkness during the tests conducted prior to sectioning of the host optic nerve, enhanced pupilloconstriction in response to transplant illumination was observed within 24 hours after lesioning. Such a rapid change in functional efficacy of the transplant is not likely due to metabolic or anatomical alterations in the transplant or its connections with the host brain resulting from loss of the host optic input. We have instead hypothesized that this enhancement results from an unmasking of the inherent responsiveness of the transplant input, perhaps by eliminating competing input from spontaneous retinal activity in the host eye. Finally, we have also shown that simultaneous illumination of a transplanted retina and the host eye elicits a greater amplitude of pupilloconstriction than that resulting from illuminating either the eye or the transplant independently (Radel et al., 1990b). It is interesting to note that the increased amplitudes do not appear to be related in a linear fashion to stepwise changes in the relative intensities with which the transplant and host eye are illuminated, suggesting that the host brain may be capable of discriminating these inputs and integrating the information they carry to produce pupilloconstriction.

These results demonstrate that, at least in this sensory response system, the host brain is capable of receiving transplant input, integrating that input with normal optic inputs, and acting upon the resulting information in an appropriate fashion. Given that a simple reflex can be driven by transplant input, two further questions arise. Although the transplant input elicits an appropriate pupilloconstrictor response, is the rat sufficiently aware of this input to make use of it? Does the rat perceive the transplant input as being similar to normal optic input?

The first of these questions was addressed in two studies. In one an unconditioned behavior, the startle response, was employed to assess transplant function. In the other a conditioned suppression response was used. The startle response is a measure of how well a novel stimulus will influence the ongoing behavior of a

A

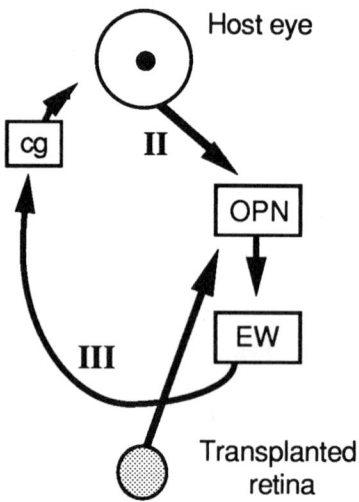

B

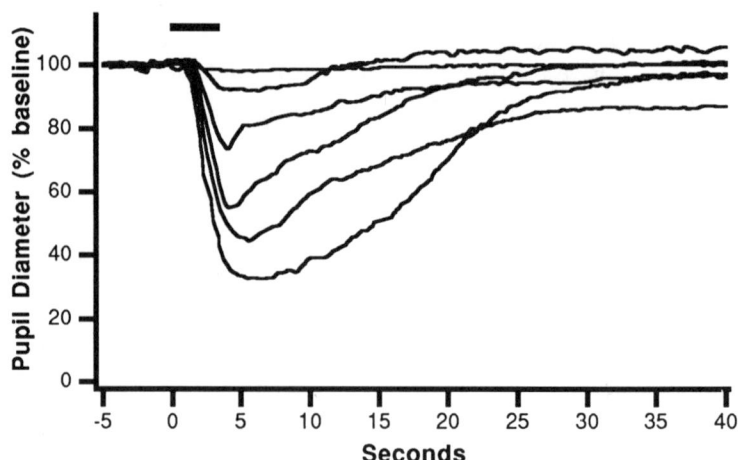

FIGURE 7.4. A: Schematic representation of the neural circuit subserving the pupillary light reflex in a host rat, with retinal input provided by one host eye or an intracranially transplanted retina. Luminance information is conveyed from the retinae to the OPN and relayed to the EW. Preganglionic motor output from the EW travels along the oculomotor

free-moving animal. A sudden change in the environment, such as the presentation of a light, will cause a normal rat to halt ongoing motor activity. Like pupillocon-striction, the startle response is reflexive and does not require training for expression. Unlike pupilloconstriction, however, voluntary activities (feeding, locomotion) are affected. A chronic indwelling fiberoptic light guide was used to deliver a burst of light directly to transplanted retinae in free-moving rats in this series of studies. When the transplant was illuminated, the host rats initially exhibited a freezing response followed by a period of random exploration (Coffey and Lund, 1991). This suggests that startle responses, which are normally mediated through the superior colliculus (Mitchell et al., 1988; Dean et al., 1989), can also be activated by transplant activity, presumably through the same neural circuit.

In the conditioned suppression approach, host rats were required to learn that light delivered to the transplanted retina is an accurate predictor of a mild foot shock. By contrast, a tone served as a less accurate predictor of the shock. Having learned the association between light and foot shock, conditioned rats showed a reduction in the rate of ongoing bar pressing for food when the transplant was illuminated, but not when the tone was presented (Coffey et al., 1989). This result demonstrates that, like normal rats, host rats ascribe a behavioral significance to visual cues in preference to auditory cues, even when the visual input is received through a transplanted retina, showing that transplant-derived information can be used to modify ongoing complex behaviors in an appropriate manner. It is of interest to note, however, that subsequent experiments have shown that there are limitations to how well this learned behavior can generalize (Coffey and Lund, 1991). Rats trained using the host eye alone fail to exhibit the light-conditioned response when the transplant is later illuminated during testing. However, when rats are initially trained using transplant illumination alone and later tested with normal vision only, the light-conditioned response initially is limited but thereafter improves rapidly. It remains to be seen whether the improvement results from a partial transfer of properties of the training and testing conditions or whether it is due to an enhanced rate of relearning the association.

This series of experiments demonstrates that light-induced transplant activity is relayed to the host brain and that the host rat is capable of responding to that information, not only at the level of a simple reflex, but also by altering ongoing behaviors in response to transplant stimulation. These results do not address the

---

nerve to the ciliary ganglion and postganglionic fibers innervate the iris musculature. OPN, olivary pretectal nucleus; EW, Edinger-Westphal nucleus; cg, ciliary ganglion; II, optic nerve; III, oculomotor nerve. **B**: Graded pupillary responses elicited by transplant illumination in an individual host rat: from upper to lower traces, stimulus intensities were 1.5, 6, 9, 10, 13, and 15×10^4 cd/m^2. The stimulus presentation periods were 3 sec in duration, and are indicated by the solid bar above the graph. Baseline (prestimulus) pupillary diameter ranged from 4.2 to 4.4 mm.

issue of whether input from a transplanted retina is perceived by the host rat as being similar to normal visual input. This question was addressed using another unconditioned response common in nocturnal rodents, light avoidance behavior. In brightly lit environments, normal rats tend to spend more time in the shadowed portions than in the light. Host rats exhibit the same response when allowed to view the test arena through their intact eye, but fail to discriminate between light and dark regions when restricted to using only the transplant as a source of photic information. Interestingly, when these same rats were then trained for the conditioned suppression study and later retested for light avoidance, they showed a preference for the dark regions of the test arena irrespective of whether only the intact eye or only the transplant was used to provide information about lighting conditions (Coffey et al., 1990). This result implies that these animals had learned to use transplant-derived information in such a way as to direct their behavior. The fact that they were initially unresponsive to transplant input suggests that transplant input may not be perceived as having qualities normally ascribed to visual input, but that appropriate training may be sufficient to lend behavioral significance to transplant activity.

SUMMARY

The experiments described here highlight one part of a large area of work that has built on the foundation provided by Gaze and his colleagues. It is clear that the conceptual approaches explored in amphibians and fish are also applicable to mammals. Furthermore, the many new techniques (such as fetal surgery, transplantation, and the capacity to label developing pathways postmortem) have made the early development of the mammalian visual system accessible to experimentation in a way that could not have been envisioned 30 years ago. The results have certainly confirmed that *Xenopus* is a good model for mammalian neural development and that nonmammalian vertebrates will continue to provide insights not easily derived from mammalian work. The unique organization of the mammalian brain and the greater complexity and diversity of response patterns to particular stimuli lend themselves to a more penetrating examination of the interactive events that play an important role in the formation of the mature visual system.

Acknowledgments

The work reviewed here represents the collaborative effort of a number of colleagues, whose contributions to the various facets are gratefully acknowledged. The studies were supported by grants EYO5283 and EYO5308 from the National Institutes of Health.

REFERENCES

Bunt SM, Lund RD, Lund PW (1983): Prenatal development of the optic projection in albino and hooded rats. *Dev Brain Res* 6: 149–168

Campbell G, Lieberman AR (1985): The olivary pretectal nucleus: Experimental anatomical studies in the rat. *Phil Trans R Soc London Ser B* 310: 573–609

Carpenter MB, Pierson RJ (1973): Pretectal region and the pupillary light reflex. An anatomical analysis in the monkey. *J Comp Neurol* 149: 271–300

Coffey PJ, Lund RD (1991): Alerting and orienting responses mediated by intracerebral retinal transplants. *Soc Neurosci* 17:557

Coffey PJ, Lund RD, Rawlins JNP (1989): Retinal transplant-mediated learning in a conditioned suppression task in rats. *Proc Nat Acad Sci USA* 86: 7248–7249

Coffey PJ, Lund RD, Rawlins JNP (1990): Detecting the world through a retinal implant. *Prog Brain Res* 82: 269–276

Constantine-Paton M, Law MI (1978): Eye-specific termination in tecta of 3-eyed frogs. *Science* 202: 639–641

Craner SL, Radel JD, Jen LS, and Lund RD (1989): Light-evoked cortical activity produced by illumination of intracranial retinal transplants: Experimental studies in rats. *Exp Neurol* 104: 93–100

Craner SL, Hoffman GE, Lund JS, Lund RD (1990): Use of c-Fos to determine the neural pathway activated by intracranial retinal transplants in rats. *Soc Neurosci Abstr* 16: 656

Cusick CG, Lund RD (1982): Modification of visual callosal projections in rats. *J Comp Neurol* 212: 385–398

Dean P, Redgrave P, Westby GWM (1989): Two response systems in the mammalian superior colliculus. *Trends Neurosci* 12: 137–147

Galli L, Maffei L (1988): Spontaneous impulse activity of retinal ganglion cells in prenatal life. *Science* 242: 90–91

Galli L, Rao K, Lund RD (1989): Transplanted rat retinae do not project in a topographic fashion on the host tectum. *Exp Brain Res* 74: 427–430

Gaze, RM (1960) Regeneration of the optic nerve in Amphibia. Int Rev Neurobiol 2:1–40 paper

Godement P, Salaün J, Imbert M (1984): Prenatal and postnatal development of retinogeniculate and retinocollicular projections in the mouse. *J Comp Neurol* 230: 551–575

Godement P, Bonhoeffer F (1989): Cross-species recognition of tectal cues by retinal fibers *in vitro*. *Development* 106: 313–320

Guillery RW, Okoro AN, Witkop CJ Jr. (1975): Abnormal visual pathways in the brain of a human albino. *Brain Res* 96: 373–377

Hankin MH, Lund RD (1987): Role of the target in directing the outgrowth of retinal axons: Transplants reveal surface-related and surface-independent cues. *J Comp Neurol* 263: 455–466

Hankin MH, Lund RD (1990): Directed early axonal outgrowth from retinal transplants into host rat brains. *J Neurobiol* 21: 1202–1218

Klassen H, Lund RD (1987): Retinal transplants can drive a pupillary reflex in host rat brains. *Proc Nat Acad Sci USA* 84: 6958–6960

Klassen H, Lund RD (1988): Anatomical and behavioral correlates of xenograft-mediated pupillary reflex. *Exp Neurol* 102: 102–108

Klassen H, Lund RD (1990): Parameters of retinal graft-mediated responses are related to underlying target innervation. *Brain Res* 533: 181–191

Land PW, Lund RD (1979): Development of the rat's uncrossed retinotectal pathway and its relation to plasticity studies. *Science* 205: 698–700

Loewy AD (1990): Autonomic control of the eye. In: Central Regulation of Autonomic Functions, Lowey AD, Spyer KM, eds.) Oxford University Press, NY

Lund RD (1964): Terminal distribution in the superior colliculus of fibers originating in the visual cortex. *Nature* 204: 1283–1285

Lund RD (1965): Uncrossed visual pathways of hooded and albino rats. *Science* 149: 1506–1507

Lund RD (1969): Synaptic patterns of the superficial layers of the superior colliculus. *J Comp Neurol* 135: 179–208

Lund RD, Bunt AH (1975): Prenatal development of central optic pathways in albino rats. *J Comp Neurol* 165: 247–264

Lund RD, Cunningham TJ (1972): Aspects of synaptic and laminar organization of the mammalian lateral geniculate body. *Invest Ophthalmol* 11: 291–302

Lund RD, Lund JS (1971): Synaptic adjustment after deafferentation of the superior colliculus of the rat. *Science* 171: 804–807

Lund RD, Lund JS (1973): Reorganization of the retinotectal pathway in rats after neonatal retinal lesions. *Exp Neurol* 40: 377–390

Lund RD, Simons DJ (1985): Retinal transplants: Structural and functional interrelations with the host brain. In: *Neural Grafting in the Mammalian CNS*, Björklund A, Steneviu, eds. Elseiver Science Publishers, B.V., Amsterdam

Lund RD, Mitchell DE, Henry GH (1978): Squint-induced modification of callosal connections in cats. *Brain Res* 144: 169–172

Lund RD, Radel JD, Das S (1990): Development of the light-activated pupillary response in rats, as mediated by optic input from normal or transplanted retinae. *Soc Neurosci Abstr* 16: 656

Lund RD, Radel RD, Horsburgh GM (1991): Intracerebral transplantation of mammalian retina. In: *Vision and Visual Dysfunction. 2. Development and Plasticity of the Visual System*, Cronly-Dillon JR, Macmillan Basingstoke

McCaffery P, Neve RL, Dräger C (1990): A dorso-ventral asymmetry in the embryonic retina defined by protein conformation. *Proc Nat Acad Sci USA* 87: 8570–8574

McLoon LK, McLoon SC, Chang F-LF, Steedman JG, Lund RD (1985): Visual system transplanted to the brain of rats. In: *Neural Grafting in the Mammalian CNS*, Bjorklund A, Stenevi U, eds. Elsevier Science Publishers, B.V., Amsterdam

McLoon S, Lund RD (1980): Specific projections of retina transplanted to rat brain. *Exp Brain Res* 40: 273–282

Meyer RL (1983): Tetrodotoxin inhibits the formation of refined retinotopography in goldfish. *Dev Brain Res* 6: 293–298

Miller BF, Lund RD (1975): The pattern of retinotectal connections in albino rats can be modified by fetal surgery. *Brain Res* 91: 119–125

Mitchell IJ, Dean P, Redgrave P (1980): The projection from superior colliculus to the cuneiform area in the rat. I. Defense-like responses to stimulation with gluatmate in cuneiform nucleus and surrounding structures. *Exp Brain Res* 72: 626–639

Perry VH, Lund RD, McLoon SC (1985): Ganglion cells in retinae transplanted to newborn rats. *J Comp Neurol* 231: 353–365

Radel JD, Das S, Lund RD: Development of light-activated pupilloconstriction in rats as mediated by normal and transplanted retinae. *Eur J Neurosoci 4*: 603–615

Radel JD, Galli-Resta L, Lund RD (1991a): Plasticity in innervation of the rat superior colliculus by transplanted retinal as a result of eye removal at maturity. *Exp Neurol* 112, 252–263

Radel J, Hankin MH, Lund RD (1990a): Proximity as a factor in the innervation of host brain regions by retinal transplants. *J Comp Neurol* 330: 211–229

Radel JD, Kustra DJ, Lund RD (1991b): Rapid enhancement of transplant-mediated pupilloconstriction after elimination of competing host optic input. *Dev Brain Res* 60: 275–278

Radel JD, Lund RD, Das S (1990b): Transplanted retinae: Assessment of functional interactions with host optic input and sensitivity to illumination. *Soc Neurosci Abstr* 16: 965

Sachs GM, Schneider GE (1984): Morphology of optic tract axons arborizing in the superior colliculus of the hamster. *J Comp Neurol* 230: 155–167

Scalia F (1972): The termination of retinal axons in the pretectal region of mammals. *J Comp Neurol* 145: 223–258

Sefton AJ, Dreher B (1985): Visual system. In: *The Rat Nervous System*, Paxinos C, ed. New York: Academic Press

Sefton AJ, Lund RD (1988): Co-transplantation of embryonic mouse retina with tectum, diencephalon or cortex to neonatal rat cortex. *J Comp Neurol* 269: 548–564

Sefton AJ, Lund RD, Perry VH (1987): Target regions enhance the outgrowth and survival of ganglion cells in embryonic retina transplanted to cerebral cortex in neonatal rats. *Dev Brain Res* 33: 145–149

Sefton AJ, Rao K, Hankin MH, Lund RD (1991): Outgrowth of transplant-derived retinal axons in cerebral cortex of neonatal rats. *Soc Neurosci Abstr* 17: 41

Simons DJ, Lund RD (1985): Fetal retinae transplanted over tecta of neonatal rats respond to light and evoke patterned neuronal discharges in the host superior colliculus. *Dev Brain Res* 21: 156–159

Simons DK, O'Leary DDM (1990): Activity dependent and independent aspects of topographic map formation in mammals. *Soc Neurosci Abstr* 16: 1287

Simons DK, O'Leary DDM (1991): Relationship of the retinotopic ordering of axons in the optic pathway of the formation of the visual maps in the control targets. *J Comp Neurol* 307: 393–404

Sperry RW (1943a): Effect of 180 degree rotation of the retinal field on visuomotor coordination. *J Exp Zool* 92: 351–361

Sperry RW (1943b): Visuomotor coordination in the newt *(Triuturus viridenscens)* after regeneration of the optic nerve. *J Comp Neurol* 79: 33–55

Sperry RW (1944): Optic nerve regeneration with return of vision in anurans. *J Neurosphysiol* 7: 57–70

Sperry RW (1948): Patterning of central synapses in regeneration of the optic nerve in teleost. *Physiol Zool* 21: 351–361

Sperry RW (1951): Mechanisms of neural maturation. In: *Handbook of ExperimentalPsychology*,S.S. Stevens (ad.), New York: Wiley, pp 236–280

Thompson ID (1979): Changes in the uncrossed retinotectal projection after removal of the other eye at birth. *Nature* 279: 63–66

Trejo LJ, Cicerone CM (1984): Cells in the pretectal olivary nucleus are in the pathway for the direct light reflex of the pupil in the rat. *Brain Res* 300: 49–62

Wise RP, Lund RD (1976): The retina and central projections of hetero-chromic rats. *Exp Neurol* 51: 68–77

Yee, KT, Lund RD (1991): Effects of reducing maturational disparity on segregation of normal and ectopic retinal inputs. *Soc Neurosci Abstr* 17: 557

Yee KT, Lund RD (1990): Transplanted neural retina expresses antigen specific for normal dorsal retina. *Soc Neurosci Abstr* 16: 1283

Young M, Klassen HK, Lund RD (1991): Effect of lesions of the olivary pretectal nucleus on direct and consensual pupillary reflexes. *Soc Neurosci* 17: 558

CHAPTER 8

Refinement of Topographic Projections in the Rodent, Avian, Amphibian, and Fish Visual Systems

JAMES W. FAWCETT

The topographic mapping of one neuronal structure onto another is a constant feature of brain architecture throughout the vertebrates. Much of our knowledge of how these topographic maps are formed has come from studying the retinotectal projection of fish and amphibia, and on the whole the developmental rules which have been deduced for the formation of the frog and fish retinotectal projection have been found to be transferrable to the formation of topographic projections in other animals and other parts of the brain. However, the lower vertebrate retinotectal projection is in many ways rather abnormal. For instance, most topographic projections are formed over a rather short period of embryonic development, yet the fish and amphibian retinotectal projection gradually enlarges throughout the animal's lifetime (Straznicky and Gaze, 1971, 1972; Gaze et al., 1979); development of most neural projections is followed by a period of naturally occurring cell death during which around half the neurons die (Cowan et al., 1984), yet an equivalent period of cell death is not seen in amphibian retinotectal development.

One way to examine the generality of the developmental mechanisms found in the frog and fish retinotectal projection is to see whether these same mechanisms perform the same functions in equivalent projections which have a developmental schedule more typical in the timing and nature of their events. The retinotectal projection of birds is one such projection; development of the whole chick retinotectal projection, which is substantially larger than that of amphibia, occupies around 12 days, including a period of retinal ganglion cell death (Cowan et al., 1961; Crossland et al., 1975). Another example is the mammalian equivalent of the retinotectal projection, the retinocollicular projection, whose developmental time course is little longer in the rat than in the chick, and which also has a period of naturally occurring retinal ganglion cell death.

The first stages of map formation in the visual system involve axonal pathfinding and recognition of appropriate positional labels in the optic tectum. Such evidence as is available would suggest that the mechanisms underlying these processes are similar in amphibia and birds. One aspect of axonal guidance, for instance, is the tendency of axons from temporal retina to grow in association with other temporal axons. This tendency was demonstrated in the chick visual system

Formation and Regeneration of Nerve Connections
Sansar C. Sharma and James W. Fawcett, Editors
© 1993 Birkhäuser Boston

in a most elegant tissue culture experiment by Bonhoeffer and colleagues (Bonhoeffer and Huf, 1985), and the same axonal behavior has been demonstrated in a similar *Xenopus* tissue culture model by Gooday and Gaze (see Chapter 4, this volume). A similar specificity is also seen in the interaction of retinal axons with their tectal target tissue; Bonhoeffer and Huf developed a tissue culture model in which axons from temporal retina could be shown to associate preferentially with rostral tectum, and subsequent advances have shown this to be due to an inhibitory cell surface glycoprotein in the caudal tectum (Bonhoeffer and Huf, 1982; Walter et al., 1987, 1990). A very similar activity is present in the goldfish optic tectum, causing axons from the temporal goldfish retina to associate more readily with rostral tectal material (Vielmetter and Stuermer, 1989). The mechanisms underlying the first phases of retinotectal map formation are therefore likely to be very similar in amphibia and birds, and therefore probably in mammals also.

Once the initial connections have been formed by retinal ganglion cell axons, a process of refinement begins, which is characterized by the growth and remodeling of terminal arbors and by gradual improvement of the accuracy of the topography of the projection (see Udin and Fawcett, 1988, for review). In fish and amphibia this refinement process is again slightly abnormal, because in addition to regulating the topography of the retinotectal projection, it is involved in the gradual shift of connections in a caudal direction as the tectal tissue is added to the posterior tectal margin, a process that does not occur in birds and mammals (Gaze et al., 1979; Easter and Stuermer, 1984; Reh and Constantine-Paton, 1984). Birds and mammals do, however, substantially remodel the terminal arbors of their retinofugal axons shortly after they have made their first connections (Sachs et al., 1986; Thanos and Bonhoeffer, 1987). One can ask, therefore, whether the processes controlling terminal arbor remodeling and map refinement are the same in the lower vertebrates as in birds and mammals.

The details of how retinal terminals and tectum interact to refine retinotectal topography in the anamniotes are discussed at length elsewhere in this volume, so a short account will suffice here. The two fundamental mechanisms are competition for terminal space on the tectal neurons, and an electrically driven process which ensures that neighboring ganglion cells project to neighboring points on the optic tectum. That tectal space is competed for is demonstrated by size mismatch experiments in which part of the retina or part of the tectum is removed, resulting in the optic projection expanding or compressing to adjust to the available synaptic space (Sharma, 1972; Schmidt et al., 1978; Yoon, 1971). At the same time as this competition for space is proceeding, there is another, more complex process going on, which uses the patterns of electrical activity in the optic axons to ensure that accurate topographic mapping is maintained. This mechanism is the main subject of this chapter.

Axonal pathfinding in the lower vertebrate retinotectal projection is exceptionally accurate, few axons appearing to go to the wrong part of the tectum (Steedman et al., 1979; Fawcett and Gaze, 1982; Stuermer and Easter, 1984a). In order, therefore, to see the electrically driven mapping mechanism in action, one has to disturb the accurate retinotopicity of the projection. The simplest way to do this is to degrade the accuracy of the retinotectal map by crushing the optic nerve and allowing axons to regenerate, which they do rather inaccurately (Gaze and Jacobson, 1963; Meyer, 1980; Schmidt and Edwards, 1983; Stuermer and Easter, 1984b). Following this somewhat random regeneration, topography gradually improves over time, unless the electrical activity mechanism is disabled. This can be done by blocking electrical activity in the retina with tetrodotoxin (TTX), modifying the response of tectal neurons by applying N-methyl-D-aspartate (NMDA) receptor blockers, or coordinating electrical activity by putting the animals under a stroboscope. Another way in which the electrically driven mechanism can be observed in action is by mixing together axons coming from neighboring and non-neighboring ganglion cells; this can be done by causing two positionally equivalent regions of retina to project to the same tectum. Thus, if two eyes project to the same tectum, initially axons from, for instance, the nasal margins of both eyes will project to the same region of tectum, directed there by axonal guidance mechanisms. Subsequently the electrically driven mechanism becomes active, and this distinguishes axons entirely on the basis of geographical proximity; since the ganglion cells in the nasal retina of the two eyes are not physical neighbors, this mechanism will tend to separate them, leading to ocular dominance stripes (Fawcett and Willshaw, 1982). Again, the formation of ocular dominance stripes under these circumstances is inhibited by TTX, by NMDA blockers, and by stroboscopic illumination (Meyer, 1982, 1983; Schmidt and Edwards, 1983; Schmidt and Eisele, 1985; Boss and Schmidt, 1984; Schmidt, 1990; Reh and Constantine-Paton, 1985, Cline and Constantine-Paton, 1989).

It can be argued that the electrically driven mechanism described above evolved so that the retinotectal projection could shift across the tectum during its prolonged development without losing its topography. Since this shift does not occur in birds and mammals, it is theoretically possible that there is no equivalent electrically driven mechanism. We therefore decided to test the higher vertebrate visual system to see whether the mechanism was active and, if so, to define its functions.

Our first experiment designed to this end was done in chicks, and involved mixing the projections from the two eyes (which normally only project contralaterally) to see whether ocular dominance stripes would form (Fawcett and Cowan, 1985). This was done by the simple expedient of damaging the region of the presumptive chiasma, so that a substantial proportion of axons growing through it would be misguided to the ipsilateral tectum, resulting in both eyes projecting to both tecta. In the chick, axons start to reach the tectum around embryonic day 7, forming an optic fiber layer superficial to the tectum. The axons remain in this superficial layer until E12, when they start to penetrate into the tectum and form synapses (Crossland et al., 1975; O'Leary et al., 1983). We looked at the distribution of axons in the optic fiber layer before synaptogenesis, and could see no signs of ocular dominance stripes at this point. It was not until axons actually started to make connections in the tectal neuropil that any eye-specific segregation was observed, and at this early stage the segregation was incomplete, with each eye-specific band having a somewhat fuzzy border. Over the following 8 days, as more retinotectal synapses became established and ganglion cell death in the retina occurred, the boundaries between stripes became extremely sharp, with no overlap between regions innervated by the left and right eyes. Our interpretation of these results was as follows: A mechanism for making ocular dominance stripes, and therefore also presumably for refining retinotectal topography by an electrically driven mechanism, exists in birds. Our disruption of the optic chiasm had comprehensively mixed up axons from the two eyes, so that axons from each eye were found throughout the optic fiber layer of each tectal lobe. When these axons started to make synapses, many of them were doing so near axons from the other eye, which therefore had different patterns of spontaneous activity. This led to a gradual separation of the different eye projections into sharply defined ocular dominance stripes. This separation could have been achieved either by a gradual shifting of axons through terminal arbor growth and pruning, by the death of ganglion cells projecting into areas in which the other eye was dominant, or by a combination of the two.

We were particularly interested in the idea that retinal ganglion cell death in birds and mammals might be a mechanism for removing axons that had made topographically inaccurate connections, and that it might be controlled by the electrical activity mechanism. This idea was suggested by our observations on the formation of ocular dominance stripes in the chick, where we saw a transition from fuzzy indistinct stripes to very sharp, well-demarcated stripes during the period of retinal ganglion cell death. We decided, therefore, to see whether we could produce any evidence that electrical activity in retinal ganglion cells might play a role in controlling their death.

To this end we first chose as an experimental system the ipsilaterally projecting ganglion cells in the rat retina. This experimental model was chosen because there was clear evidence that the death of these ipsilaterally projecting cells during the period of naturally occurring cell death shortly after birth was controlled, at least in part, by competitive interactions with other axons on the surface of the superior colliculus. In rats there is a projection from the retina to the ipsilateral colliculus at birth, but most of this projection is lost due to the death of ganglion cells by 10 days after birth. However, if one eye is removed at birth to remove competing axons, the ipsilaterally projecting cells in the remaining eye mostly survive (Jeffery and Perry, 1982; Land and Lund, 1979; Bunt et al., 1983; Insausti et al., 1984). We wished to see if the competition between ipsilateral and contralateral axons was in some way mediated by an activity-related mechanism (Fawcett et al., 1984). We therefore gave newborn animals repeated injections of TTX into one eye for the first 12 days of life, after which we retrogradely labeled the surviving ganglion cells by injections of the fluorochrome fast blue to the colliculus, or orthogradely labeled the surviving axons by injecting Wheat germ agglutinin-horse radish peroxidase (WGA-HRP) into the eye. This experiment showed a clear increase, by approximately a factor of two, in the number of surviving ipsilaterally projecting ganglion cells when the other eye was treated with TTX; however, this still meant that around 70% of the ipsilaterally projecting cells were dying. We had to conclude that this particular form of interaxonal competition was only partly mediated by an electrically activated mechanism. This was, however, the first direct evidence of a link between activity and the control of retinal ganglion cell death.

In the next experiment we looked directly for a link between retinal ganglion cell death and the removal of axons that had made topographic targeting errors (Fawcett and O'Leary, 1985; O'Leary et al., 1986). This would be extremely difficult to do in a frog or fish, in which initial axon targeting appears to be very accurate. However, many axons growing onto the chick or rat tectum appear to grow in a most disorderly way, initially going to the wrong region of the tectum. This has been shown directly in recent studies by O'Leary and his colleagues, in which axons originating from small regions of the retina have been anterogradely labeled (Nakamura and O'Leary, 1989; Simon and O'Leary, 1990). These experiments show that before the period of ganglion cell death, axons from all parts of the retina can be seen to have grown almost randomly on the tectum, although there is a concentration of axons which have grown to the topographically correct area. Axons from peripheral temporal retina, for instance, are found in some profusion in the caudal tectum, as well as in the rostral tectum, which is their correct target. If the label is applied to an adult animal, axons labeled from a small region of retina almost all follow a path directly to the correct region of the tectum and have their terminals concentrated in a very small region of the tectum; a few axons have acute bends in them, suggesting pathway corrections. The axons that had grown randomly must therefore be removed, either by death of the whole axon or by removal of the part of it that projects inappropriately. We were able to show that at least some of these axons which had made targeting errors were removed by ganglion cell death. To do this, we made small injections of fast blue into the posterior margin of the rat superior colliculus. If this is done in an adult animal, all the ganglion cells labeled by retrograde transport from this small injection are in the nasal margin of the retina. However, if a newborn animal is similarly injected and killed before the period of retinal cell death, ganglion cells all over the retina are labeled as well as a concentration in nasal retina, indicating that many ganglion cells all over the retina have sent their axons incorrectly to the caudal superior colliculus. The transition between these two labeling patterns involves ganglion cell death. If an animal is injected with label at birth, so that all the erroneously projecting cells are labeled, but then allowed to survive until day 12, after the period of ganglion cell death, then it can be seen that very few of the erroneously projecting ganglion cells have survived. Almost all the label is concentrated in a group of cells in the nasal retina, just as if an adult animal had been labeled. Thus almost all the ganglion cells scattered throughout the retina whose axons had made targeting errors died during the period of retinal ganglion cell death. Since around half the total number of ganglion cells die during this period, the implication is that many of the cells that die during the period of cell death will be ones whose axons had grown to the wrong region of their target; there is preferential elimination of erroneously projecting ganglion cells.

The mechanism for correcting topographic targeting errors in the lower vertebrate retinotectal projection is electrically driven. The question therefore arises as to whether the mechanism that decides which ganglion cells have connected to the wrong area of the tectum, and will therefore die, is a similar activity-driven mechanism. Before answering this question, however, it is necessary to consider whether there is any electrical activity in retinal ganglion cells during the period of ganglion cell death. During the first 10 days of life, when retinal ganglion cells are dying, the eyes of rats are still closed and the retinal circuitry itself is not fully connected up. If there is electrical activity in retinal ganglion cells during this period, therefore, it must be spontaneous. Spontaneous activity in immature neurons is the rule rather than the exception, and there is now direct evidence that electrical activity in rodent retinal ganglion cells is established well before the onset of cell death and continues on into this period (Masland, 1977; Meister et al., 1991). Moreover, the activity is far from random. Waves of excitation have been recorded sweeping across the retina, activating all the neighboring ganglion cells on the crest of the wave simultaneously. To determine whether this activity is involved in controlling the pattern of ganglion cell death, we blocked nerve transmission in the retina during the period of ganglion cell death by giving daily TTX injections, having first given a fast blue injection to the caudal margin of the colliculus on the day of birth to label up the erroneously projecting ganglion cells. On killing these animals after 12 days, the pattern of retinal labeling was very similar to that seen at birth: there were many labeled cells spread throughout the retina, and the ratio of the density of labeled cells in the correctly projecting (nasal) retina to the density of labeled cells in temporal retina was the same as is found at birth. TTX blockade had therefore prevented the elimination of the incorrectly projecting ganglion cells. This could in theory have been either because all cell death had been prevented, or because cell death had occurred randomly rather than being directed toward erroneously projecting cells. To distinguish between these two possibilities, we counted the number of axons in the optic nerves of the TTX-treated animals and found that there was no significant inhibition of cell death. Cell death had therefore occurred at the same rate as in normal animals, but it was random rather than being directed toward erroneously projecting cells. From this evidence it is reasonable to suppose that an electrically driven mechanism, similar to that which rearranges axon terminals in the frog and fish retinotectal projection, is also responsible in the higher vertebrates for marking out for eventual death ganglion cells whose axons have made topographic targeting errors.

Many types of long-term synaptic plasticity are mediated via the NMDA type of glutamate receptor, which, when activated, allows calcium to enter the cell as long as it has already been depolarized by other excitatory inputs (Morris et al., 1986; Morris, 1989). In particular, the mechanism responsible for ocular dominance stripe formation and map refinement in fish and frogs can be inactivated by NMDA receptor blockers (Schmidt, 1990; Cline et al., 1987). The circuitry here is reasonably straightforward, since glutamate is the transmitter of retinotectal axons: the postsynaptic tectal neurons are excited by glutamate released from retinofugal axons, and at the same time glutamate activates NMDA channels, allowing calcium to enter the postsynaptic neurons, provided they have been sufficiently depolarized by their retinal and other inputs. The calcium entry must put into train the events which stabilize those synapses which are active while the cell is strongly depolarized. It is reasonable to ask whether the control of retinal ganglion cell death in mammals via the electrical activity mechanism also involves NMDA channels, but here the exact way in which this might work is less clear. The transmitter of the retinocollicular axons is not glutamate, and it is not entirely clear which transmitter these axons secrete, although there is reasonably good evidence that the transmitter is N-acetylaspartlyglutamate, which does have activity on NMDA channels. On the other hand, there are intrinsic collicular glutaminergic circuits, and there is also a major glutaminergic input from the visual cortex to the retinorecipient layers of the superior colliculus. It is reasonable to suppose, therefore, that the activity of retinorecipient neurons in the colliculus might be influenced by glutaminergic inputs, and that an NMDA receptor blocker might also have effects. Indeed, there is some direct evidence for an influence of NMDA receptor blockers on long-term potentiation resulting from the stimulation of optic fibers (Tsai et al., 1990; Miyamoto et al., 1990; Fosse et al., 1984, 1989; Huettner and Baughman, 1988). There is also evidence that the remodeling of terminal arbors in the lateral geniculate, the axons of which are branches of those projecting to the superior colliculus, is inhibited by NMDA blockade (Hahm et al., 1991).

Given this information, it seemed reasonable to look for an effect of NMDA blockade on the control of retinal ganglion cell death. We tried two methods of administering the blockade: APV was administered locally by impregnating it into a polymer which was implanted over the colliculus, and MK801 was administered systemically by subcutaneous injection. Both methods had disadvantages. Systemic MK801 made the newborn animals drowsy and disinclined to suckle, while the APV-containing polymer tended to get dislodged. Our clearest results were obtained with the animals that had daily MK801 injections for the first 12 days of life. These animals had a fast blue injection to the posterior colliculus on the day of birth, as in the TTX-treated animals described previously, and were then killed on day 12 and their retinae examined. The pattern of labeling in the retinae was almost identical to that seen following TTX blockade. The relative densities of correctly and erroneously projecting cells were the same as at birth, indicating that retinal ganglion cell death had been either randomized or prevented. Optic nerve axon counts showed that retinal ganglion cell death had occurred at the normal time in MK801-treated animals, and in fact around 10% more cells than normal had died during this period. Thus the effects of NMDA blockade on retinal ganglion cell death are very similar to those of TTX blockade. Retinal ganglion cell death occurs on schedule and in roughly normal amounts, but the cells that die are selected at random rather than being predominantly those whose axons have made targeting errors.

The experiments described above allow one to make some detailed comparisons of the mechanisms active in the formation of the topographic retinotectal map in frogs and fish, and the equivalent mechanisms involved in the much faster development of birds and mammals. Axonal pathfinding is clearly much less precise in the higher vertebrates, where many retinofugal axons grow to the wrong region of the target structure. This disorder is considerably more pronounced in albino than in pigmented rats (Bunch and Fawcett, 1990). Albinos tend to misroute axons at the chiasma (Guillery, 1986), and it may be that axons become topographically disordered at the same point, this disorder persisting until axons reach their target. However, even pigmented animals have many more aberrant axons than frogs or fish. It is possible that the very much greater speed of development might be a reason for this. In frogs and fish axons grow a few at a time onto the tectum, and these few axons are probably guided by axons that have already reached their targets (Gaze and Fawcett, 1983; Fawcett, 1985). In birds and mammals the number of axons growing at any one time is very much greater, and many of the axons arrive as a large group. This might allow fewer possibilities for interactions with pioneer axons. Having reached their targets, frog and fish axons have the topography of their connections refined by an activity-related mechanism. In mammals this activity-related mechanism probably still has the same function, driving the reordering of axons that have made initial connections near their correct target area. However, the same mechanism is also used for a second purpose. This purpose is to determine which ganglion cells have projected to completely the wrong region of their target, and to mark them out for death during the period of naturally occurring ganglion cell death during the first few days of life. In comparing the mechanisms responsible for map formation in amniotes and anamniotes, one can say, therefore, that the same basic mechanisms are doing roughly the same job, but with changes of detail and emphasis. Despite its atypical features, the lower vertebrate retinotectal projection has proven an excellent model system in which to work out developmental mechanisms that are applicable in different species and in different neuronal projections.

REFERENCES

Bonhoeffer F, Huf J (1982): In vitro experiments on axon guidance demonstrating anterior-posterior gradient on the tectum. *EMBO J* 1: 427–431

Bonhoeffer F, Huf J (1985): Position dependent properties of retinal axons and their growth cones. *Nature* 315: 409–410

Boss VC, Schmidt JT (1984): Activity and the formation of ocular dominance patches in dually innervated tectum of goldfish. *J Neurosci* 4: 2891–2905

Bunch ST, Fawcett JW (1990): A comparison of the initial retinal ganglion cell projection to the contralateral superior colliculus in albino and pigmented rats. *Dev Brain Res* 52: 259–264

Bunt SM, Lund RD, Land PW (1983): Prenatal development of the optic projections in albino and hooded rats. *Dev Brain Res* 6: 149–168

Cline HT, Constantine-Paton M (1989): NMDA receptor antagonists disrupt the retinotectal topographic map. *Neuron* 3: 413–426

Cline HT, Debski EA, Constantine-Paton M (1987): N-Methyl-D-aspartate receptor antagonist desegrates eye-specific stripes. *Proc Natl Acad Sci USA* 84: 4342–4345

Cowan WM, Adamson L, Powell TPS (1961): An experimental study of the avian visual system. *J Anat* 95: 545–563

Cowan WM, Fawcett JW, O'Leary DDM, Stanfield BB (1984): Regressive events in neurogenesis. *Science* 225: 1258–1265

Crossland WJ, Cowan WM, Rogers LA (1975): Studies on the development of the chick optic tectum. IV. An autoradiographic study of the development of the retinotectal connections. *Brain Res* 91: 1–23

Easter SS Jr., Stuermer CAO (1984): An evaluation of the hypothesis of shifting terminals in the goldfish optic tectum. *J Neurosci* 4: 1052–1063

Fawcett JW (1985): Factors guiding regenerating retinotectal fibres in the frog *Xenopus laevis*. *J Embryol Exp Morphol* 90: 233–250

Fawcett JW, Cowan WM (1985): On the formation of eye dominance stripes and patches in the doubly innervated optic tectum of the chick. *Dev Brain Res* 17: 149–163

Fawcett JW, Gaze RM (1982): The retinotectal fibre pathways from normal and compound eyes in *Xenopus*. *J Embryol Exp Morphol* 72: 19–37

Fawcett JW, O'Leary DDM (1985): The role of electrical activity in the formation of topographic maps in the nervous system. *TINS* 8: 201–206

Fawcett JW, Willshaw DJ (1982): Compound eyes project stripes on the optic tectum in *Xenopus*. *Nature* 296: 350–352

Fawcett JW, O'Leary DDM, Cowan WM (1984): Activity and the control of ganglion cell death in the rat retina. *Proc Natl Acad Sci USA* 81: 5589–5593

Fosse VM, Heggelund P, Fonnum F (1989): Postnatal development of glutamatergic, GABAergic, and cholinergic neurotransmitter phenotypes in the visual cortex, lateral geniculate nucleus, pulvinar, and superior colliculus in cats. *J Neurosci* 9: 426–435

Fosse VM, Heggelund P, Iversen E, Fonnum F (1984): Effects of area 17 ablation on neurotransmitter parameters in efferents to area 18, the lateral geniculate body, pulvinar and superior colliculus in the cat. *Neurosci Lett* 52: 323–328

Gaze RM, Fawcett JW (1983): Pathways of *Xenopus* optic fibres regenerating from normal and compound eyes under various conditions. *J Embryol Exp Morphol* 73: 17–38

Gaze RM, Jacobson M (1963): A study of the retinotectal projection during regeneration of the optic nerve in the frog. *Proc R Soc Lond (Biol)* 57: 420–448

Gaze RM, Keating MJ, Ostberg A, Chung SH (1979): The relationship between retinal and tectal growth in larval *Xenopus*. Implications for the development of the retinotectal projection. *J Embryol Exp Morphol* 53: 103–143

Guillery RW (1986): Neural abnormalities of albinos. *TINS* 9: 364–367

Hahm J-O, Langdon RB, Sur M (1991): Disruption of retinogeniculate afferent segregation by antagonists to NMDA receptors. *Nature* 351: 568–570

Huettner JE, Baughman RW (1988): The pharmacology of synapses formed by identified corticocollicular neurons in primary cultures of rat visual cortex. *J Neurosci* 8: 160–175

Insausti R, Blakemore C, Cowan WM (1984): Ganglion cell death during development of ipsilateral retino-collicular projection in golden hamster. *Nature* 308: 362–365

Jeffrey G, Perry VH (1982): Evidence for ganglion cell death during development of the ipsilateral retinal projection in the rat. *Dev Brain Res* 2: 176–190

Land PW, Lund RD (1979): Development of the uncrossed retinotectal pathway, and its relation to plasticity studies. *Science* 205: 698–700

Masland RH (1977): Maturation of function in the development rabbit retina. *J Comp Neurol* 175: 275–286

Meister M, Wong ROL, Baylor DA, Shatz CJ (1991): Synchronous bursts of action potentials in ganglion cells of the developing mammalian retina. *Science* 252: 939–943

Meyer RL (1980): Mapping the normal and regenerated retinotectal projection of goldfish with autoradiographic methods. *J Comp Neurol* 189: 273–289

Meyer RL (1982): Tetrodotoxin blocks the formation of ocular dominance columns in goldfish. *Science* 218: 589–591

Meyer RL (1983): Tetrodotoxin inhibits the formation of refined retinotopography in the goldfish. *Dev Brain Res* 6:293–298

Miyamoto T, Sakurai T, Okada Y (1990): Masking effect of NMDA receptor antagonists on the formation of long-term potentiation (LTP) in superior colliculus slices from the guinea pig. *Brain Res* 518: 166–172

Morris RGM (1989): Synaptic plasticity and learning: Selective impairment of learning in rats and blockade of long-term potentiation *in vivo* by the *N*-methyl-D-aspartate receptor antagonist AP5. *J Neurosci* 9: 3040–3057

Morris RGM, Anderson E, Lynch GS, Baudry M (1986): Selective impairment of learning and blockade of long-term potentiation by an N-methyl-D-aspartate receptor antagonist, AP5. *Nature* 319: 774–776

Nakamura H, O'Leary DDM (1989): Inaccuracies in initial growth and arborization of chick retinotectal axons followed by course corrections and axon remodeling to develop topographic order. *J Neurosci* 9: 3776–3795

O'Leary DDM, Fawcett JW, Cowan WM (1986): Topographic targeting errors in the retinocollicular projection and their elimination by selective ganglion cell death. *J Neurosci* 6: 3692–3705

O'Leary DDM, Gerfen C, Cowan WM (1983): The development and restriction of the ipsilateral retinofugal projections in the chick. *Dev Brain Res* 10: 93–109

Reh TA, Constantine-Paton M (1984): Retinal ganglion cell terminals change their projection sites during larval development of *Rana pipiens*. *J Neurosci* 4: 442–457

Reh TA, Constantine-Paton M (1985): Eye specific segregation requires neural activity in 3 eyed *Rana pipiens*. *J Neurosci* 5: 1132–1143

Sachs GM, Jacobson M, Caviness VS (1986): Postnatal changes in arborization patterns of murine retinocollicular axons. *J Comp Neurol* 246: 395–408

Schmidt JT (1990): Long-term potentiation and activity-dependent retinotopic sharpening in the regenerating retinotectal projection of goldfish: Common sensitive period and sensitivity to NMDA blockers. *J Neurosci* 10: 233–246

Schmidt JR, Edwards DL (1983): Activity sharpens the map during the regeneration of the retinotectal projection in goldfish. *Brain Res* 268: 28–39

Schmidt JT, Eisele LE (1985): Stroboscopic illumination and dark rearing block the sharpening of the regenerated retinotectal map in goldfish. *Neuroscience* 14: 535–546

Schmidt JT, Cicerone CM, Easter SS Jr. (1978): Expansion of the half retinal projection to the tectum in goldfish: An electrophysiological and anatomical study. *J Comp Neurol* 177: 257–278

Sharma SC (1972): Reformation of the retinotectal projections after various tectal ablations in adult goldfish. *Exp Neurol* 34: 171–182

Simon DK, O'Leary DDM (1990): Limited topographic specificity in the targeting and branching of mammalian retinal axons. *Dev Biol* 137: 125–134

Steedman JG, Stirling RV, Gaze RM (1979): The central pathways of optic fibres in *Xenopus* tadpoles. *J Embryol Exp Morphol* 50: 199–215

Straznicky C, Gaze RM (1971): The growth of the retina in *Xenopus laevis*: An autoradiographic study. *J Embryol Exp Morphol* 26: 67–79

Straznicky C, Gaze RM (1972): The development of the tectum in *Xenopus laevis*: An autoradiographic study. *J Embryol Exp Morphol* 28: 87–115

Stuermer CA, Easter SS Jr. (1984a): Rules of order in the retinotectal fascicles of goldfish. *J Neurosci* 4: 1045–1051

Stuermer CAO, Easter SS Jr. (1984a): A comparison of the normal and regenerated retinotectal pathways of goldfish. *J Comp Neurol* 223: 57–76

Thanos S, Bonhoeffer F (1987): Axonal arborization in the developing chick retinotectal system. *J Comp Neurol* 261: 155–164

Tsai G, Stauch BL, Vornov JJ, Deshpande JK, Coyle JT (1990): Selective release of N-acetylaspartylglutamate from rat optic nerve terminals in vivo. *Brain Res* 518: 313–316

Udin SB, Fawcett JW (1988): Formation of topographic maps. *Ann Rev Neurosci* 11: 289–328

Vielmetter J, Stuermer CAO (1989): Goldfish retinal axons respond to position-specific properties of tectal cell membranes in vitro. *Neuron* 2: 1331–1339

Walter J, Allsopp TE, Bonhoeffer F (1990): A common denominator of growth cone guidance and collapse. *TINS* 13: 447–452

Walter J, Henke-Fahle S, Bonhoeffer F (1987): Avoidance of posterior tectal membranes by temporal retinal axons. *Development* 101: 909–913

Yoon MG (1971): Reorganization of retinotectal projection following surgical operations on the optic tectum in goldfish. *Exp Neurol* 33: 395–411

CHAPTER 9

Pyramidal Cell Modules in Rat Visual Cortex: Their Structure and Development

ALAN PETERS

INTRODUCTION

Several years ago it was found that the apical dendrites of some pyramidal cells in the neocortex do not ascend through the cortex in a random fashion, as might be deduced from an examination of Golgi-impregnated material in which only a random sample of neurons is visible. Instead, when all of the apical dendrites are visible, as they are in semithick "plastic" sections of material prepared for electron microscopy, it is found that the apical dendrites ascend through the cortex in aggregates that we have termed "clusters" (Peters and Walsh, 1972). Such clustering of apical dendrites was demonstrated almost concurrently in primate cortex (von Bonin and Mehler, 1971), rat somatosensory cortex (Peters and Walsh, 1972), and rabbit and cat sensorimotor cortex (Fleischhauer et al., 1972). Subsequent studies (e.g., Feldman and Peters, 1974; Fleischhauer, 1974; Fleischhauer and Detzer, 1975; Winkelman et al., 1975) suggested that the clustering of apical dendrites is probably a basic arrangement in the neocortices of all mammals, with the sizes and number of the apical dendrites within the clusters, and the spacing between them, depending upon the particular cortical area being examined and upon the species of animal involved.

For several years, this arrangement of the apical dendrites remained little more than an interesting fact and was not pursued further. One reason for this was that interest shifted away from studies of arrangements of populations of cortical neurons toward studies of individual neurons, which could be examined in detail by electron microscopy after they had been labeled either by Golgi impregnations (Fairén et al., 1977) or by intracellular filling with horseradish peroxidase (Martin, 1988). These techniques allowed the characteristics and synaptic connections of specific types of neurons to be examined, and particular attention was paid to the nonpyramidal inhibitory GABAergic neurons, whose role in the cerebral cortex had hitherto been only poorly understood. Our own particular studies of the inhibitory neurons concentrated on the cells in rat visual cortex (Peters, 1985), which was chosen because the layering is simple (Figure 9.1) and there are

Formation and Regeneration of Nerve Connections
Sansar C. Sharma and James W. Fawcett, Editors
© 1993 Birkhäuser Boston

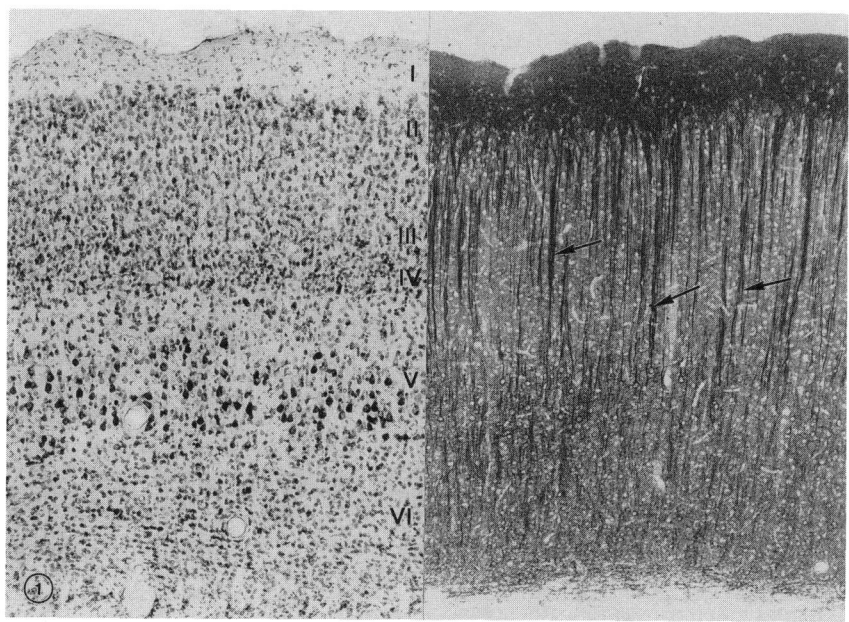

FIGURE 9.1. Nissl- (left) and MAP2- (right) stained preparations of rat visual cortex. In the MAP2-stained section the apical dendrites of the large and medium-sized pyramidal cells of layer V form clusters (arrows) that ascend toward the upper layers of the cortex. There they are joined by the apical dendrites of the layer II/III pyramids. These apical dendrites form their terminal tufts in layer I, which is densely stained in the MAP2-stained section. ×54.

relatively few types of neurons contained within it. These types of studies were carried out because it was felt that by understanding the synaptic relations of the various neuronal types it would be possible to understand how the physiological response properties of neurons are generated. However, this goal remains largely unfulfilled (Martin, 1988). One reason is that although the types of synapses made by the axons of individually labeled neurons can be determined, and the form of their postsynaptic partners recognized, the sources of the postsynaptic elements cannot usually be specifically identified. Consequently, the input to individual neurons cannot yet be determined. The difficulty of this task can be imagined if it is remembered that the average neurons in rat visual cortex, for example, can receive about 10,000 synapsing axon terminals (Peters, 1987).

Difficulties such as this led us to consider that further progress in understanding the functioning of the cerebral cortex might be better achieved by examining the neuronal composition of the cerebral cortex and how its neurons are arranged. Our first attempt at achieving this new goal was made in a study of the rat visual cortex (Peters and Kara, 1985a, 1985b, 1987; Peters et al., 1985). These studies led us to the realization that the neocortex is composed of repeating modules of pyramidal cells, and that the axes of these modules are the dendritic clusters that we had first encountered almost 20 years ago.

Since the arrangement of the pyramidal cells is particularly evident in rat visual cortex, this piece of cortex can serve as a model for understanding the similar arrangements in other cortices.

RAT VISUAL CORTEX

Neuronal Composition

Nissl-stained preparations (Figure 9.1) show that rat primary visual cortex (area 17) contains six layers (Schober and Winkelman, 1975; Peters and Kara, 1985a, 1985b). Layer I on the outside of the cortex contains few neurons, and its border with layer II/III is well marked because the outer portion of layer II/III contains closely packed, small neurons. With increasing depth the neurons of layer II/III become gradually larger, but there is no obvious boundary that allows layer II to be distinguished from layer III. Both layers contain a preponderance of pyramidal cells, whose apical dendrites usually reach as far as upper layer II before forming their terminal tufts that spread into layer I. Layer IV in this cortex also contains a preponderance of cells with apical dendrites, and these neurons can be regarded as star-pyramids, because unlike the pyramidal cells present in layer II/III and in layer V, they have dendrites radiating out in all directions from their cell bodies and have only a flimsy apical dendrite. No clearly defined small spiny stellate cells like those that characterize layer IV in cat and monkey visual cortex have been encountered in the rat. In Nissl preparations the cell bodies of the small neurons in layer IV are closely packed, in marked contrast to the situation in layer V, in which the neurons are sparser and have larger cell bodies. Layer V contains both large and medium-sized pyramidal cells, and the former are more prevalent in the lower half of layer V. This allows the border with layer VIa to be easily discerned, because in layer VIa the cell bodies of the small pyramidal cells are closely packed and many of them are arranged in vertical rows. Unlike those in other layers, the apical dendrites of the small pyramidal cells in layer VIa only rarely reach as far as layer I. Instead, most of the layer VI pyramids form their apical tufts in layer IV. Finally, the deepest layer of the cortex is layer VIb. This also contains rather small neurons, but instead of being rounded, the cell bodies of these neurons are usually elongate in the horizontal plane. None of the neurons in layer VIb have the features of pyramidal cells. Instead they are either multipolar or bitufted, horizontally aligned cells.

At this point it should be stated that in Golgi-impregnated preparations it is readily apparent that the most common type of neuron in rat visual cortex is the pyramidal cell (Parnavelas et al., 1977; Werner et al., 1979; Peters and Kara, 1985a). Pyramidal cells occur in all layers except I and VIb, and they account for 85% to 90% of all neurons (Peters et al., 1985; Gabbott and Stewart, 1987). The other types of neurons are nonpyramidal neurons, which mostly contain GABA and are inhibitory in function (Meinecke and Peters, 1987)

Dendritic Arrangements

As stated earlier, the majority of the apical dendrites of the pyramidal cells in layer V of rat visual cortex are aggregated into clusters that pass through the cortex to reach layer I, where they form their terminal tufts. As the clusters reach layer II/III, the apical dendrites of the pyramidal cells in that layer are added to them, so that the clusters become progressively larger. In the first instance, this information had to be gained by following apical dendrites in serial semithick plastic sections (Peters and Kara, 1987), but recently a more efficient and less painstaking way to examine the arrangement of dendrites has been developed through use of antibodies produced against microtubule-associated protein 2 (MAP2), a protein which occurs in the perikarya and dendrites of neurons (de Camilli et al., 1984). We have recently used a MAP2 antibody (anti-MAP2; 5F9; Kosik et al., 1984), whose binding sites can be revealed with the horseradish peroxidase reaction, to reexamine area 17 of rat visual cortex. An illustration of a vertical section of this cortex stained with the MAP2 antibody is shown on the right side of Figure 9.1.

The MAP2 antibody clearly shows the darkly staining clusters of apical dendrites of layer V pyramidal cells passing through the cortex, and after they reach layer III it can be seen that the clusters become increasingly thicker as the apical dendrites of the layer III, and then the layer II pyramidal cells, are added to them. As they approach layer I, all of these apical dendrites form their apical tufts, and this produces a dense concentration of dendritic branches in layer I, causing it to stain darkly with the MAP2 antibody. This arrangement of apical dendrites can be seen to better advantage in Figure 9.2. The clusters are also well defined in horizontal sections taken at the level of layer IV (Figure 9.3) in which the thick apical dendrites of the layer V pyramidal cells are in cross section. If the clusters containing the thick dendrites are counted, then it is found that there are some 360 clusters/mm^2 of section area, so that their average center-to-center spacing is about 60 μm. In our earlier analysis (Peters and Kara, 1987) it was determined that on average each cluster contains the apical dendrites of 8 large and 38 medium-sized pyramids. However, not all of the apical dendrites of the layer V pyramidal cells are contained within the clusters. Others ascend either singly or in pairs as they traverse layer IV, but whether these dendrites join the clusters at upper levels is not known.

Interestingly, the apical dendrites of the layer VIa pyramidal cells do not join the clusters. Instead, they form their own independent groups, which will be referred

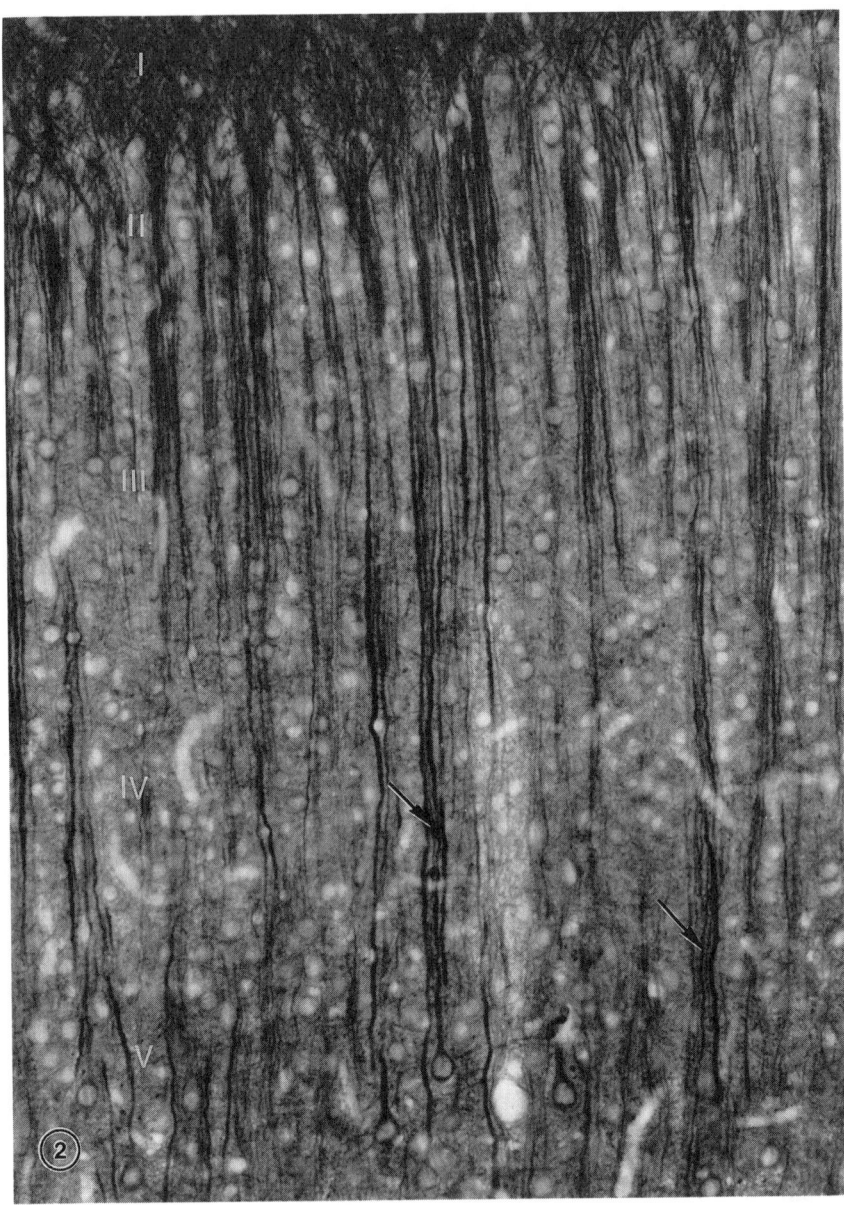

FIGURE 9.2. MAP2-stained vertical section of area 17. At this higher magnification the darkly stained apical dendrites of the layer V pyramidal cells can be better seen to form clusters (arrows). In layers III and II other pyramidal cells add apical dendrites to the clusters, which splay out into their terminal tufts just below layer I. ×175.

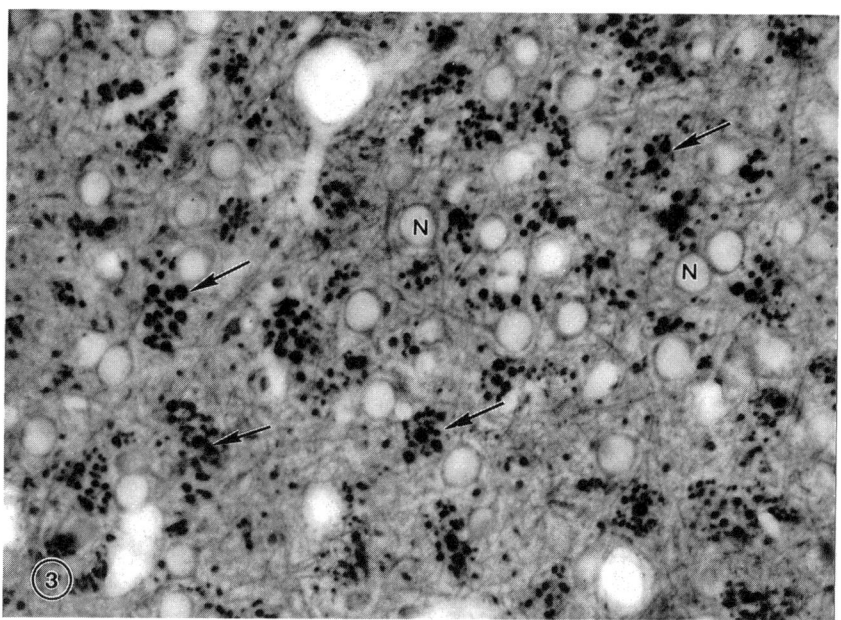

FIGURE 9.3. MAP2-stained tangential section at the level of layer IV. The clusters of darkly stained and transversely sectioned apical dendrites are readily apparent (arrows). In the clusters the thicker dendrites are from the large layer V pyramids, and the thinner ones are from the medium-sized pyramids. Note the faintly stained layer IV neurons *(N)* with dendrites extending into the neuropil. ×350.

to as "bundles." These bundles are numerous, and because the pyramidal cells of layer VIa are rather small, their apical dendrites are thin. To add to the bundles, many of the apical dendrites of the layer VIa pyramids first pass obliquely, so that their parent cell bodies are often tilted. Upon reaching a bundle, however, the apical dendrites bend and ascend toward layer V by following a more vertical, but often sinuous, trajectory. Reconstructions suggest that the bundles are formed so that the apical dendrites of the layer VIa pyramids can pass through the gaps between the groups of neuronal cell bodies in layer V. Thus, they attain layer IV, in which most of the apical dendrites in the bundles form their terminal tufts. Consequently, in lower layer IV both the clusters and the bundles of apical dendrites are present, but they can be readily distinguished from each other by the differences in the thickness of the apical dendrites that they contain.

The apical dendrites of the small layer IV pyramidal cells in the rat visual cortex also ascend independently of the clusters. Because they are so thin, these apical dendrites are difficult to follow, but as far as can be determined they follow a similar pattern to the layer VIa pyramids. Their apical dendrites aggregate into groups, but in this instance the groups are formed so that they can pass between the cell bodies of the lower layer III pyramidal cells.

In summary, the apical dendrites of the layer V and layer II/III pyramidal cells largely ascend through the cortex in well-organized, rather evenly spaced clusters. In contrast, the apical dendrites of the smaller layer VIa and layer IV pyramidal cells ascend in a more random fashion, only grouping together as a means to pass between the cell bodies of the neurons in the layer above them. Because the layer V and layer II/III pyramidal cells are so well organized, it is suggested that they are assembled into vertical neuronal modules centered upon the clusters of their apical dendrites, and that these neuronal modules represent the fundamental neuronal units of the cerebral cortex. In rat visual cortex the diameters of the modules would be 60 μm, which is the center-to-center spacing of the clusters. Each module of this kind would have an average of 290 neurons associated with it, of which about 140 are the layer V and layer II/III pyramids, and the others the nonclustered pyramidal cells of layers VIa and IV, as well as the nonpyramidal cells of the cortex (Peters et al., 1985). A diagram representing such a module is shown in Figure 9.4.

DEVELOPMENT OF NEURONAL MODULES

If the cerebral cortex is composed of repeating neuronal units, or modules, then it is likely that these modules are formed during development. As is now well known, during development of the cortex, cells destined to be pyramidal neurons undergo their final divisions in the generative epithelium of the ventricular zone, which borders the cerebral ventricles. The postmitotic cells then ascend through the intermediate zone and move toward the outside of the developing cortex or cortical plate. As the cells migrate, they have a bipolar shape (Rakic, 1972, 1977; Peters and Feldman, 1973), with a stout leading process extended toward the pial surface and a thinner and shorter trailing process, so that the migrating neurons are radially oriented. This migration of the neuroblasts toward the pial surface is guided by radial glial fibers that extend from the ventricle to the glial limiting membrane, and a number of recent studies have examined the dramatic changes that occur in the radial glial fiber system during development. One interesting point is that early in development the radial glial fibers are gathered into fascicles that extend across the entire depth of the developing hemispheres, but by embryonic day 16 (E16), in the mouse at least, the fascicles begin to break down and only single radial glial fibers with a relatively constant spacing between them persist (Gadisseux et al., 1989, 1990). This event seems to coincide with the time when the majority of cells are entering the cortical plate. This means that late in embryonic development, as migrating cells pass through the intermediate zone (Figure 9.5: IZ), they are in contact with a number of radial glial fibers. By once

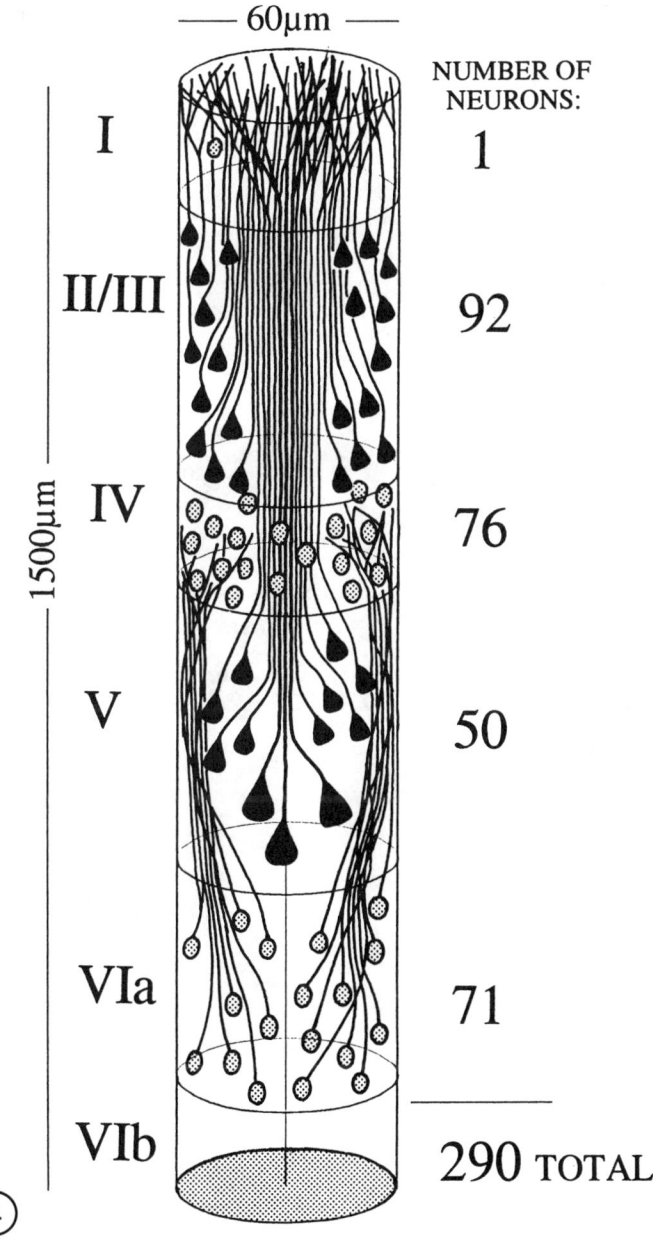

FIGURE 9.4. Diagrammatic representation of the pyramidal cell module in area 17 of the rat. The module, containing 290 neurons, is 60 μm in diameter, and extends through the depth of the cortex. While the apical dendrites of the layer V and layer II/III pyramids form the cluster that is the axis of the module, the apical dendrites of the smaller layer VIa and layer IV pyramids form bundles.

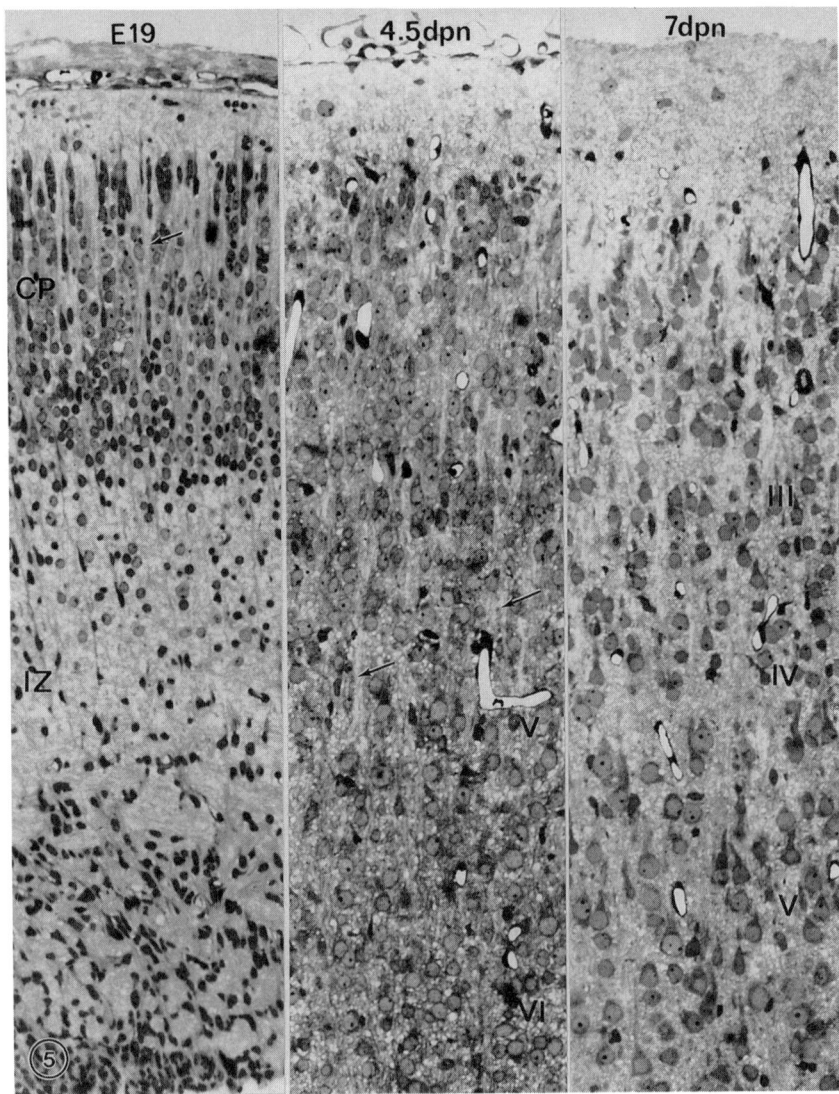

FIGURE 9.5. Vertical semithick plastic sections through the developing cerebral cortex. These sections are through the cortex of a 19-day embryo (E19), at 4.5 days postnatal (4.5dpn), and at 7 days postnatal (7dpn). At E19 the cortical plate (CP) lies above the intermediate zone (IZ), and consists of vertical stacks of cells separated by neuronal processes (arrow). By 4.5dpn the cortex has become thicker. Layers VIa and V have differentiated and the clusters (arrows) are visible. By 7dpn all of the cortical layers can be discerned and the neurons have become more separated from each other. ×190.

they enter the cortical plate (Figure 9.5: CP), where the density of radial glial fibers is reduced, a number of migrating neurons are grouped around a single radial glial fiber (Gadisseux et al., 1990).

The first neurons to migrate into the cortical plate are those of layer VIa, and as successive waves of migrating neuroblasts enter the cortical plate, they pass between the neurons that have already entered the plate and come to rest above them. Thus, the cortex is laid down in an inside-to-outside sequence (Berry and Rogers, 1965; Hicks and D'Amato, 1968). Interestingly, Hirst et al. (1991) have shown that even at E16 in the developing rat cortex, when the cortical plate is only about 6 cells deep, there is an aggregation of some apical dendrites. At E16 the only cells in the cortical plate are the presumptive layer VI pyramidal cells, and so perhaps Hirst et al. (1991) are seeing the beginnings of the organization of the apical dendrites of the layer VI pyramids into the bundles present in the mature cortex. By E19 the cortical plate is much better developed (Figure 9.5), and it is about 20 cells thick. At this stage the neurons within the plate are very orderly, and they are arranged into vertical stacks that are separated from each other by intervening pale streaks. These pale streaks are aggregates of vertically oriented processes of the developing neurons, most of which are still essentially bipolar in form at this stage. And in thin sections it is evident that these processes are organized into groups, each of which is associated with a radial glial fiber (Peters and Feldman, 1973; Gadisseux et al., 1990). The arrangement of the processes of the neurons within the cortical plate at this stage is best seen in thin sections taken in the tangential plane, or parallel to the pial surface (Figure 9.6). In such sections it is evident that the processes of the neurons are arranged into tightly packed clusters in which thicker processes predominate. These thicker processes, most of which are 2 to 3 μm in diameter, represent the apical processes of the bipolar-shaped developing neurons, and in the case of the pyramidal cells, these apical processes will become their apical dendrites. Most of these processes have a rather pale cytoplasm (Figure 9.6: P_2), but some of the larger ones have a darker cytoplasm, containing frequent ribosomes and profiles of the Golgi apparatus (Figure 9.6: P_1), which indicates that they are the bases of the apical processes of the neurons. Yet other processes in the clusters are thin, less than 0.5 μm thick. These are probably the profiles of the trailing processes of the neurons and will eventually produce their axons. Unfortunately, it is not usually possible to discern which of the profiles in the clusters belong to the radial glial fibers along which the developing neurons migrate, because the only reliable way to identify them in normal tissue is by their content of glycogen, which can easily be lost during the preparation of tissue (Peters and Feldman, 1973). But in general, it seems that each cluster of neuronal processes in the cortical plate has one or more radial glial fiber associated with it.

After birth the developing cerebral cortex gradually becomes thicker, but it is still immature, so that in the 4.5-day postnatal rat, for example, the pyramidal cells of layer V have begun to assume recognizable shapes, and in semithick plastic sections clusters of apical dendrites can be seen to arise from them (Figure 9.5). At about this time the layer V pyramidal cells begin to sprout basal dendrites and their apical dendrites also begin to form side branches (Miller, 1981). However, above these neurons there is still little differentiation of the neurons of the cortical plate into layers. The cells are more loosely packed and larger than in the late embryo, but as determined by the timed series of autoradiographs of Berry and Rogers (1965) and Hicks and D'Amato (1968), migration is still taking place along the radial glial fibers as the neurons destined for the supragranular layers are passing along them. This migration of neurons is still taking place, even though the generation of neurons destined to form area 17 of the rat is completed by birth (Miller, 1988).

In horizontal thin sections taken through the middle of the cortical plate at postnatal day 4.5 (Figure 9.7), the clusters of neuronal processes are still apparent, but as compared with the late prenatal cortex the cytoplasm of the more distal apical dendrites has become more filled with microtubules and polyribosomes

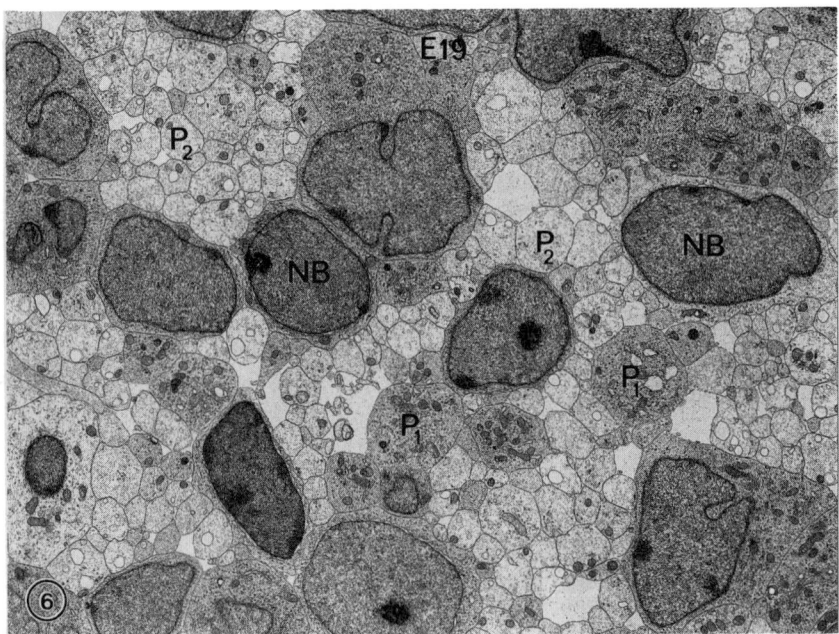

FIGURE 9.6. Electron micrograph of a tangential section through the cortical plate of a 19-day rat fetus. The neuroblasts (NB) have dark nuclei, and between the neuroblasts are the closely packed apical processes of other neurons. Some of these processes (P_1) are sectioned close to their cell bodies, while others (P_2) are sectioned more distally. $\times 3,700$.

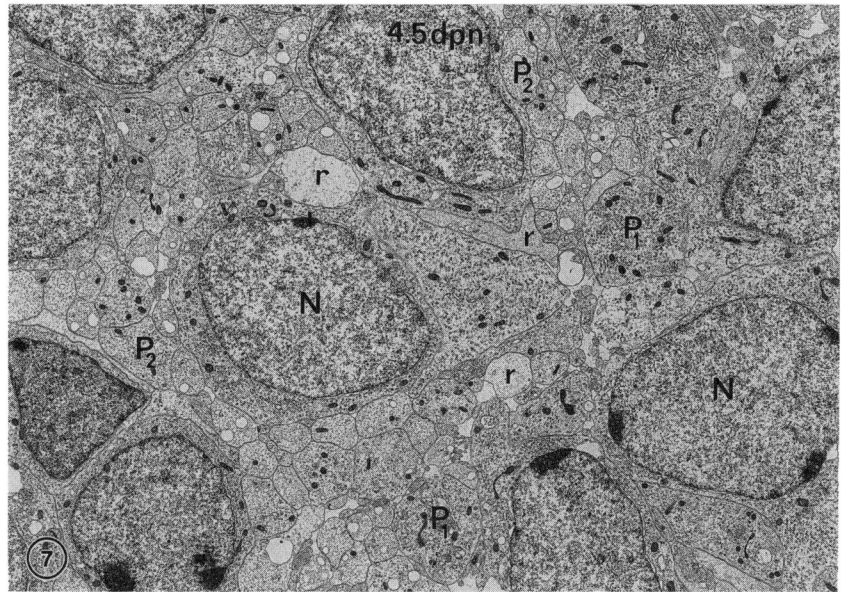

FIGURE 9.7. Electron micrograph of a tangential section through the upper third of the developing cortex of a 4.5-day postnatal rat. The neurons (N) have become paler. Between the neurons are the processes of neurons, some (P_1) sectioned close to the cell body, and others (P_2) sectioned more distally. Some processes of radial glial fibers (r) are still present. ×3,700.

(Figure 9.7: P_2), and some of the dendrites have begun to form spines. In addition, a few immature synapses are present and more laterally extending processes have made an appearance, but pale radial glial fibers (Figure 9.7: r) are still in evidence.

In subsequent days the maturation of the rat cerebral cortex is more rapid. At postnatal days 7 to 8, for example, all of the layers evident in the mature cortex can be distinguished (Figure 9.5), and by about postnatal day 9, when neuronal migration has been completed, the radial glial fibers disappear (Schmechal and Rakic, 1979; Miller, 1981; Pixley and De Villis, 1984).

If the cortex of a 7- to 8-day postnatal rat is examined in thin sections, it is evident that there has been an extensive remodeling of the neuropil between the cell bodies of the neurons (Figure 9.8). The obvious difference compared to the cortex of the 4.5-day rat is that the neuropil has become looser, and it has been invaded by numerous thin axonal process that cross in all directions. This invasion of the neuropil by axons leads to the formation of synapses with dendritic

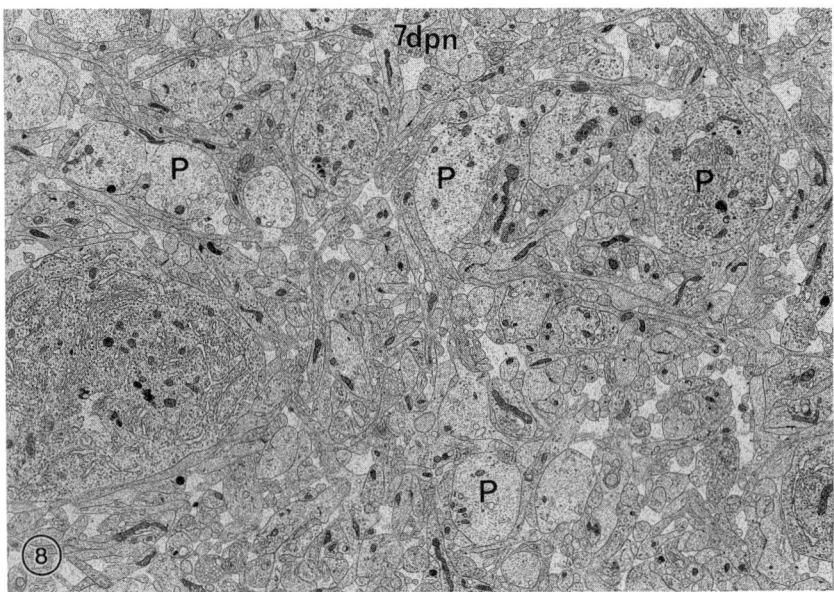

FIGURE 9.8. Electron micrograph of a tangential section through the cortex of a 7-day postnatal rat. The section is at about the level of layer III. The apical dendrites (P) of pyramidal cells are now separated into clusters by the ramifying processes in the maturing neuropil. ×3,700.

processes and cell bodies, but it also leads to the isolation of apical dendrites into groups. In effect, the dendritic clusters seen in the mature cerebral cortex have been formed and segregated from each other by other components of the neuropil, by cell bodies of neurons, and by the processes of maturing neuroglia.

This pattern of maturation of the cerebral cortex strongly suggests that the neuronal modules present in the mature cerebral cortex represent groups of pyramidal cells that have migrated into the cortical plate along the same radial glial fiber, and that their ascending processes aggregate into vertically oriented bundles that will eventually become segregated by the developing neuropil to produce the apical dendritic clusters. It is interesting to note that in the developing cortex of the mouse, Gadisseux et al. (1989) found that the single radial glial fibers that extend across the depth of the cortical plate at E16 are regularly spaced and are 8 to 10 μm apart. In mature rat area 17, the clusters are also regularly spaced, but at about 60 μm apart, and the spacing can be so regular that in favorable horizontal sections the dendritic clusters show a hexagonal packing pattern (Peters and Kara, 1987).

Similar steps in the formation of the dendritic clusters have been described by Schmolke (1989) during the development of the visual cortex in the rabbit. She also describes the formation of bundles of ascending dendrites that are subsequently separated from each other by the maturing neuropil, and points out that in the early phases of development, when the dendritic clusters first appear, there are numerous puncta adherentia between adjacent dendrites. Presumably, these puncta play a role in holding together the dendrites in the forming clusters, because the puncta have largely disappeared from the developing rat cortex just after birth.

MECHANISM OF PYRAMIDAL CELL MODULE FORMATION

From the description of the events taking place during the formation of the rat cerebral cortex, and from what is known of the neuronal arrangements in the mature cortex, it can be proposed that the layer VI pyramidal cells reach the cortical plate at about the time, or before, the radial fibers within it defasciculate. Hirst et al. (1991) showed that bundles of radial glial processes are present at E16 in the rat, but they are not regularly spaced. The dendritic bundles formed by the apical dendrites of the layer VI pyramidal cells in the mature cortex are also not spaced evenly.

If the layer V pyramidal cells arrived at the cortical plate just after the radial glial fibers had defasciculated, they would migrate into the plate along regularly spaced, individual radial glial fibers. This would mark the beginnings of the formation of the dendritic clusters, whose presence is clearly evident in the late fetal cortex (Figure 9.5). Subsequently, layer IV, then layers III and II pyramidal cells arise in sequence and pass through the layer V pyramids. Of these more superficial neurons, study of the mature cortex shows that only the layers III and II pyramids contribute to the dendritic clusters (Figure 9.2). But in the mature cortex the apical dendrites of these superficial pyramids are located at the peripheries of the clusters, surrounding the central core of layer V apical dendrites (Peters and Kara, 1987). How this arrangement is achieved is not yet apparent. If the radial glial fibers lie at the centers of the developing clusters of apical dendrites, then the migration of the layer III pyramids along them might be expected to push the large layer V apical dendrites apart. In this situation it would be expected that the apical dendrites of the layer III pyramids would come to lie in the center of a cluster. Examination of the cortical plate in thin sections suggests, in fact, that the radial glial fibers lie at the peripheries of the clusters of neuronal processes in the developing cortical plate (Figure 9.7). Consequently, the migration of the more superficial neurons whose apical dendrites contribute to the mature clusters probably takes place on the peripheries of the clusters. During this stage the integrity of the clusters may be maintained by the puncta adhaerentia between the neuronal processes.

It is suggested that those neurons that migrate into the cortical plate along the same radial glial fiber constitute a neuronal module. These modules are proposed as the basic units of organization of the cortex, and their integrity seems to be recognized as the neuropil of the cortex begins to form. In rat visual cortex this occurs by postnatal day 7, for at this time the clusters of apical dendrites have become separated from each other as the cortex expands laterally. This requires that the neuropil develop differentially, leaving the neurons of the modules, and especially their apical dendrites, as discrete units. This should not be taken to suggest that the neuronal modules remain well defined in the mature cortex. They do not, but the dendritic clusters to maintain their integrity and serve to mark the axes of the neuronal modules.

GENERALIZATIONS AND DISCUSSION

This pattern of development of the cerebral cortex in the rat, which leads to the formation of pyramidal cell modules that are based upon the clustering of the apical dendrites of the pyramidal cells of layers V, III, and II, seems to be a basic one, since it is also present in other animals. For example, in mouse cerebral cortex there is a similar arrangement of apical dendrites, with the apical dendrites of the layer VI pyramidal cells again forming a system of bundles that are independent of the clusters (Escobar et al., 1986), and Gadisseaux et al. (1989, 1990) also suggest that the clusters result from the migration of pyramidal cells along the radial glia into the cortical plate. Fleischhauer et al. (1972) have described dendritic clusters in the visual cortex of the rabbit, in which their spacing is given as 40–45 μm, and during the development of that cortex Schmolke (1989) has shown the presence of groups of apical dendrites that eventually become separated by the developing neuropil into well-defined clusters. The extensive studies of Rakic (e.g., 1972, 1977, 1990) on rhesus monkey visual cortex have been seminal in understanding the migration of neurons along the radial glia, and we have recently shown that this visual cortex also contains modules of pyramidal cells (Peters and Sethares, 1991). These modules in monkey visual cortex are different from those in other cortices, however, because the neurons are small and their packing density is about twice that in other cortices. Consequently, the modules are only about 30 μm in diameter (Peters and Sethares, 1991), and the thin apical dendrites of the layer V pyramidal cells pass for a long distance through several sublayers of the complex layer IV in which spiny stellate cells prevail.

It is of interest that in area 17 of rhesus monkey visual cortex, we have calculated that each 30-μm-wide neuronal module contains an average of 142 neurons (Peters and Sethares, 1991). In his studies of the development of this cortex, Rakic (1988) has estimated that an "ontogenetic" or "embryonic" column, which he defines as the neurons that migrate along the same radial glial fiber, contains about 120 neurons. Thus there is close correspondence of numbers in these two estimates. The differences between them might be accounted for by the nonpyramidal, or inhibitory, neurons that enter the cortex. In his studies of the generation and maturation of nonpyramidal neurons in rat visual cortex, Miller (1981, 1988) has shown that the nonpyramidal cells and the pyramidal cells that are destined to have their cell bodies in the same cortical layer are generated and migrate at the same time. But whether these nonpyramidal cells can be considered to belong to the pyramidal cell modules, or whether their distribution in the cortex is independent of the modules needs to be investigated.

Because of its position within a particular cortical area, each neuronal module will receive a unique input, one that is slightly different from that of its neighbors. In the visual cortex of the rhesus monkey, for example, each module will differ in the input that it receives in terms of such variables as the part of the visual field to which it is responsive, the eye that dominates that input, and the color and orientation of the image. The neurons in the module would then convey their responses to other structures, so that in effect the neuronal modules can be regarded as output units. In this respect it is of interest that the pyramidal cells of layers V, III, and II are all projection neurons that convey their responses to other structures or targets. Because of their close relationships, it is likely that the neurons in the module have a common physiological response. It may also be that a number of neurons in the same module project to the same target, so that the information conveyed is reliable and not subject to the vagaries of individual neurons, whose response properties are unlikely to be completely predictable in view of the large number of synapses that each one receives.

In many respects the neuronal modules possess the properties that Mountcastle (1957, 1978) predicted for the small, basic and irreducible, vertically oriented neuronal units that he called "minicolumns." In cortices in which various kinds of functional columns have been described, such as the eye preference and orientation columns in cat and monkey visual cortex (Hubel and Wiesel, 1977), it seems likely that the neuronal modules can be recruited in different combinations to produce these columns. Since the columns occupy the same cortical space, this also implies that a given module can at various times be a component of different columns, depending upon which cortical property is being examined.

After finding that the neocortex can be considered to consist of vertical modules of neurons, the next step will be to examine the interrelations and synaptic input to the neurons within them. Then it may become possible to generate a neuronal model that can explain the physiological response properties recorded from neurons in various cortical areas.

Acknowledgments

I wish to thank Claire Sethares and Karen Josephson for their help in these studies, which were supported by a Javits Award (NS 07016) from the National Institute of Neurological Disorders and Stroke of the U.S. National Institutes of Health.

REFERENCES

Berry M, Rogers AW (1965): The migration of neuroblasts in the developing cerebral cortex. *J Anat* 99: 691–709

de Camilli P, Miller PE, Navone F, Theurkauf WE, Vallee RB (1984): Distribution of microtubule-associated protein 2 in the nervous system of the rat studied by immunofluorescence. *Neuroscience* 11: 819–846

Escobar MI, Pimienta H, Caviness VS, Jacobson M, Crandall JE, Kosik KS (1986): Architecture of apical dendrites in the murine neocortex: Dual apical dendritic systems. *Neuroscience* 17: 975–989

Fairén A, Peters A, Saldanha J (1977): A new procedure for examining Golgi impregnated neurons by light and electron microscopy. *J Neurocytol* 6: 311–337

Feldman ML, Peters A (1974): A study of barrels and pyramidal dendritic clusters in the cerebral cortex. *Brain Res* 77: 55–76

Fleischhauer K (1974): On different patterns of dendritic bundling in the cerebral cortex of the cat. *Z Anat Entwickl-Gesch* 143: 115–126

Fleischhauer K, Detzer K (1975): Dendritic bundling in the cerebral cortex. In: *Physiology and Pathology of Dendrites: Advances in Neurology, Volume 12*, Kreutzberg GW, ed. New York: Raven Press

Fleischhauer K, Petsche H, Wittkowski W (1972): Vertical bundles of dendrites in the neocortex. *Z Anat Entwickl-Gesch* 136: 213–223

Gabbott PLA, Stewart MG (1987): Distribution of neurons and glia in the visual cortex (area 17) of the adult albino rat: A quantitative description. *Neuroscience* 21: 833–845

Gadisseux JF, Evrard P, Misson JP, Caviness VS (1989): Dynamic structure of the radial glial fiber system of the developing murine cerebral wall. An immunocytochemical analysis. *Dev Brain Res* 50: 55–67

Gadisseux JF, Kadhim HJ, van den Bosch de Aguilar P, Caviness VS, Evrard P (1990): Neuron migration within the radial glial fiber system of the developing murine cerebrum: An electron microscopic autoradiographic analysis. *Dev Brain Res* 52: 39–56

Hicks SP, D'Amato CJ (1968): Cell migration to the isocortex in the rat. *Anat Rec* 160: 619–634

Hirst E, Asante J, Price J (1991): Clustering of dendrites in the cerebral cortex begins in the embryonic cortical plate. *J Neurocytol* 20: 431–438

Hubel DH, Wiesel TN (1977): Functional architecture of macaque monkey visual cortex. *Proc Roy Soc Lond B* 198: 1–59

Kosik KS, Duffy LK, Dowling MM, Abraham C, McClusky A, Selkoe DJ (1984): Monoclonal antibody to microtubule-associated protein 2 (MAP2) labels Alzheimer neurofibrillary tangles. *Proc Natl Acad Sci USA* 81: 7941–7945

Martin KAC (1988): From single cells to simple circuits in the cerebral cortex. *Q J Exp Physiol* 73: 637–702

Meinecke DL, Peters A (1987): GABA immunoreactive neurons in rat visual cortex. *J Comp Neurol* 261: 388–404

Miller M (1981): Maturation of rat visual cortex. I. A quantitative study of Golgi-impregnated pyramidal neurons. *J Neurocytol* 10: 859–878

Miller MW (1988): Development of projection and local circuit neurons in neocortex. In: *Cerebral Cortex: Development and Maturation of Cerebral Cortex, Volume 7*, Peters A, Jones EG, eds. New York: Plenum Press

Mountcastle VB (1957): Modality and topographic properties of single neurons of cat's somatic sensory cortex. *J Neurophysiol* 20: 408–434

Mountcastle VB (1978): An organizing principle for cerebral function. The unit module and the distributed system. In: *The Mindful Brain*, Edelman GM, Mountcastle VB, eds. Cambridge, MA: MIT Press

Parnavelas JG, Lieberman AR, Webster KE (1977): Organization of neurons in the visual cortex, area 17, of the rat. *J Anat* 124: 305–322

Peters A (1985): The visual cortex of the rat. In: *Visual Cortex, Volume 3*, Peters A, Jones EG eds. New York: Plenum Press

Peters A (1987): Number of neurons and synapses in primary visual cortex. In: *Cerebral Cortex. Further Aspects of Cortical Function, Including Hippocampus, Volume 6*, Jones EG, Peters A, eds. New York: Plenum Press

Peters A, Feldman M (1973): The cortical plate and molecular player of the late rat fetus. *Z Anat Entwickl-Gesch* 141: 3–27

Peters A, Kara DA (1985a): The neuronal composition of area 17 of rat visual cortex. I. The pyramidal cells. *J Comp Neurol* 234: 213–241

Peters A, Kara DA (1985b): The neuronal composition of area 17 of rat visual cortex. II. The nonpyramidal cells. *J Comp Neurol* 234: 242–263

Peters A, Kara DA (1987): The neuronal composition of area 17 of rat visual cortex. IV. The organization of pyramidal cells. J Comp Neurol 26: 573–590

Peters A, Sethares C (1991): Organization of pyramidal neurons in area 17 of monkey visual cortex. *J Comp Neurol* 306: 1–23

Peters A, Walsh TM (1972): A study of the organization of apical dendrites in the somatic sensory cortex of the rat. *J Comp Neurol* 144: 253–268

Peters A, Kara DA, Harriman KM (1985): The neuronal composition of area 17 of rat visual cortex. III. Numerical considerations. *J Comp Neurol* 238: 263–274

Pixley SR, De Vellis J (1984): Transition between immature radial glia and mature astroyctes studied with a monoclonal antibody to vimentin. *Dev Brain Res* 15: 201–209

Rakic P (1972): Mode of cell migration to the superficial layer of fetal monkey cortex. *J Comp Neurol* 145: 61–84

Rakic P (1977): Prenatal development of the visual system in rhesus monkey. *Philos Trans R Soc Lond B* 278: 245–260

Rakic P (1988): Specification of cerebral cortical areas. *Science* 241: 170–176

Rakic P (1990): Principles of neural cell migration. *Experientia* 46: 882–891

Schmechel DE, Rakic P (1979): A Golgi study of radial glial cells in developing monkey telencephalon: Morphogenesis and transformation into astrocytes. *Anat Embryol* 156: 15–152

Schmolke C (1989): The ontogeny of dendrite bundles in rabbit visual cortex. *Anat Embryol* 180: 371–381

Schober W, Winkelmann E (1975): Der visuelle Kortex der Ratte. Cytoarchitektonic und Stereostaktische Parameter. *Z Mikrosk Anat Forsch* 89: 431–446

Von Bonin G, Mehler WR (1971): On columnar arrangement of nerve cells in cerebral cortex. *Brain Res* 27: 1–10

Werner L, Hedlich A, Winkelmann E, Brauer K (1979): Versuch einer Identifizierung von Nervenzellen der visuellen Kortex der Ratte nach Nissl-und Golgi-Kopsch-Darstellung. *J Hirnforsch* 20: 121–139

Winkelmann E, Brauer N, Berger U (1975): Zur columnaren Organisation von Pyramidenzellen in visuellen Cortex der Albinoratte. *Z Mikroskop Anat Forsch* 89: 239–256

CHAPTER 10

Architecture of Sensory Fiber Projections: Implications for Neuronal Specificity in the Spinal Cord

LORNE M. MENDELL AND H. RICHARD KOERBER

INTRODUCTION

The notion of neuronal specificity implies that a neuron has an identity, conferred, for example, by its surface chemistry, that permits it to contact a defined subset of postsynaptic neurons. Schematic diagrams of connections between regions of the nervous system generally contain an implicit notion of a precise point-to-point projection. Similarly, the concept of a nucleus to which such a neuron might project is generally considered to encompass a compact group of cells.

Although the receptive fields of individual sensory afferent fibers are quite discrete, their zones of termination in the spinal cord are relatively widespread. This is particularly true of large myelinated afferents whether they innervate individual muscle stretch receptors or localized clusters of receptors in the skin. These fibers enter the spinal cord via the dorsal root, and then typically bifurcate to run rostrally and caudally in the dorsal columns. Recent reconstructions from single horseradish peroxidase (HRP)- injected afferents in the cat show very clearly that they give off branches at intervals of about 1 mm which enter the gray matter to terminate in specific laminae which differ depending on the receptor type in the periphery (reviewed in Brown, 1981). For example, hair afferents typically terminate in laminae III and IV, whereas muscle spindle afferents terminate in laminae VI, VII, and IX.

This architecture is interesting because it gives each individual afferent access to a large number of postsynaptic target cells in a given lamina, both in the segment of entry and in neighboring segments. This divergence might at first glance appear to violate the neuronal specificity doctrine and to be functionally counterproductive, since it would disseminate the effects of stimulation of a small area in the periphery very widely in the spinal cord. This would result in considerable overlap in the spinal representation of peripheral structures. As we shall see, this deficit is only apparent, since spinal neurons of similar function (common receptive field or common target muscle) are arranged in loose columns, so that in effect this arrangement provides individual afferents access to a large number of cells of similar, but not necessarily identical, function. The fact that cells in a nucleus can be arranged as a loose column does not minimize the existence or importance of

Formation and Regeneration of Nerve Connections
Sansar C. Sharma and James W. Fawcett, Editors
© 1993 Birkhäuser Boston

specificity. However, it raises questions concerning the functional significance of such an arrangement. For example, schematically reducing the divergent connections made by an individual afferent to a single connection makes the implicit assumption that these connections all function identically. Must we reconsider our notions of specificity, at least in its functional aspects, in the light of these anatomical facts?

In his monograph *The Formation of Nerve Connections* (1970, p. 217), Dr. R. M. Gaze stated:

> The conception of a nervous system that is engendered by the hypothesis of neuronal specificity is one in which the system is extensively prewired. . . . A nervous system built in this fashion would have many switchboard-like characteristics. . . . Our difficulty in envisaging how such a system could work stems . . . from our inability (or unwillingness; it amounts to the same thing) to bring our ideas on switchboards up to date.

He then pointed out the fact, perhaps less familiar to the reader of the early 1970s than the 1990s, that "with a morphologically rigid system of sufficient complexity, it is possible to obtain an almost unlimited variety of different responses from the same stimulus, merely by shifting the distribution of central facilitation and inhibition" (p. 218).

Spinal circuits have provided excellent model systems with which to analyze this balance between excitation and inhibition. There are a number of ways in which a change in this balance can be achieved. For example, segmental and descending control of interneuronal activity (reviewed in Baldissera et al., 1981) and presynaptic control of transmitter release (reviewed in Rudomin, 1990) can affect specific excitatory or inhibitory inputs, thereby affecting the balance of input to individual spinal cells.

These physiological mechanisms undoubtedly play some role in modulating the response to input in a given afferent system. However, it is now clear that despite the morphological rigidity of spinal afferent inputs (at least at the level of the light microscope), they are not functionally rigid in the sense that the efficacy of the synaptic drive they provide to spinal neurons is subject to modulation during activity. From the viewpoint of neuronal specificity, a fundamental question is whether all the divergent connections made by an individual afferent function in an identical manner during natural stimulation. A given afferent may synapse on a distributed array of spinal neurons of apparently identical function, e.g., motoneurons. However, as we shall see below, the amplitude modulation of excitatory postsynaptic potential (EPSP) amplitude in the different motoneurons contacted by the *same* afferent fiber can differ substantially in response to the same program of stimulation. Thus the specificity dictating these highly stereotyped connections on cells innervating a common target does not necessarily imply that each of these connections functions in an identical manner.

An additional implication of this plasticity is that each afferent may not have an identical central action in response to each mode of activation (i.e., discharge pattern), nor will different afferents necessarily evoke identical central actions in response to identical patterns of input. We shall see that the parameters of this plasticity appear to be determined by the afferent "type," i.e., by the receptor type that is innervated by the afferent. In addition, it will be shown that the different connections made by an individual afferent can exhibit considerable diversity in this plasticity which is determined by the spinal target cell. Furthermore, this plasticity may be of considerable importance from a functional viewpoint. The present essay explores some of these issues in the context of the physiological plasticity which acts to temper the action of a (portion of the) nervous system hard-wired according to the rules of specificity.

PROJECTION FROM Ia AFFERENTS TO MOTONEURONS

Group Ia fibers supplying muscle spindles are known to make monosynaptic projections to alpha-motoneurons in the ventral horn (reviewed in Henneman and Mendell, 1981). Physiological analysis using the spike-triggered averaging method (Mendell and Henneman, 1971) has revealed that each of the 60 spindle afferents from the cat medial gastrocnemius muscle can evoke an EPSP in virtually all of the 300 alpha-motoneurons in the homonymous motoneuron pool, and in about 60% of the motoneurons projecting to close synergists such as lateral gastrocnemius and soleus (Scott and Mendell, 1976). This organization is a general one, having also been demonstrated in the knee flexor muscles (Nelson and Mendell, 1978). The motoneuron pools are organized in long rostrocaudal columns (Burke et al., 1977). Each afferent enters the spinal cord, and after bifurcating to run rostrocaudally in the dorsal columns, gives off collaterals at about 1-mm intervals which descend into the ventral horn (Burke et al., 1979; Brown and Fyffe, 1981). This arrangement provides each afferent access to virtually all of the motoneurons in the elongated motoneuron pool, and thus is the morphological substrate for the physiological finding that each afferent can elicit EPSPs in virtually every homonymous motoneuron.

The fact that a population of motoneurons innervating a given muscle can receive a projection from a particular afferent irrespective of their positions in a 10-mm long motoneuron pool represents excellent evidence that neuronal specificity operates in this system, since there are many other neurons interspersed in this region which receive no connections from these afferents. Furthermore, there are other motoneurons interdigitating with these that innervate synergist muscles. Connections to these heteronymous motoneurons have been described by many authors (reviewed in Henneman and Mendell, 1981), but they differ from those made on homonymous motoneurons by virtue of their smaller EPSPs and the lower probability of these connections being made. This represents another example of neural specificity in that the motoneurons supplying different muscles can be distinguished by afferents terminating in the ventral horn. Scott and Mendell (1976) used the term "species specificity" to describe this situation whereby motoneurons receive connections whose strength, on the average, is correlated with the muscle they innervate. This conclusion is not without controversy, since others (Luscher and Vardar, 1989) have concluded that the position of a motoneuron within the spinal cord is a more important determinant of the strength of connectivity from a particular group Ia fiber than the muscle innervated.

Even within a pool of motoneurons there are systematic differences in the strength of monosynaptic projection according to the properties of the motoneuron and its motor unit. Individual Ia axons make more potent connections on motoneurons innervating type S (slowly contracting, fatigue-resistant) motor units, as evidenced by the large amplitude of the single-fiber EPSPs evoked in these motoneurons. On the average, the same population of sensory axons elicits smaller EPSPs in motoneurons innervating type FR (rapidly contracting, fatigue-resistant) motor units, and the smallest EPSPs are developed in the motoneurons innervating the type FF (rapidly contracting, fatigable) motor units. These differences in EPSP amplitude have a postsynaptic component in that they are correlated with the input resistance of the motoneurons, which in turn is related to two major factors, motoneuron size (input resistance: Burke, 1968) and specific membrane resistance (Gustafsson and Pinter, 1984) which vary systematically in these populations of motoneurons (motoneuron size: S< FR< FF; specific resistance; S> FR > FF).

Recent work has suggested that other factors more intimately related to the synapse also differ systematically across the motoneuron pool. The steady-state synaptic currents induced by repetitive activation of spindle afferent fibers are systematically largest in motoneurons with the highest values of input resistance and decrease as motoneuron resistance decreases (Heckman and Binder, 1988). An additional finding is that large EPSPs in small motoneurons (as measured by small values of rheobase: Zengel et al., 1985) exhibit depression during high-frequency stimulation, whereas small EPSPs in large motoneurons (large values of rheobase) are prone to facilitation (Collins et al., 1984, 1988; Figure 10.1). This suggests that EPSPs produced by large synaptic currents depress during high-frequency activation, whereas those with small synaptic currents facilitate. One

FIGURE 10.1. EPSPs averaged in medial gastrocnemius motoneurons in response to bursts of 32 shocks (maximal group Ia stimulation) to the lateral gastrocnemius nerve in an anesthetized cat (Nembutal). Stimulus bursts consist of 32 shocks with interstimulus interval of 6 msec. In each case a burst was presented every 2 sec for a total of 32 bursts, with the 32 EPSPs being averaged in register (for further details see Collins et al., 1984). Note that for the motoneuron with a small value of rheobase, the EPSP amplitude diminished during the burst, whereas the EPSP amplitude increased for the same stimulus program in the motoneuron with the large value of rheobase.

possible explanation is that synapses generating large synaptic currents are prone to transmitter depletion during high-frequency stimulation, which would prevent any effects of residual Ca^{2+} from being expressed, whereas at synapses with small synaptic current, the lack of transmitter depletion would allow residual Ca^{2+} to elevate transmitter release during high-frequency stimulation (see Collins et al., 1984, for discussion). However, at present postsynaptic differences among these motoneurons cannot be ruled out as a cause for these differences in behavior in response to high-frequency stimulation.

Differences in the EPSP amplitude modulation described above have been shown to be correlated with the postsynaptic motoneuron rather than the presynaptic spindle afferent (Koerber and Mendell, 1991a), since patterns of amplitude modulation recorded in a motoneuron in response to high-frequency stimulation of different afferents tend to be similar. In contrast, these patterns of EPSP amplitude modulation tend to differ when elicited in different motoneurons by the same single afferent. Although the motoneuron rather than the afferent appears from these studies to *determine* the pattern of EPSP amplitude modulation, it is not clear that the motoneuron itself is the *cause* of these differences. For example, properties of presynaptic terminals on motoneurons (e.g., the amount of depletion: see above) might be determined in a trophic manner by the motoneuron, much as these synapses have been shown to change when the motoneuron is axotomized (Gustaffson et al., 1986).

These findings indicate that the Ia/motoneuron connection, normally considered to have a highly stereotyped central action as well as a highly specific central connectivity (reviewed in Henneman and Mendell, 1981), can exhibit considerable plasticity in its synaptic action during stimulation at frequencies similar to that reported during naturally evoked activity. The modulation of EPSP amplitude differs according to the target. This systematic difference in the synaptic action of terminals of a given Ia afferent fiber in accordance with the target motoneuron permits the single afferent to distribute appropriate action to the diverse motoneurons that it activates despite the highly stereotyped connections, i.e., the single afferent sending terminals to virtually all homonymous motoneurons in the homonymous motoneuron pool.

The projection of single Ia fibers to virtually all motoneurons in the homonymous pool represents an extraordinary example of specificity on the part of individual spindle afferent fibers. The major advantage conferred by this specificity is that it represents a relatively simple way to ensure an equal distribution of spindle afferent input to the motoneuron pool (Henneman et al., 1965; see discussion in Mendell et al., 1990). Initially, this was felt to ensure greater depolarization of small motoneurons that are more susceptible to recruitment during the myotatic reflex. These motoneurons have the higher input resistance that would subject them to a greater level of depolarization than large motoneurons in response to a given input, thereby ensuring their earlier recruitment in response to a given input, i.e., the Size Principle.

This equalized input ensuring orderly recruitment due to different EPSP amplitudes could be disadvantageous under other circumstances. For example, during the high frequency of inputs characteristic of Ia inputs, temporal summation would cause the small motoneurons to depolarize disproportionately compared with the large ones. One consequence of this might be to make small motoneurons disproportionately *insensitive* to inhibition during high-frequency stimulation. Thus the anatomical arrangement that is idea to ensure orderly recruitment is probably inappropriate to maintain the proper relative levels of depolarization of motoneurons in the pool during maintained activation. The modification of synaptic transmission that occurs during high-frequency activation would act to minimize these disparities in depolarization that are the result of temporal summation: the large EPSPs in small motoneurons would tend to diminish in amplitude and the small EPSPs in large motoneurons would tend to increase in amplitude.

Another aspect to the flexibility in the physiology of this highly stereotyped anatomical connection is that it may contribute to altering the recruitment scheme. When the cycle time of a repetitive movement is decreased, such as when going from a slow walk to a gallop, or to take a more extreme example, to shaking the paw vigorously, the differences in recruitment threshold of small and large motoneurons may be diminished (Chanaud et al., 1991). In accordance with this, we have found that the distribution of potentiation of the initial EPSP following high-frequency bursts differs according to the cycle time of the task (Koerber and Mendell, 1991b). For tasks involving brief cycle times (e.g., paw shaking), EPSPs in large motoneurons tend to be potentiated to a greater extent than those in small motoneurons, in effect working against the "usual" order of recruitment. This does not occur in tasks with longer cycle times (e.g., walking). This then is another aspect of the plasticity of the Ia synapse on motoneurons: not only can it adjust the excitatory input to the motoneuron during high-frequency stimulation, but it can also help to shape the input according to the task that the muscle is carrying out to influence recruitment order.

Another factor of potential importance in determining the strength of the projection of an afferent is the relative position of the motoneuron with respect to its entry zone into the spinal cord. Several authors have commented that projections of medial gastrocnemius afferents entering the spinal cord become weaker as one examines connections further and further rostral from the entry zone (Scott and Mendell, 1976; Luscher and Vardar, 1989). One could ask whether these smaller EPSPs are more subject to facilitation than the large ones recorded in motoneurons located in the vicinity of the afferent's entry zone. At present no definitive answer is available on this point. However, it seems likely that at least some of the decrement in amplitude results from a decreased number of boutons available for synapses. If one assumes that the individual synapses have similar strength at these different locations (the differences in amplitude being accounted for by numbers of boutons on motoneurons), one might not expect systematic differences in facilitation/depression as a function of location.

Group II afferents make relatively weak monosynaptic projections to motoneurons (Stauffer et al., 1976; Munson et al., 1982), as demonstrated by their very small EPSPs. It is not known whether these are particularly susceptible to facilitation, which in the context of the hypotheses advanced above would be interpreted in terms of each of their individual synapses being "weak." An alternative is that there are fewer synapses made by each of these afferents on motoneurons than in the case of Group Ia fibers. This would result in smaller EPSPs, but no difference in their modulation during high-frequency activity.

PROJECTION OF CUTANEOUS AFFERENTS

It is interesting that the spinal projections of individual cutaneous afferents appear very similar in their organization in the sagittal plane to those of muscle spindle afferents. Individual fibers enter the spinal cord via the dorsal root, bifurcate (in many cases) to run rostrally and caudally along the cord, and give off collaterals that enter the gray matter at intervals of about 1 mm. This organization has been studied for many different kinds of cutaneous afferents (reviewed in Brown, 1981), and it has been shown to be true for both large (A-beta) and small (A-delta) myelinated afferents. Each afferent type, determined according to sensory receptor (e.g., guard hair, muscle spindle), has the focus of its projection in specific laminae of Rexed, often with a characteristic anatomy (e.g., flame-shaped arborization patterns of guard hairs).

It is the rostrocaudal distribution of terminals from individual cutaneous afferents that is the focus of the present analysis. The anatomy, based on visualization of single filled axons, suggests a widespread distribution of boutons, up to 12 mm. The functional capacity of boutons over all these regions has been revealed by stimulating single afferents and demonstrating that cord dorsum field potentials can be measured over an extensive rostrocaudal region (Koerber and Mendell, 1988; Figure 10.2). The correspondence between the physiology and the HRP anatomy has been confirmed by filling individual afferents with HRP and comparing the resultant distribution of boutons in the dorsal horn with the distribution of cord dorsum potentials (CDPs) produced by stimulation of the same afferent (Koerber et al., 1990). Although the absolute length of the projection is relatively great, there is a central region of relatively great concentration of boutons with relatively sparse concentrations of boutons both rostral and caudal of this central region. The fact that the total extent of individual boutons is generally much greater than the rostrocaudal spinal representation of the skin area innervated by the single afferent, as measured by extracellular recording, suggests that some of these boutons might be nonfunctional (Meyers and Snow, 1984) or at least not invaded by the impulse. However, boutons in these regions relatively distant from their fiber's root entry zone are functional, since they contribute to the CDP to the same extent as the more centrally located boutons (Koerber et al., 1990). This raises the possibility that total spinal representations of specific areas of the skin may not be as compact as indicated by extracellular recording. More specifically, the architecture of individual afferents suggests that any region of the cord receives a heterogeneous mixture of afferents in terms of their receptive field location, but that most of the boutons come from afferents innervating a particular patch of skin.

HAIR FOLLICLE AFFERENT

HIGH THRESHOLD
MECHANORECEPTOR

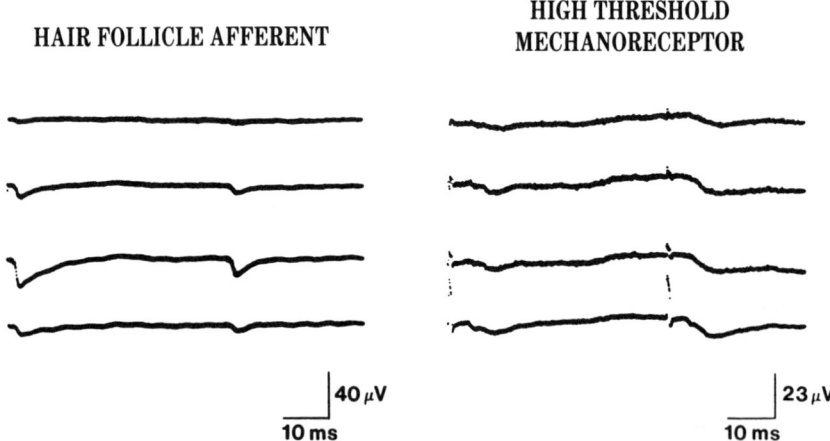

40 μV

10 ms

23 μV

10 ms

FIGURE 10.2. Cord dorsum potentials (CDPs) evoked simultaneously at four locations on the dorsal surface of the spinal cord separated by about 5 mm (total extent about 15 mm). The top trace in each group of four is the rostral electrode placed close to the L5/L6 border, while the bottom trace is the CDP recording from the most caudal electrode in rostral S1. Middle two traces from intermediate positions. For further details of methods see Koerber and Mendell (1988). In each case a single dorsal root ganglion cell in L7 was penetrated, identified as to receptive field modality and location, and then stimulated with pairs of shocks separated by 50 msec. This pair of shocks was repeated every 1,500 msec for a total of 128 times, and the resultant CDPs were averaged in register. Note that when the hair follicle afferent was stimulated (left), the test CDP (evoked 50 msec after the initial conditioning one) was depressed in both amplitude and duration compared with the conditioning one. When the high-threshold mechanoreceptor was stimulated (right), the test CDP was facilitated in amplitude. In both cases the effects, whether depression or facilitation, were observed at all four electrode positions.

Thus any given region of the dorsal horn probably contains cells strongly activated by one portion of the skin, but interdigitated among these are other cells more weakly (and perhaps only subliminally) activated by surrounding skin regions.

In the case of spindle projections to motoneurons, we have noted that the projections are systematically different according to the motor unit type supplied by the motoneuron. In the case of the population of dorsal horn interneurons activated by a single afferent, the evidence is much less direct on this issue. Relay neurons receive much stronger monosynaptic connections from afferents innervating skin in the center of their receptive field than from afferents innervating the peripheral portion of their receptive field (Hongo and Koike, 1975; Brown et al., 1987). But this may not reflect differences in the postsynaptic cells, since individual spinal neurons can presumably have inputs from afferents innervating either the center or the periphery of their receptive field. In terms of group Ia connections to motoneurons discussed above, this might be more analogous to the differences between projections to homonymous and heteronymous motoneurons rather than those to type S and type F motoneurons.

When individual cutaneous afferents are stimulated with pairs or more complicated trains of pulses, one often observes modulation of the amplitude of the CDPs (Figure 10.2). These modulations vary systematically for afferents innervating different types of receptors (Koerber and Mendell, 1988; Koerber et al., 1991). For example, afferents innervating slowly adapting receptors tend to evoke CDPs that do not depress during repetitive stimulation, whereas afferents innervating rapidly adapting receptors evoke CDPs that depress during the same program of stimulation (Figure 10.2). CDPs evoked by afferents innervating nociceptors tend to exhibit facilitation (Figure 10.2). Thus transmission properties differ according to the afferent (i.e., peripheral receptor) type, although at present it is not known whether this is a consequence of differences in the synapses made by these individual afferent types or the properties of the spinal networks they activate. These different response patterns can also be observed in individual target cells contacted by a particular afferent type, suggesting that the CDP is an accurate reflection of the underlying cellular activity, at least on the average (Koerber et al., 1991).

Since recordings of CDPs are generally made simultaneously from four positions along the spinal cord, it has been possible to observe whether the modulation differs systematically at the different locations. It is generally found that modulation is similar despite large differences in the amplitude of the CDPs examined at relatively low frequencies of stimulation (Figure 10.2). Thus it appears that there are no systematic differences in the amplitude modulation of the CDPs produced by topographically disparate terminals (e.g., caudal vs. rostral). However, since the CDP represents an average response produced by the activity of many different neurons activated by the same afferent, it cannot be said that there is not some diversity in the response evoked in different spinal neurons by the same afferent. In other words, there may be some differences among the cells in their response to a particular pattern of stimulation of a particular afferent, but these differences are not spatially localized across the column of cells activated by that afferent. Such differences could correspond to the postsynaptic targets of these cells, much as has been reported for Ia fiber connections on motoneurons, where facilitation/depression behavior depends on the type of motor unit supplied by the motoneuron.

DISCUSSION

The tendency for spinal afferents to project widely along the rostrocaudal axis appears to be an adaptation to the axial structure of the spinal cord, particularly the fact that the neural representation of peripheral structures is generally elongated. Thus the column of motoneurons innervating a muscle typically stretches over two segments within the cord, while the neural representation of even a single toe covers several millimeters. The ability of individual afferents to access this entire representation helps to provide an even input to all regions of this representation. Although this scheme ensures a uniform innervation of the spinal representation of specific structures, it should not be inferred that individual afferents project equally to all the neurons in the columnar representation. Thus it has been shown by the use of the intracellular HRP method that boutons are most highly concentrated in some central region and that the bouton numbers fall off steeply on either side of this central region (see above).

Excitatory synapses such as those made by group Ia fibers on alpha-motoneurons are subject to modification of their efficacy in the course of normal physiological activation. This modulation of synaptic efficacy is highly dependent on the postsynaptic target, in particular the identity of the motoneuron on which the connection is made. Thus the apparent rigidity of the Ia fiber connectivity to the homonymous motoneuron pool implied by the architecture of this projection is tempered by the plasticity of the connections, which enable them to act quasi-independently even though the stimulation pattern at each of the connections is identical.

In the case of cutaneous afferents, the different terminals made by a given afferent appear to result in similar modulation patterns during physiologically patterned stimulation. However, the different afferent types induce very different modulation patterns in response to the same program of stimulation, suggesting that the networks that they activate and/or their synaptic properties differ systematically.

Thus the appropriate spinal output from these highly specified, morphologically rigid systems is determined not only by the balance between fixed excitatory and inhibitory inputs but also by the ability of the afferent input to be modulated during stimulation at physiological frequencies. Several factors can influence the nature of this modulation, including the receptor supplied by the afferent and the type of postsynaptic cell contacted by the afferent (in the case of muscle spindle afferents, at least). These mechanisms are of prime importance in determining the functional synaptic input from the periphery.

In conclusion, it is clear that cells in the nervous system are precisely interconnected according to the cues implied in the term "neuronal specificity." The progress of the 20 years since the publication of Gaze's monograph does not relieve us of the necessity to continue our search for the nature of these cues. However, it is clear that once we have discovered what makes one cell attract input from another, or itself synapse with a third cell, we will still be left with numerous functional questions. It is not unlikely that the answers to these questions, such as why EPSPs facilitate or depress on different target cells, will reside in another set of specificity markers that determine function. Of course, it is always possible that connectivity and functional determinants will be one and the same. Whatever the answer, investigation of the linkage between molecules and function remains an important priority for the neurosciences.

Acknowledgments

The experimental work described in this paper was supported by NIH (NS 16996, NS 14899-LMM, and NS 23275-HRK). L.M.M. was the recipient of a Javits Neuroscience Award.

REFERENCES

Baldiserra F, Hultborn H, Illert M (1981): Integration in spinal neuronal systems. In: *Handbook of Physiology. Section 1. The Nervous System. Vol. II. Motor Control. Part 2*, Brooks VB, ed. Bethesda, MD: American Physiological Society

Brown AG (1981): *Organization in the Spinal Cord*. Berlin, Heidelberg, New York: Springer-Verlag

Brown AG, Fyffe REW (1981): Direct observations on the contacts made between Ia afferent fibers and α-motoneurons in the cat's lumbosacral spinal cord. *J Physiol* 313: 121–140

Brown AG, Koerber HR, Noble R (1987): An intracellular study of spinocervical tract cell responses to natural stimuli and single hair afferent fibres in cats. *J Physiol* 382: 331–354

Burke RE (1968): Group Ia synaptic input to fast and slow twitch motor units of cat triceps surae. *J Physiol* 196: 605–630

Burke RE, Strick PL, Kanda K, Kim CC, Walmsley B (1977): Anatomy of medial gastrocnemius and soleus motor nuclei in cat spinal cord. *J Neurophysiol* 40: 667–680

Burke RE, Walmsley B, Hodgson JA (1979): Structural functional relations in monosynaptic action on spinal motoneurons. In: *Integration in the Nervous System*, Brooks VB, Asanuma H, eds. Tokyo: Igaku, Shoin

Chanaud CM, Pratt CM, Loeb GE (1991): Compartmentalized muscles of the cat hindlimb. V. The roles of histochemical fiber-type regionalization and mechanical heterogeneity in differential muscle activation. *Exp Brain Res* 85: 300–313

Collins WF III, Davis BM, Mendell LM (1988): Modulation of EPSP amplitude during high frequency stimulation depends on the correlation between potentiation, depression and facilitation. *Brain Res* 442: 161–165

Collins WF III, Honig MG, Mendell LM (1984): Heterogeneity of group Ia synapses on homonymous α-motoneurons as revealed by high frequency stimulation of Ia afferent fibers. *J Neurophysiol* 52: 980–993

Gaze RM (1970): *The Formation of Nerve Connections*. New York: Academic Press

Gustafsson B, Pinter MJ (1984): Relations among passive electrical properties of lumbar α-motoneurones of the cat. *J Physiol* 356: 401–431

Gustafsson B, Pinter MJ, Wigstrom H (1986): The effect of axotomy on posttetanic potentiation of group Ia synapses in the cat. *J Neurophysiol* 56: 1174–1184

Heckman CJ, Binder MD (1988): Analysis of steady-state synaptic currents generated by homonymous Ia afferent fibers in motoneurons of the cat. *J Neurophysiol* 60: 1946–1966

Henneman E, Mendell LM (1981): Functional organization of the motoneuron pool and its inputs. In: *Handbook of Physiology. The Nervous System. Motor Control*, Brookhart JM, Mountcastle VB, eds. Bethesda, MD: American Physiological Society

Henneman E, Somjen GG, Carpenter DO (1965): Excitability and inhibitability of motoneurons of different sizes. *J Neurophysiol* 28: 599–620

Hongo T, Koike H (1975): Some aspects of synaptic organization in the spinocervical tract cell of the cat. In: *The Somatosensory System,* Kornhuber HH, and Stuttgart: Thieme

Koerber HR, Mendell LM (1991a): Modulation of synaptic transmission at Ia afferent fiber connections on motoneurons during high frequency stimulation: Role of postsynaptic target. *J Neurophysiol* 65: 590–597

Koerber HR, Mendell LM (1991b): Modulation of synaptic transmission at Ia afferent fiber connections on motoneurons during high frequency stimulation: Dependence on motor task. *J Neurophysiol* 65: 1313–1320

Koerber HR, Seymour AW, Mendell LM (1991): Tuning of spinal networks to frequency components of spike trains in individual afferents. *J Neurosci 11*: 3178–3187

Koerber HR, Brown PB, Mendell LM (1990): Correlation of monosynaptic field potentials evoked by single action potentials in single primary afferent axons and their bouton distributions in the dorsal horn. *J Comp Neurol* 294: 133–144

Koerber HR, Mendell LM (1988): Functional specialization of central projections from identified primary afferent fibers. *J Neurophysiol* 60: 1597–1614

Luscher HR, Vardar U (1989): A comparison of homonymous and heteronymous connectivity in the spinal monosynaptic reflex arc of the cat. *Exp Brain Res* 74: 480–492

Mendell LM, Collins WF III, Koerber HR (1990): How are synapses distributed on spinal motoneurons to permit orderly recruitment? In: *The Segmental Motor System,* Binder MD, Mendell LM, eds. New York: Oxford University Press

Mendell LM, Henneman E (1971): Terminals of single Ia fibers: Location, density and distribution within a pool of 300 homonymous motoneurons. *J Neurophysiol* 34: 171–187

Meyers DER, Snow PJ (1984): Somatotopically inappropriate projections of single hair follicle afferents to the cat spinal cord. *J Physiol* 347: 59–73

Munson JB, Sypert GW, Zengel JE, Lofton SA, Fleshman JW (1982): Monosynaptic projections of individual spindle group II afferents to type-identified medial gastrocnemius motoneurons in the cat. *J Neurophysiol* 48: 1164–1174

Nelson SG, Mendell LM (1978) Projection of single knee flexor Ia fiben to homonymous and heteronymous motoneurons. *J Neurophysiol* 42: 778–787

Rudomin P (1990): Presynaptic control of synaptic effectiveness of muscle spindle and tendon organ afferents in the mammalian spinal cord: In: *The Segmental Motor System,* Binder MD, Mendell LM, eds. New York: Oxford University Press

Scott JG, Mendell LM (1976): Individual EPSPs produced by single triceps surae Ia afferents in homonymous and heteronymous motoneurons. *J Neurophysiol* 39: 679–692

Stauffer EK, Watt DGD, Taylor A, Reinking RM, Stuart DG (1976): Analysis of muscle receptor connections by spike triggered averaging. 2. Spindle group II afferents. *J Neurophysiol* 39: 1393–1402

Zengel JE, Reid SA, Sypert GW, Munson JB (1985): Membrane electrical properties and prediction of motor-unit type of medial gastrocnemius motoneurons in the cat. *J Neurophysiol* 53: 1323–1344

CHAPTER 11

The Conditioned Goldfish Retinal Explant as an Experimental Model of Nerve Regeneration

B.W. Agranoff and A.M. Heacock

INTRODUCTION

Some years ago, we developed a simple procedure that permitted us to demonstrate neurite outgrowth from cultured explants of adult goldfish retina. It has since enjoyed considerable application in our laboratory as well as that of others, principally because it provides a vigorous outgrowth of neurites into an environment in which both the growth medium and the substratum can be manipulated experimentally. We describe here in some detail the experimental system, its properties, and several applications. It is particularly appropriate to review its origin in this volume honoring the many contributions of R.M. Gaze to developmental biology.

In 1974 one of us (B.W.A.) was a visiting investigator in the Laboratory of Developmental Biology at the National Medical Research Institute at Mill Hill. We encountered technical difficulties in explanting retinae from early-stage *Xenopus* tadpoles, principally because of the minute size and friability of the tissue. Those retinal bits that could be transferred successfully to a collagen substratum in an explant culture dish supported outgrowth of neurites after a few days *in vitro*. While it was difficult to tease out the retinal fragments in a reproducible fashion, late-stage retinae could be removed and explanted easily. Unlike our experience with earlier-stage retinal explants, neurite outgrowth from mature tadpole retina was not seen. We surmised that the rapidly growing neurites present in immature tissue continue to grow in vitro, whereas severed mature axons are much less likely to initiate in vitro growth. Gaze suggested we try a conditioning lesion, i.e., crush the optic nerve of mature tadpoles in vivo, wait a few days, and then explant the retinae. The concept of a conditioning lesion can be traced to clinical literature on nerve trauma, much of it military (for review see Ducker et al., 1969), which indicated that peripheral nerves often grew back more rapidly if they were cut proximally several days after the initial injury. This strategy proved successful for our *Xenopus* retinal explants (Agranoff et al., 1976). We subsequently reasoned that since the injured optic nerve has been demonstrated to regrow in both frogs and fishes, the conditioning lesion should

Formation and Regeneration of Nerve Connections
Sansar C. Sharma and James W. Fawcett, Editors
© 1993 Birkhäuser Boston

also work when applied to adult goldfish retinal explants, and this proved to be the case (Landreth and Agranoff, 1976).

The cellular basis for conditioning effects on nerve regeneration has been much discussed in the literature and has been the subject of studies in experimental animals. As is detailed below, the robust effect of a prior *in vivo* conditioning lesion on neurite outgrowth from explants *in vitro* has furnished additional useful information on the cellular basis of conditioning lesions. We do not include a number of additional applications of the conditioned goldfish retinal explant that also demonstrate the convenience and versatility of the preparation. For example, Freeman et al. (1985) have used the conditioned goldfish retinal explant to examine the possible generation of, or influence by, electrical currents at the growth cone. Schechter and his colleagues have demonstrated the dependence of neurites on supporting tissues for the expression of ON_1 and ON_2 neurofilament proteins and their induction upon addition of retinoic acid to the explant medium (Hall et al., 1990). The preparation was also used to reveal the non-neuronal nature of plasminogen activator activity associated with regeneration (Sallés et al., 1990). Koenig's laboratory has used the goldfish retinal explant for studies on nonperikaryal protein synthesis (Koenig and Adams, 1982; Koenig, 1989) and on the nature of neurite-associated varicosities (Koenig et al., 1985; Edmonds and Koenig, 1990). Bastmeyer et al. (1990) demonstrated the presence of N-CAM immunoreactivity on goldfish explant neurites and growth cones, employing a monoclonal antibody against chick neuronal cell adhesion molecule (N-CAM).

EXPERIMENTAL PROCEDURES

The following procedures are those used routinely in our laboratory, and are modified somewhat from previous descriptions (Landreth and Agranoff, 1976, 1979; Schwartz and Agranoff, 1989).

Optic Nerve Crush

Goldfish weighing 7–11 gm (147–17 cm in body length), maintained at a fixed temperature in the range 21–30°C, are anesthetized with tricaine methane-sulfonate. The fish are then placed on the stage of a low-power dissecting microscope, and the extraocular muscles of one eye (routinely the right, in our laboratory) are cut with a fine scissors over the posterodorsal aspect of the eye, after which the eye is gently pulled forward, exposing the optic nerve as it leaves the eyeball. Variable amounts of fat are present and may obscure the nerve. This can easily be removed with a blunt-ended 16-gauge needle attached to a suction apparatus. The optic nerve is then crushed through the intact dura with fine forceps. The forceps are held closed in place for 3 sec, then opened, and the crush procedure is repeated one additional time. Care is taken not to injure the blood supply to the eye. Fish that bleed retro-orbitally are discarded.

Preparation of Explants

Several days (generally 6–14) following the unilateral optic nerve crush, goldfish are dark-adapted for 30–60 min, anesthetized as above, then sacrificed by spinal cord transection. All subsequent procedures prior to retinal removal are carried out in subdued light. The conditioned and unoperated (control) eyes are removed from the orbit by blunt dissection with forceps and dipped briefly in several changes of sterile water, placed in 70% aqueous ethanol for 10 sec, then rinsed three more times in sterile saline. The front of each eye is cut away and the lens removed, exposing the retina. Stainless steel tubing (2 mm i.d.) with a beveled edge is used as a punch to remove the central retina, including the optic disk. The thinner nature of the peripheral retina makes it more suitable for explantation. The remainder of the retina is rinsed, then gently placed on the platform of a McIlwain tissue chopper, the surface and blade of which have been first sterilized with ethanol, and cut into 500-μm ribbons. The retina is repositioned by rotating the cutting plate 90° and is rechopped, yielding 500-μm squares of full-thickness retinal tissue. The pooled retinal pieces are rinsed in saline followed by Leibowitz L-15 nutrient medium containing 20 mM Na HEPES (pH 7.4), 0.1 mg/ml gentamicin sulfate, 0.1 mM 5'-fluorodeoxyuridine, 0.2 mM uridine, and 1% fetal calf serum (FCS). Retinal explants are then arranged as a 3 × 3 or 4 × 4 matrix on a 35-mm Nunclon tissue culture dish (9 or 16 explants per dish) previously coated with poly-L-lysine, laminin, or a combination of both. No regard is given to whether the explants lie with photoreceptor layer up or down. Sufficient culture medium is added to just cover the explants, but not enough to permit them to float free. After explants become firmly attached, additional medium may be added. The time required for attachment varies with culture conditions, particularly the nature of the substratum. The poly-L-lysine-coated substrata are prepared by adding to each culture dish 1 ml of a 1 mg/ml sterile solution of poly-L-lysine (Sigma) in 0.2 M sodium borate buffer (pH 8.4) for 6–24 hr at room temperature. For laminin substrata, each dish is coated with a solution of laminin (Collaborative Research) 100 μg/ml in the borate buffer for 18–24 hr. For poly-L-lysine + laminin substrata, poly-L-lysine-treated dishes are rinsed three times with sterile distilled water, then exposed to 1 ml of a 10 μg/ml solution of laminin in the borate buffer for 18–24 hr. Dishes (for all three substrata) are then rinsed three times in water, once in phosphate-buffered saline, then equilibrated with culture medium at least 30 min prior to plating the explants. Explants are grown in a humidified chamber at 20°C. It should be noted that for studies in *Xenopus*, the procedure is similar, but adjustment must be made for the lower osmolarity of the species (Agranoff et al., 1976).

Computerized Assay

The extent of neurite outgrowth in goldfish retinal explants was initially determined somewhat subjectively, by scoring the explants for both length and density and multiplying these two scores to obtain a neurite growth index (NGI; Landreth and Agranoff, 1979). The measure has been utilized in determining the effects of various agents such as difluoromethylornithine (Kohsaka et al., 1982), tunicamycin (Heacock, 1982), and diazacholesterol (Heacock et al., 1984) on neuritic outgrowth. An automated assay based on a video image microcomputer analysis, which bears the potential advantages of reduced scoring time, increased reproducibility, and elimination of subjectivity, was subsequently developed (Ford-Holevinski et al., 1986a). We have recently improved this assay by substituting the Apple II computer used in our initial automation procedure with the faster Macintosh computer. We now generate neuritic growth values at an average rate of 8 sec per explant. The lower-power, pseudo-dark-field image provided by the Wild M420 Makroskop permits visualization of an entire explant and its neurites within the field. The use of the technique is illustrated in Figure 11.1.

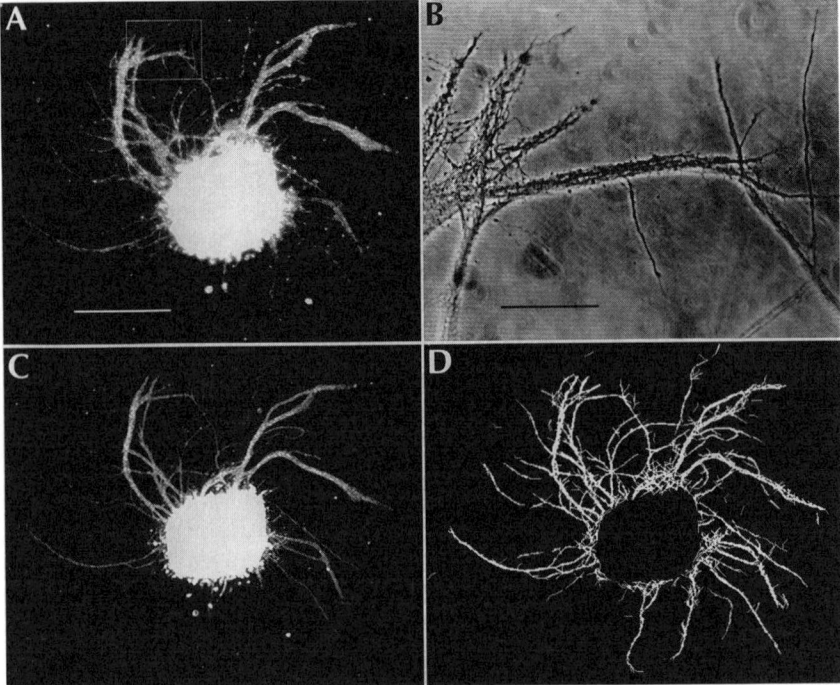

FIGURE 11.1. Imaging of a goldfish retinal explant after 5 days *in vitro*. **A:** Pseudo-dark-field photomicrograph of an explant; scale bar = 600 μm. **B:** Phase contrast photomicrograph of the boxed area in **A**; scale bar = 100 μm. **C:** Video image of same explant as in **A**. **D:** Image generated by Macintosh neurite recognition program currently in use in our laboratory.

GROWTH FROM CONDITIONED EXPLANTS

Origin of Neurites

Of the many cell types within the retina, the ganglion cells would seem the most probable source of the neurites, since optic nerve crush selectively injures the ganglion cell axons and leads to greatly enhanced neurite outgrowth (Landreth and Agranoff, 1976). Evidence that this is indeed the case was obtained by histological examination of explants after neurite outgrowth had occurred (Johns et al., 1978a). The explants were observed to undergo a progressive necrosis, starting with the photoreceptor nuclei, then progressing to the inner nuclear layer until, by 3 weeks *in vitro*, only the ganglion cells could be identified. Since extensive neuritic outgrowth is still apparent after 3 weeks *in vitro*, these observations point to the ganglion cells as the source of the neurites. Additional proof was provided by silver staining of the explants and retrograde axonal tracing with horseradish peroxidase (HRP) (Johns et al., 1978a). Silver staining of the explants also revealed that an area of dense asymmetric neurite outgrowth was often aligned with an array of parallel fibers in a region of the explant corresponding to the retinal optic fiber layer. We subsequently demonstrated that these parallel fibers were indeed ganglion cell axons. The earliest neurites which grew out of the explant were in fact those that coursed along the explant margin that had been nearest the optic disc *in vivo*, i.e., along the degenerated distal fiber track, and could thus be considered to be growing toward a phantom optic disc (Johns et al., 1978b).

Outgrowth as a Function of the Substratum

For most growth assays, we have used an inverted microscope and low-power, pseudo-dark-field illumination. For in vivo morphological examination, phase contrast or interference contrast optics are suitable. For high-power examination of neurites, a hole can be made in the plastic bottom of a Nunclon culture dish, which is then resealed by gluing a thin glass coverslip to the dish. The morphology and appearance of the goldfish retinal explant neurites vary as a function of the tissue culture substratum employed. In initial studies, we used a collagen gel substratum, on which we found the neurites to radiate out randomly from the explant in sparse fascicles (Landreth and Agranoff, 1976, 1979). We subsequently turned to a thin coating on the plastic surface for the inherent improvement in optical properties and overall convenience. On a polycation-coated substratum, a striking clockwise directionality of the neurite fascicles was observed (Heacock and Agranoff, 1977; see also Figure 11.1). This clockwise outgrowth pattern did not vary with the chirality of the substratum (i.e., poly-L- or poly-D-lysine), nor did it depend on the anatomical frame of reference of the explant (i.e., right or left eye, or whether the retinal explant's vitreal or photoreceptor layer had been placed against the substratum). An influence of external factors such as electromagnetic, gravitational, or even Coriolis force fields was ruled out by an experiment in which explants were grown upside down while attached to a poly-L-lysine-coated coverslip. They still exhibited clockwise neuritic outgrowth with respect to the substratum, although to an observer looking down at an explant culture from above, the growth would appear to be counterclockwise. These observations suggest that there is an inherent right-handed helicity within the neurites and neurite fascicles, which when given a planar substratum of the appropriate degree of adhesiveness, is expressed as a clockwise curvature of outgrowth. Subsequent experiments led to the suggestion that the degree of curvature of outgrowth was a function of the relative degrees of fiber–fiber vs. fiber–substratum affinity. Outgrowth on the extracellular matrix component laminin led to a very different morphology (Hopkins et al., 1985). On laminin, neurites grew out as straight spokes in a radial pattern with little or no curvature and much reduced fasciculation. This pattern may reflect a high affinity of the neurites for laminin, a substratum for which the rate of neurite outgrowth was enhanced fivefold over that seen for poly-L-lysine. On a combination poly-L-lysine + laminin substratum, fasciculation remained reduced, yet there was a marked accentuation of the clockwise curvature over that seen with poly-L-lysine alone. We explain this as the combined effects of the acceleration of outgrowth by laminin mediated by the fiber–substratum interaction and the expression of curvature by remaining fiber–fiber interaction. The compound nature of neurites is not apparent from phase microscopic examination but can be readily visualized by use of the vital stain dioctadecylindocarbocyanine (DiI; Agranoff et al., 1980; Feldman et al., 1982). These were, to our knowledge, the first reports of the use of this useful dye in the nervous system. The possibility that the high affinity of the neurites for laminin was mediated by a heparin sulfate proteoglycan (HSP) binding site was explored. The addition of HSP was found to have no effect on neurite outgrowth on a laminin substratum; however, growth on a poly-L-lysine substratum was completely blocked (Hopkins and Agranoff, 1987). Thus the neurite-promoting activity of laminin in the goldfish retinal explant system does not depend on the HSP binding site.

Agents Affecting Neuritic Outgrowth

While a laminin substratum clearly increases the rate of neurite outgrowth, thus far no agent that we have added to the retinal explant culture medium has elicited measurable enhancement of neuritic outgrowth from either conditioned or control goldfish retinal explants. Lima et al. (1988) report a stimulatory effect of taurine on neurite outgrowth. Addition of gangliosides, a glycolipid class that has been reported to promote regeneration, has not been reported in the goldfish explant system, although antiganglioside antibodies appear to block neurite outgrowth from explanted conditioned retinas (Spirman et al., 1982). The nature of membrane synthesis, transport, and assembly is complex. This was explored by labeling goldfish retinal explant cultures with the lectin concanavalin A (Con A), followed by antibodies to Con A, which immobilized the Con A binding sites but still permitted neurite outgrowth. After 24 hr, treatment with fluorescent goat anti-rabbit antibodies revealed that labeling was confined to the region of old growth, indicating that new membrane is added in the region of the growing tip (Feldman et al., 1981).

Conditions that permit neurite outgrowth from conditioned goldfish retinal explants in the absence of added fetal calf serum (FCS) have been reported (Johnson and Turner, 1982), although FCS nevertheless stimulates outgrowth. These authors reported that goldfish tectal extracts from optic nerve-lesioned goldfish can replace the requirement for FCS, but when added together with FCS, may slightly inhibit outgrowth. Schwartz and her collaborators (Mizrachi and Schwartz, 1982; Schwartz et al., 1982) have reported the stimulatory effect of extracts of control goldfish brain on neurite outgrowth, and their potentiating effect on outgrowth when added with FCS to conditioned explants. The stimulatory activity present in the brain extracts is reportedly heat-labile and nondialyzable.

It should be noted that no *in vivo* or *in vitro* treatment has thus far been reported that will lead to outgrowth from control, i.e., unconditioned, explants. Nerve growth factor (NGF) has not been reported to stimulate outgrowth when added to goldfish retinal explants, but it is claimed that it accelerates and enhances the stimulatory effects of conditioning lesions if injected intraocularly at the time of optic nerve crush (Turner et al., 1980).

The Role of Intracellular Messengers

We have recently explored the possible involvement of second messengers in the neuritic outgrowth process. Participation of the phosphoinositide signal transduction pathway was probed by analysis of the effects of lithium ion, since chronic exposure to Li^+ might be expected to compromise this second messenger system. Effects of Li^+ on developing organisms have been described (for review, see Berridge et al., 1989). Of more direct relevance is a report of inhibition of NGF-induced neurite outgrowth by Li^+ in PC-12 cells (Burstein et al., 1985). Neurite outgrowth in goldfish retinal explants was found to be inhibited by relatively high concentrations of Li^+ (Figure 11.2), with an IC_{50} of $\sim$ 5 mM. This effect of Li^+ was found, however, not to be due to perturbation of the phosphoinositide signal transduction pathway, as indicated both by the failure of inositol (10 or 30 mM) to reverse the Li^+ inhibition, and by a similar sensitivity of neurite outgrowth to cesium ion (data not shown). Li^+-induced alterations in cation transport (Wood and Goodwin, 1987) rather than a block of stimulated phosphoinositide turnover may underlie the observed effects on neurite outgrowth.

An alternative approach to investigating the role of the second messenger systems in neurite outgrowth is the perturbation of protein kinase C (PKC) activity. Diacylglycerol, the major intracellular lipid modulator of PKC, can arise via the stimulated breakdown of phosphoinositides or of other phospholipids. Both inhibitory and stimulatory effects of PKC activators and blockers on neurite outgrowth in various neuronal culture systems have been described (see Bixby,

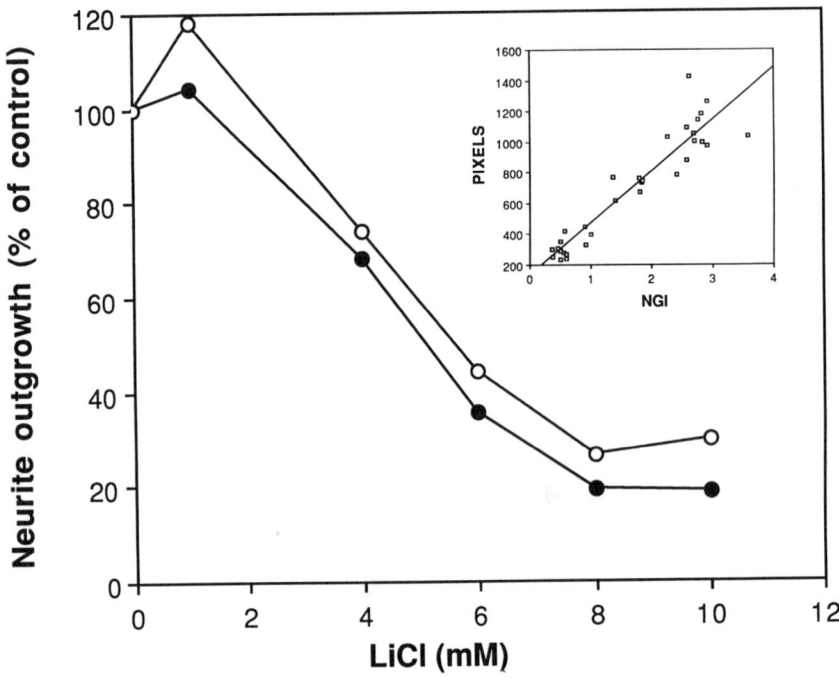

FIGURE 11.2. Inhibition of neurite outgrowth by lithium chloride. Postcrush retinal explants were grown for 7 days in the presence of the indicated concentrations of LiCl. The extent of neurite outgrowth was then determined by measurement of the neurite growth index (●) or by the computerized growth assay (○). Inset shows good correlation between the two methods of measurement of neurite outgrowth ($r = 0.88$).

1989). The effect of several protein kinase inhibitors on neurite outgrowth from goldfish retinal explants onto a poly-L-lysine substratum was examined. Staurosporine, sphingosine, K_{252a}, and H7 all blocked neurite outgrowth in a reversible manner with IC_{50} values compatible with their inhibitory effects on PKC. In contrast, HA1004, which is 20 times more potent against protein kinase A than against PKC, was ineffective. The phorbol ester PKC activator (teradecanolyphorbol acetate (TPA) blocks neurite outgrowth by 50–60%, possibly a reflection of the ability of chronic treatment with TPA to down-regulate PKC. In addition, the protein phosphatase inhibitor okadaic acid proved to be a very potent inhibitor of neurite outgrowth from the retinal explants. Taken together, these results suggest a role for PKC-mediated protein phosphorylation/dephosphorylation in the regulation of neurite outgrowth in this system (Heacock and Agranoff, in preparation).

Neurite Outgrowth from Conditioned Retinal Explants of Other Species

Conditioned retinal explants from *Xenopus* prepared as previously described (Agranoff et al., 1976) have been used to examine the distribution of monoclonal antibodies directed against cytoskeletal elements (Steen et al., 1989). The clockwise outgrowth pattern seen in goldfish explants (Heacock and Agranoff, 1977) has been demonstrated in older (stages 40–50), but not in earlier (stage 25) *Xenopus* tadpole retinal explants (Grant and Tseng, 1989). Gooday (1990) compared neurite outgrowth in conditioned *Xenopus* explants from various retinal regions to establish possible differences in response to glia. Since regeneration in the central nervous system (CNS) appears to be a general property of the poikilotherm brain, one may anticipate success in eliciting neurite outgrowth from conditioned retina of fish species other than the goldfish. We have in fact demonstrated such outgrowth from the zebrafish (Figure 11.3), a species that has generated much interest because of its potential for molecular genetic approaches. Outgrowth from mammalian retinal explants has also been investigated. We found neurite outgrowth from a conditioned rat retinal explant onto a laminin-fortified substratum much more likely to occur than from a control explant (Ford-Holevinski et al., 1986b). The conditioning effect was confirmed by Johnson et al. (1989) under culture conditions in which laminin was not stimulatory to outgrowth. Conditions have been established for long-term survival of unconditioned mammalian retinal explants, and successful neurite outgrowth from postmortem human retina onto a Schwann cell substratum has recently been reported (Hopkins and Bunge, 1991).

DISCUSSION

It was early noted, on the basis of clinical surgical experience, that the probability that a damaged nerve will successfully regenerate is increased by waiting for some days after the initial trauma, and then axotomizing the injured nerve proximal to the original lesion. An initial delay in new outgrowth from the proximal stump can be attributed first to the retrograde transport of a signal from the site of injury to the cell body, and second to the gearing up of the cell's macromolecular synthetic machinery to support new axonal growth. Old histochemical observations of the axon reaction (Lieberman, 1974) were shown to reflect an increased rate of RNA and protein synthesis (Brattgård et al., 1957; Smith et al., 1984), which persists until nerve regeneration is complete. Experimental evidence for conditioning effects in nerve regeneration *in vivo* in the mammalian peripheral nervous system (Young et al., 1940; Gutmann, 1942) confirms clinical experience. It can be shown in rats that nerve regrowth in a conditioned nerve proceeds at a more rapid rate than does primary regeneration (McQuarrie and Grafstein, 1973; McQuarrie, 1979). It is generally believed that, since the initial trauma initiates the perikaryal regenerative response, there will be a much briefer lag period in initiation of regrowth from the fresh neural stump proximal to the initial injury. Most such studies have been performed in long motor nerves such as the hypoglossal or sciatic nerve. In nerves with long axons, the volume of the cell body is a fraction of that of the long axon, which is under the support of the macromolecular synthetic machinery in the perikaryon, so it is not surprising that such neurons

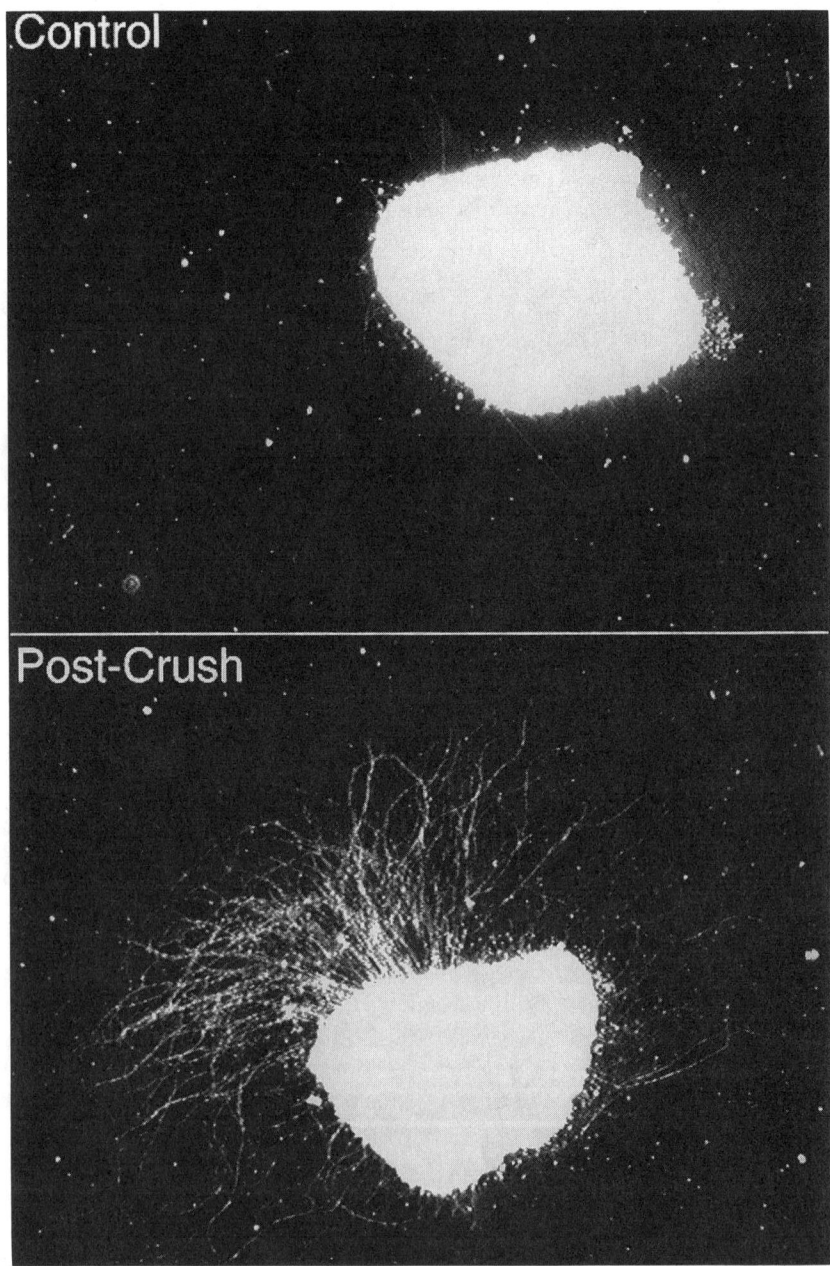

FIGURE 11.3. Conditioning lesion effect in explanted zebrafish retina. Control and 8-day postcrush zebrafish retina explants after 4 days outgrowth *in vitro* on a poly-L-lysine + laminin substratum.

have an enormous capacity for RNA and protein synthesis. Both of these measures are increased during the period of nerve regrowth and reconnection. One might speculate that given a long enough nerve, a conditioning lesion could actually lead to functional recovery earlier than a singly lesioned nerve, i.e., the enhanced rate of synthesis would be maintained so high that the twice-cut nerve could overtake the rate of recovery of a singly crushed one. Such a finding might indicate simply that two retrograde signals to the perikaryon to regenerate are better than one, or alternatively, that the rapid rate of new growth from the cut conditioned nerve stump overcomes some peripheral process that might otherwise block regrowth. For example, glial scar formation at the site of injury is known in some instances to block regrowth. Rapid neurite outgrowth in conditioned nerve could traverse a given region before glial replication blocked regrowth. Our conditioned explant studies indicate that the conditioning lesion has profound stimulating effects on neurite outgrowth in the absence of any interaction with non-neuronal cells at the site of outgrowth. Furthermore, although there may be extracellular growth factors that stimulate regeneration, the conditioning effect that we have studied does not rely on them. When control and conditioned retinal explants are grown in the same incubation medium in a single culture dish, we find no evidence that conditioned explants stimulate outgrowth from control explants by a humoral effect or, conversely, that control explants inhibit neurite outgrowth from conditioned explants. If we remove half of the tectum, we produce a goldfish retina that is in effect half control and half conditioned.

In that instance, we observe ganglion cell hypertrophy only in the denervated portion of the retina, as well as neurite outgrowth only from the denervated side upon explanation (Landreth and Agranoff, 1979). The unaxotomized retinal region shows no histological evidence of nerve crush, nor are neurites extended from such explants, again arguing for the absence of a humoral effect. Furthermore, if a systemically mediated process were to account for the conditioning effect (e.g., a circulating growth factor or hormone), one might expect that the contralateral retina would also evince outgrowth. This is not the case. In fact, we routinely use the contralateral retina for control explants. They do not differ from retina explanted from previously untreated fish in regard to their inability to extend neurites, even after many days in culture.

One might predict that since ganglion cell axons are cut upon explanation, neurites would grow out from control retinae after several days *in vitro*. In the original *Xenopus* studies there was an indication that such was the case (Agranoff et al., 1976). In goldfish studies we found no indication of such "catch up" *in vitro*. These results indicate that *in vivo* conditions are in some ways more conducive to the various biochemical events that lead to the conditioning effect seen *in vivo*. Could it be that explants simply run downhill in culture, and by the time the ganglion cell perikarya are conditioned, the preparation is unable to support neurite outgrowth? This does not appear to be the case, since conditioned explants can be kept in the culture medium for several days without contact with the substratum. They may then be explanted and will demonstrate excellent neurite extension. On the other hand, it should be noted from studies in mammalian retina in which long-term survival conditions are optimized that neurite outgrowth can be elicited in the absence of a prior conditioning lesion (see, for example, Hopkins and Bunge, 1991).

Explant studies have addressed but have not answered, the questions of why successful regeneration occurs in the CNS of cold-blooded vertebrates but not in the avian or mammalian nervous system, and why the peripheral nervous system (PNS) of warm-blooded animals, unlike the CNS, regenerates. It is clear from such studies, however, that explanted CNS neurons will support a sustained extension of neurites in the absence of supporting cells at the growth cone. It is also apparent that a conditioning lesion and a favorable substratum, while not obligatory, enhance outgrowth.

We have not addressed in this chapter an extensive literature on biochemical studies on retinal regeneration in the goldfish (for reviews see Agranoff et al., 1980; Benowitz and Routtenberg, 1987; Grafstein, 1986). Increased axonal transport of newly synthesized proteins from the retinal ganglion cells has been studied in some detail. In addition to tubulin (Heacock and Agranoff, 1976) and other structural cytoskeletal elements, GAP-43 (Benowitz et al., 1981; Heacock and Agranoff, 1982) as well as other growth-related proteins appear to be selectively produced and transported down the regrowing fibers in vivo. A growth-related protein doublet termed p68-70 (Heacock and Agranoff, 1982; Agranoff and Ford-Holevinski, 1984) has been the subject of considerable interest in this laboratory. Its synthesis is greatly enhanced in regenerating retinal ganglion cells and it is axonally transported at a slow rate, but does not appear to be associated with known cytoskeletal components. A polyclonal antibody to the doublet indicates that their amounts are increased in regenerating ganglion cells and their axons (G. Wilmot, in press). Recent studies employing in situ hybridization techniques indicate that mRNA for tubulin remains elevated until reconnection in the tectum occurs. In contrast, mRNA for a cholinergic receptor subunit is apparently enhanced only after reconnection takes place (Hieber et al., 1991). One may safely predict such molecular biological approaches will shed new light on the nature of the signal that initiates and regulates regeneration.

REFERENCES

Agranoff BW, Ford-Holevinski TS (1984): Biochemical aspects of the regenerating goldfish visual system. In: *Advances in Neurochemistry, Vol 6: Axonal Transport in Neuronal Growth and Regeneration*, Elam J, Cancalon P, eds. New York: Plenum Press

Agranoff BW, Feldman EL, Heacock AM, Schwartz M (1980): The retina as a biochemical model of central nervous system regeneration. *Neurochem Int* 1: 487–500

Agranoff BW, Field P, Gaze RM (1976): Neurite outgrowth from explanted *Xenopus* retina: An effect of prior optic nerve section. *Brain Res* 113: 225–234

Benowitz LI, Routtenberg A (1987): A membrane phosphoprotein associated with neural development, axonal regeneration, phosopholipid metabolism, and synaptic plasticity. *Trends Neurosci* 10: 527–532

Bastmeyer M, Schlosshauer B, Stuermer CAO (1990): The spatiotemporal distribution of N-CAM in the retinotectal pathway of adult goldfish detected by the monoclonal antibody D3. *Development* 108: 299–311

Benowitz LI, Shashoua VE, Yoon MG (1981): Specific changes in rapidly transported proteins during regeneration of the goldfish optic nerve. *J Neurosci* 1: 300–307

Berridge MJ, Downes CP, Hanley MR (1989): Neural and developmental actions of lithium: A unifying hypothesis. *Cell* 59: 411–419

Bixby JL (1989): Protein kinase C is involved in laminin stimulation of neurite outgrowth. *Neuron* 3: 287–297

Brattgård S-O, Edström J-E, Hyden H (1957): The chemical changes in regenerating neurons. *J Neurochem* 1: 316–325

Burstein DE, Seeley PJ, Greene LA (1985): Lithium ion inhibits nerve growth factor-induced neurite outgrowth and phosphorylation of nerve growth factor-modulated microtubule-associated proteins. *J Cell Biol* 101: 862–870

Ducker TB, Kempe LG, Hayes GJ (1969): The metabolic background for peripheral nerve surgery. *J Neurosurg* 30: 270–280

Edmonds PT, Koenig E (1990): ATP and calmodulin dependent actomyosin aggregates induced by cytochalasin D in goldfish retinal ganglion cell axons *in vitro*. *J Neurobiol* 21: 555–566

Feldman EL, Axelrod D, Schwartz M, Heacock AM, Agranoff BW (1981): Studies on the localization of newly added membrane in growing neurites. *J Neurobiol* 12: 591–598

Feldman EL, Heacock AM, Agranoff BW (1982): Lectin binding to neurites in goldfish retinal explants. *Brain Res* 248: 347–354

Ford-Holevinski TS, Dahlberg TA, Agranoff BW (1986a): A microcomputer-based image analyzer for quantitating neurite outgrowth. *Brain Res* 368: 339–346

Ford-Holevinski TS, Hopkins JM, McCoy JP, Agranoff BW (1986b): Laminin supports neurite outgrowth from explants of axotomized adult rat retinal neurons. *Dev Brain Res* 28: 121–126

Freeman JA, Manis PB, Snipes GJ, Mayer BN, Samson PC, Wikswo JP Jr, Freeman DB (1985): Steady growth cone currents revealed by a novel circularly vibrating probe: A possible mechanism underlying neurite growth. *J Neurosci Res* 13: 257–283

Gooday DJ (1990): Retinal axons in *Xenopus laevis* recognize differences between tectal and diencephalic glial cells in vitro. *Cell Tissue Res* 259: 595–598

Grafstein B (1986): The retina as a regenerating organ. In: *The Retina: A Model for Cell Biology Studies, Part II*, Adler R, Farber DB, eds. New York: Academic Press

Grant P, Tseng Y (1989): *In vitro* growth properties of *Xenopus* retinal neurons undergo developmental modulation. *Dev Biol* 133: 502–514

Gutmann E (1942): Factors affecting recovery of motor function after nerve lesions. *J Neurol Psychiatry* 5: 81–95

Hall CM, Else C, Schechter N (1990): Neuronal intermediate filament expression during neurite outgrowth from explanted goldfish retina: Effect of retinoic acid. *J Neurochem* 55: 1671–1682

Heacock AM (1982): Glycoprotein requirement for neurite outgrowth in goldfish retina explants: Effects of tunicamycin. *Brain Res* 241: 307–315

Heacock AM, Agranoff BW (1976): Enhanced labeling of a retinal protein during regeneration of optic nerve in goldfish. *Proc Natl Acad Sci USA* 73: 828–832

Heacock AM, Agranoff BW (1977): Clockwise growth of neurites from retinal explants. *Science* 198: 64–66

Heacock AM, Agranoff BW (1982): Protein synthesis and transport in the regenerating goldfish visual system. *Neurochem Res* 7: 771–788

Heacock AM, Klinger PD, Seguin EB, Agranoff BW (1984): Cholesterol synthesis and nerve regeneration. *J Neurochem* 42: 987–993

Hieber V, Agranoff BW, Goldman D (1992): Target-dependent regulation of retinal nicotinic acetylcholine receptor and tubulin RNAs during optic nerve regeneration, *J Neurochem* 58: 1009–1015

Hopkins JM, Agranoff BW (1987): Neurite outgrowth of laminin does not require the heparin-binding site. *Neurosci Res Commun* 1: 57–63

Hopkins JM, Bunge RP (1991): Regeneration of axons from adult human retina *in vitro*. *Exp Neurol* 112: 243–251

Hopkins JM, Ford-Holevinski TS, McCoy JP, Agranoff BW (1985): Laminin and optic nerve regeneration in the goldfish. *J Neurosci* 5: 3030–3038

Johns PR, Heacock AM, Agranoff BW (1978a): Neurites in explant cultures of adult goldfish retina derive from ganglionc cells. *Brain Res* 142: 531–537

Johns PR, Yoon MG, Agranoff BW (1978b): Directed outgrowth of optic fibres regenerating *in vitro*. *Nature* 271: 360–362

Johnson AR, Gregson NA, Wigley CB, Berry M (1989): The conditioning effect of optic nerve injury upon axonal regrowth from adult rat retinal ganglion cells explanted in vitro. *Neurosci Letts* 97: 63–68

Johnson JE, Turner JE (1982): Growth from regenerating goldfish retinal cultures in the absence of serum or hormonal supplements: Tissue extract effects. *J Neurosci Res* 8: 315–329

Koenig E (1989): Cycloheximide-sensitive [^{35}S]methionine labeling of proteins in goldfish retinal ganglion cell axons in vitro. *Brain Res* 481: 119–123

Koenig E, Adams P (1982): Local protein synthesizing activity in axonal fields regenerating *in vitro*. *J Neurochem* 39: 386–400

Koenig E, Kinsman S, Repasky E, Sultz L (1985): Rapid mobility of motile varicosities and inclusions containing α-Spectrin, actin, and calmodulin in regenerating axons *in vitro*. *J Neurosci* 5: 715–729

Kohsaka S, Heacock AM, Klinger PD, Porta R, Agranoff BW (1982): Dissociation of enhanced ornithine decarboxylase activity and optic nerve regeneration in goldfish. *Dev Brain Res* 4: 149–156

Landreth GE, Agranoff BW (1976): Explant culture of adult goldfish retina: Effect of prior optic nerve crush. *Brain Res* 118: 299–303

Landreth GE, Agranoff BW (1979): Explant culture of adult goldfish retina: A model for the study of CNS regeneration. *Brain Res* 161: 39–53

Lieberman A (1974): The axon reaction: A review of the principal features of perikaryal responses to axon injury. *Int Rev Neurobiol* 14: 49–123

Lima L, Matus P, Drujan B (1988): Taurine effect on neuritic growth from goldfish retinal explants. *Int J Dev Neurosci* 6: 417–424

McQuarrie IG (1979): Accelerated axonal sprouting after nerve transection. *Brain Res* 167: 185–188

McQuarrie IG, Grafstein B (1973): Axon outgrowth enhanced by a previous nerve injury. *Arch Neurol* 29: 53–55

Mizrachi Y, Schwartz M (1982): Goldfish tectal explants have a growth-promoting effect on neurites emerging from co-cultured regenerating retinal explants. *Dev Brain Res* 3: 502–505

Sallés FJ, Schecther N, Stickland S (1990): A plasminogen activator is induced during goldfish optic nerve regeneration. *EMBO J* 9: 2471–2477

Schwartz M, Agranoff BW (1989): Retinal explants as in *in vitro* model for optic nerve regeneration in lower vertebrates. In: *A Discussion and Tissue Culture Manual of the Nervous System*, Shahar A, de Vellis J, Vernadakis A, Haber B, eds. New York: Alan R. Liss

Schwartz M, Mizrachi Y, Eshhar N (1982): Factor(s) from goldfish brain induce neuritic outgrowth from explanted regenerating retinas. *Dev Brain Res* 3: 29–35

Smith CB, Crane AM, Kadekaro M, Agranoff BW, Sokoloff L (1984): Stimulation of protein synthesis and glucose utilization in the hypoglossal nucleus induced by axotomy. *J Neurosci* 4: 2489–2496

Spirman N, Sela B-A, Schwartz M (1982): Antiganglioside antibodies inhibit neuritic outgrowth from regenerating goldfish retinal explants. *J Neurochem* 39: 874–877

Steen P, Kalghatgi L, Constantine-Paton M (1989): Monoclonal antibody markers for amphibian oligodendrocytes and neurons. *J Comp Neurol* 289: 467–480

Turner JE, Delaney RK, Johnson JE (1980): Retinal ganglion cell response to nerve growth factor in the regenerating and intact visual system of the goldfish *(Carassius auratus)*. *Brain Res* 197: 319–330

Wilmot GR, Raymond PA, Agranoff BW (1992): The expression of the protein p68/70 within the goldfish visual system suggests a role in both regeneration and neurogenesis, *J Neurosci*, in press

Wood AJ, Goodwin GM (1987): A review of the biochemical and neuropharmacological actions of lithium. *Psychol Med* 17: 579–600

Young JZ, Holmes W, Sanders FK (1940): Nerve regeneration. *Lancet* 3: 128–130

CHAPTER 12

The Morphology of Developing and Regenerating Retinal Ganglion Cells

Sarah Dunlop, Sandra Fraley, and Lyn Beazley

SUMMARY

In this chapter we describe aspects of the changing morphology of retinal ganglion cells during both normal development and regeneration. We labeled cells with the neuronal tracer horseradish peroxidase (HRP) to visualize the dendritic trees or the axons of individual cells. This approach has allowed us to gain insights into the morphology of cell processes beyond that afforded by an examination of the cell population as a whole. We have examined the development of dendrites in the quokka (a type of wallaby or small kangaroo) and of axons in the developing optic nerve of the chick, quokka, rabbit, and cat. We have also examined the morphology of optic axons in frog as they regenerate through the site of an optic nerve crush.

We describe a possible sequence by which dendritic trees develop from immature cells lacking dendrites. The first dendrites are always seen to arise on the side of the soma opposite the axon hillock. As the first dendrites elongate and branch, new dendrites are added sequentially between this site and the axon. We suggest that the overall polarity of the cell is determined by the point at which the axon exits the soma. Spines are a prominent feature of developing dendrites, and the branching pattern of the dendritic tree is more complex than in the adult.

Developing and regenerating optic axons have several features in common. Both are tipped with growth cones; long and numerous filopodia are a hallmark of growth cones on developing axons in the chick. Axonal spines are also present in the developing and regenerating systems. Furthermore, we saw axons with highly aberrant trajectories. These are in a minority during development but are numerous during axon regeneration. We suggest that some ganglion cells die, both during development and optic nerve regeneration, because their axons become misdirected. In addition, we propose that features such as growth cone filopodia, axonal spines, and axons with aberrant trajectories may have contributed to the high counts of axonal profiles reported in optic nerves during development and regeneration.

Formation and Regeneration of Nerve Connections
Sansar C. Sharma and James W. Fawcett, Editors
© 1993 Birkhäuser Boston

INTRODUCTION

The formation of nerve connections is fundamental to brain development and to the restoration of function after damage. In the last three decades, we have gained a much greater understanding of the mechanisms whereby populations of neurons navigate to their target tissue and choose their partner cells within it during both development and regeneration. It is the work of Michael Gaze, honored at this meeting, that has pioneered much of our knowledge in this field.

However, for all the advances gained by studying populations of neurons, we know very little of the changes that take place in the morphology of individual cells during either normal development or axonal regeneration. Such knowledge is basic to our understanding of the processes that underlie the formation of functional circuitry in the central nervous system. In this chapter we report some recent studies in which retinal ganglion cells were labeled *in vitro* with HRP. We describe the morphology of dendrites (Dunlop, 1990) and of axons in the optic nerve during development (Dunlop et al., 1990; Fraley et al., 1991; Dunlop et al., 1992); in addition, we have examined optic axons regenerating after axotomy in the frog.

DEVELOPMENT OF DENDRITIC TREES

We have traced the emergence of dendritic trees from newly postmitotic ganglion cells in the mammalian retina (Figure 12.1A-H). This approach contrasts with studies that have concentrated on later stages of dendritic elaboration (Morest, 1970; Perry and Walker, 1980; Maslim et al., 1986; Dann et al., 1987, 1988; Ramoa et al., 1987, 1988; Wong, 1990). Our study was conducted using the quokka, *Setonix brachyurus,* a small wallaby.

The most immature ganglion cells lack dendrites (Figure 12.1A-C,D2), and some have an endfoot still attached to the ventricular zone (Figure 12.1A). The somata are elongate and are presumably in the process of migrating from the cytoblastic to the ganglion cell layer (Figure 12.1A-C). Our method of placing HRP on the optic nerve allows us to deduce that these immature retinal ganglion cells nevertheless possess an axon which has penetrated the optic nerve. Once in the ganglion cell layer, the somata become rounded and the first dendrites emerge as short processes (Figure 12.1D1) or as lamellipodia.

We found that all of the cells are strongly polarized with respect to the emergence of their axons and dendrites. That is, the axons emerge from one side of the soma and dendrites from the opposite side. No exceptions to this polarity rule are seen. In the majority of cases, axons emerge from the side of the soma closest to the optic disc, the dendrites from the side closest to the retinal margin (Figure 12.1D1,E1,F,G). However, in a minority of cells, axons emerge from the side of the soma toward the retinal margin (Figure 12.1D2,E2). In all such cases, the dendrites are seen to emerge on the side nearest the disc (Figure 12.1E2).

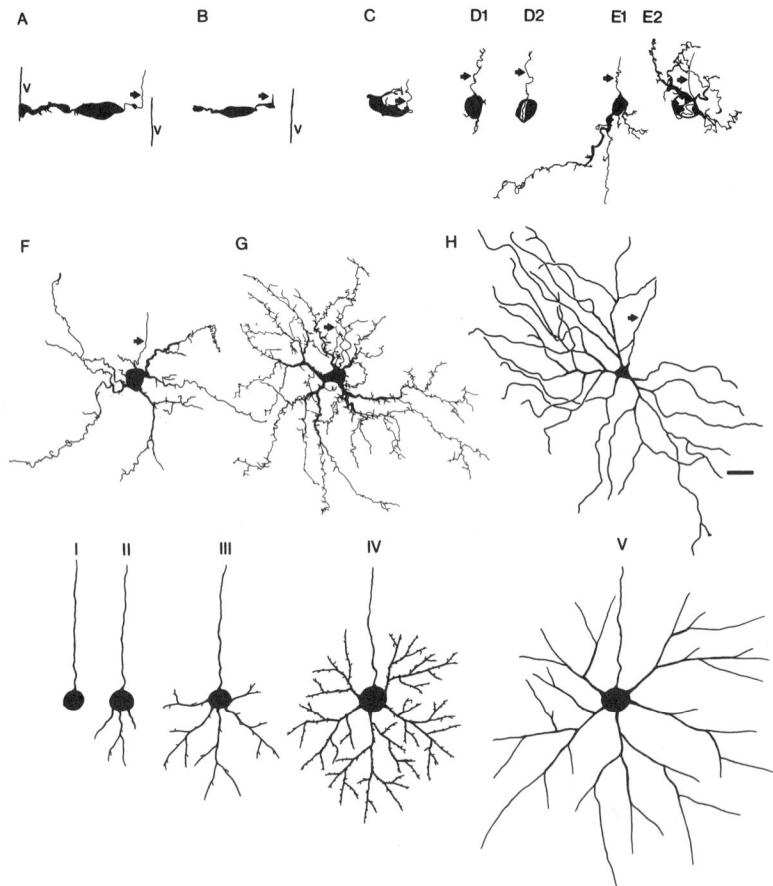

FIGURE 12.1. Dendritic trees of retinal ganglion cells in the quokka during development **(A-G)** and in the adult **(H)**. Cells were drawn with a camera lucida using Nomarski optics at 1,500×. The optic disc is toward the top of the page. **A-C**: Very immature ganglion cells span the depth of the retina; the ventricular surface is indicated by a line to the left of each cell and the positions of the axon mark the location of the nerve fiber layer. **D-H**: Ganglion cells in the ganglion cell layer seen in whole mount preparations. Axons (arrows) exit the soma from the side toward the optic disk in A-D1, E1, F-H but toward the retinal margin in D2 and E2. The HRP was applied to the severed optic nerve of isolated eye cups, resulting in the dendritic labeling of only a small percentage of ganglion cells scattered across the retina. The procedure labels only cells with an axon projecting into the optic nerve (i.e., retinal ganglion cells); displaced amacrine cells (Perry, 1982) and any ganglion cells with axons yet to enter the optic nerve remain unlabeled. Adult ganglion cells were labeled by applying HRP to severed axons in the nerve fiber layer and identified by the presence of an axon. For details of the procedures see Dunlop (1990). **I-V**: Summary diagram of the development of dendritic trees. Scale = 190 μm for **A-F**, 135 μm for **G**, and 50 μm for **H**. Reprinted with permission of Wiley-Liss, a division of John WIley & Sons, Inc. from Dunlop SA (1990): Early development of retinal ganglion cell dendrites in the marsupial *Setonix brachyurus*, quokka. *J Comp Neurol* 293: 425–447. © 1990 S. Dunlop.

Our observations support the concept (Maffei and Perry, 1988) of an intrinsic determinant for the initial emergence of an axon and dendrites from opposite sides of the soma. The concept is also supported by observations of cells in the developing cerebral cortex of the mouse. Disoriented cells are seen in normal animals (Van der Loos, 1965) and, in greater numbers, in the reeler mutant (Goffinet, 1984). Nevertheless, as we found for retinal ganglion cells, the disoriented cortical neurons have a polarized morphology with an axonal and a dendritic pole.

As development proceeds, dendritic trees become elaborated by the addition of new dendrites sequentially between the established ones and the axon hillock. We saw examples of cells with dendrites emerging from a 90° (Figure 12.1E1 and 12.1E2) or a 180° sector (Figure 12.1F) opposite the axon hillock and yet others arising from around the circumference of the cell (Figure 12.1G). This sequence is shown diagrammatically in Figure 12.1I-V. In adult rat, the bulk of the dendritic tree is oriented away from the axon hillock (Maffei and Perry, 1988). We suggest that this arrangement is determined by the sequence in which dendrites arise during development.

Furthermore, among retinal ganglion cells of the adult cat and monkey, the overall shape of the dendritic tree tends to be related to the cell's retinal location (Leventhal and Schall, 1983; Schall et al., 1986; Schall and Leventhal, 1987). The asymmetry displayed by the majority of dendritic trees is considered to be the morphological substrate for the physiological property of orientation selectivity (Levick and Thibos, 1982). We suggest that the initial axodendritic polarity of ganglion cells may influence the symmetry of the developing dendritic tree and thereby determine the functional properties of the cell.

At later stages, dendritic trees become longer and a clear developmental sequence is seen in the maturation of cell classes. In the adult quokka, ganglion cells can be classified according to their dendritic trees into large- (Figure 12.1H), small-, medium-, and wide-field cells, which resemble respectively the alpha, beta, delta, and gamma classes in the cat (Boycott and Wassle, 1974; Stone and Clarke, 1980; Piechl and Wassle, 1981; Wassle et al., 1981; Rowe and Dreher, 1982; Saito, 1983; Leventhal et al., 1985). As development proceeds in the quokka, wide-field cells are distinguishable first, and small-field cells last. We do not know whether this sequence reflects the timing of ganglion cell birth. The sequence of dendritic maturation has yet to be related to ganglion cell classes in species other than the quokka. However, in the cat, different ganglion classes, identified by their soma size (Walsh et al., 1983; Walsh and Polley, 1985) are known to be born at different times.

As dendrites grow, short (up to 5 μm), unbranched processes emerge from the dendritic shaft (Figure 12.1E1-G). These processes, termed spines, were first described in developing tissue using Golgi preparations (Morest, 1970). We observed dendritic spines on all four classes of developing ganglion cell in the quokka. By maturity, these structures have been lost from all but the small-field cells. In addition to spines, dendrites transiently develop a more complex branching pattern (Figure 12.1G) than that of the adult (Figure 12.1H). The development and subsequent loss of both spines and excess dendritic branches reported here have also been seen in other species (Dann et al., 1988; Wong, 1990).

The roles played by spines and by excess branches in development are unknown. In the cat the emergence of spines coincides with synaptogenesis in the inner plexiform layer (Maslim and Stone, 1986), suggesting that these structures may serve to increase postsynaptic area. However, neither the formation nor the loss of spines or excess branches requires visual input. The structures appear well before eye opening (Morest, 1970; Maslim et al., 1986; Ramoa et al., 1987, 1988; Dunlop, 1990) and are lost even when animals are deprived of normal visual experience (Lau et al., 1990).

In summary, the location of the first dendrites and the sequence in which subsequent dendrites arise are probably determined by the point at which the axon exits the soma. Spines are a prominent feature of developing dendrites but are lost from most classes of ganglion cell by adulthood. In addition, dendrites bear more branches during development than in the adult.

DEVELOPMENT AND REGENERATION OF AXONS

We have examined the morphology of individual axons as they grow along the optic nerve during development in the chick, quokka, rabbit, and cat; in addition we have studied axons regenerating after axotomy in the frog. This approach has been applied previously to study the terminations of developing (Sretevan and Shatz, 1984) and regenerating (Fujisawa et al., 1982; Schmidt et al., 1988) optic axons in target tissue, but not, to our knowledge, to study the axons as they navigate their way along the optic pathway. Our studies extend those of populations of pioneer axons during development (Horsburgh and Sefton, 1986; Silver and Rutishauser, 1984) and of total numbers of axonal profiles in the optic nerve both during development (Ng and Stone, 1982; Crespo et al., 1985; Perry et al., 1983; Rakic and Riley, 1983; Braekevelt et al., 1986; Williams et al., 1986; Robinson et al., 1987; Kirby et al., 1988) and regeneration (Murray, 1982; Stelzner and Strauss, 1986).

Development

We describe here developing optic axons (Figure 12.2A-J) at stages when numbers of axonal profiles in the nerve are at their highest. There are embryonic days (E)9–11 (Rager and Rager, 1978), postnatal day (P)40–45 (Braekevelt et al., 1986), E24 (Robinson et al., 1987), and E40 (Williams et al., 1986) for the chick, quokka, rabbit, and cat, respectively. To gain a more comprehensive picture, future studies will encompass other developmental stages and different regions of the optic pathway, such as the optic nerve fiber layer, chiasm, and optic tract.

Growth cones (Figure 12.2A-C)) are seen in all the developing optic nerves but are particularly prominent in the chick (Figure 12.2A,B). The observation of growth cones at the stage of peak numbers of axonal profiles confirms previous ultrastructural observations (Williams et al., 1986; Beazley et al., 1989). Furthermore, at these stages, dying ganglion cells are known to be present (Stone et al., 1984; Dunlop and Beazley, 1987), suggesting that there is a significant overlap between the outgrowth of axons and the loss of those axons destined to die.

In the chick, growth cones are complex, being either elongate or round and bearing a large number of filopodia, some of which are 40-50 μm long (Figure 12.2A,B). Most filopodia have a similar caliber to the main axon shaft and are aligned parallel to the surrounding axons. By contrast, growth cones in the mammalian optic nerves have a very simple lobulate shape and lack filopodia (Figure 12.2C), matching a previous description in embryonic mice (Bovolenta and Mason, 1987).

It has been suggested that growth cones elaborate filopodia only when axons reach decision points along their pathway (Harris et al., 1987; Bovolenta and Mason, 1987; Holt, 1988; Godemont et al., 1990). In the chick optic nerve, the complex morphology of growth cones may indicate that they actively navigate the confines of the optic nerve. By contrast, if our method has revealed the mammalian growth cones in their entirety, it seems likely that at this stage growing optic axons are not required to search actively within the nerve.

Another feature of developing axons in the species studied is that a minority follow highly aberrant trajectories (Figure 12.2E-G). Whereas most axons course directly toward the brain (Figure 12.2H), some are seen to loop back on themselves and track retrogradely toward the eye. Some of these turn again and resume their course toward the brain (Figure 12.2E). Other axons take sinuous trajectories across the nerve (Figure 12.2F), while yet others make dogleg (Figure 12.2G) or hairpin turns often to join another axon fascicle.

We suggest that axons with aberrant trajectories tend to be removed during development, presumably by ganglion cell death (Sengelaub and Finlay, 1982; Stone et al., 1984; Dunlop and Beazley, 1987), since such axons are seen only very rarely in the adult. The death of misdirected axons presumably results from their becoming lost and failing to reach target tissue. An alternative is that, as a result of their increased length, these axons would have asynchronous firing patterns compared with their neighbors, a property that is thought to result in cell death (Mastronarde, 1989).

As with developing dendrites, spines are seen on immature axons. In the chick, numerous spines, which themselves are often branched, are seen on axonal shafts for several hundred microns behind a growth cone (Figure 12.2D). Occasionally, in both chick and mammalian optic nerves, spines are seen further back on axons. At this level, spines are seen on axons coursing directly toward the brain (Figure 12.2I) but are more often associated with axons that follow aberrant trajectories (Figure 12.2J). The observation that spines occur particularly at points where axons change direction suggests that axonal spines may be involved in pathfind-

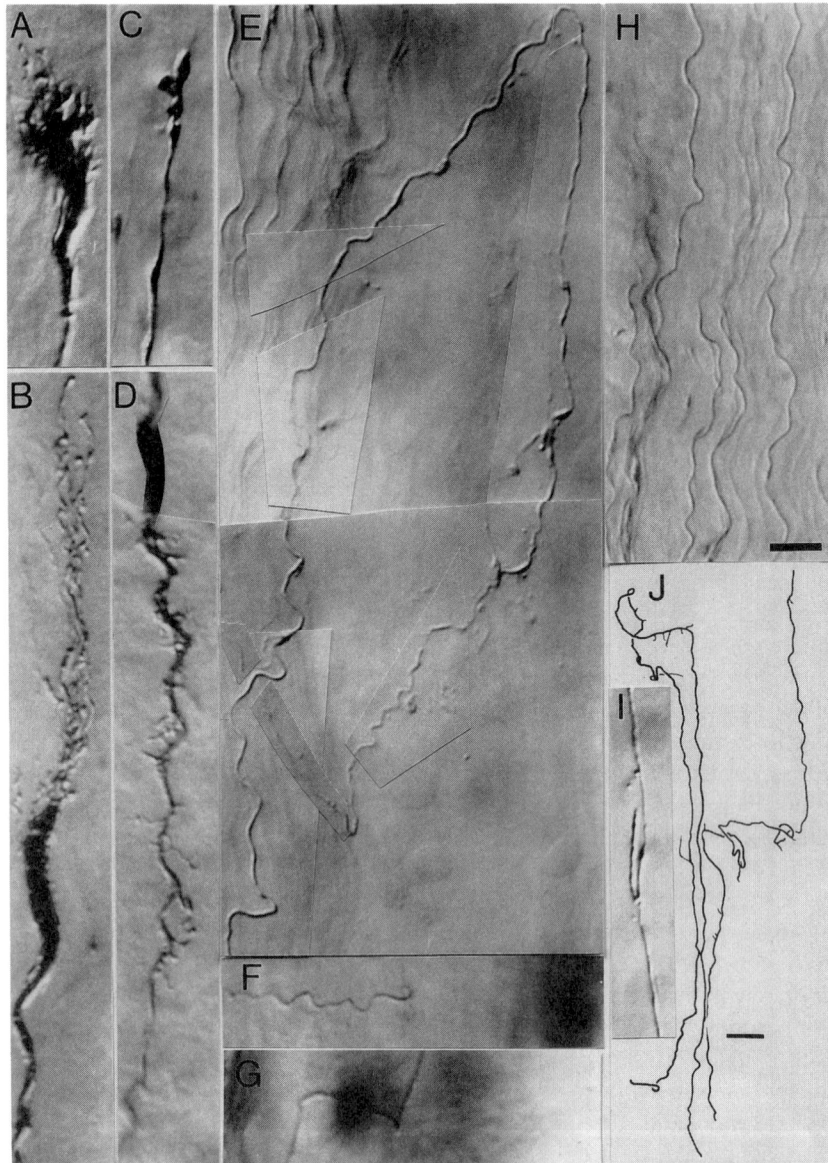

FIGURE 12.2. Developing axons in the optic nerves of chick (**A,B,D**), quokka (**C,E,I,J**), rabbit (**F**), and cat (**G**). The eye is to the bottom and the brain to the top of the page. **A-C:** Growth cones; **D:** the region of an axon immediately behind a growth cone; **E-G:** misdirected axons; **H:** appropriately oriented axons; **I,J:** axons with spines. Photographed or drawn with a camera lucida using Nomarski optics at 1,500×. **D** and **E** are montages of photographs at more than one plane of focus. In **G** a red blood cell lies beneath the axon. In **H** the undulating axon trajectories may partly be due to shrinkage but may also relate to the

ing. In agreement with this interpretation, spines have been observed on axons as they form terminal arbors in target tissue (Reh and Constantine-Paton, 1985; Sagaguchi and Murphey, 1985).

The observation of axonal spines at the level of the light microscope confirms our previous report of axonal spines reconstructed from serial electron micrographs of optic nerves (Beazley et al., 1989). Our analysis indicated that axonal spines do not form synapses or other contacts with the parent or adjacent axons. This observation suggests that axonal spines do not fulfil the synaptogenic role proposed for their dendritic counterparts.

Our results suggest that not every developing axon is a single shaft progressing directly to the brain. Axonal spines have not been previously reported on axons within the developing optic nerve, although they have been seen in the retinal nerve fiber layer (Ramoa et al., 1988; Dunlop, 1990). Misdirected axons must also be added to our picture of the morphology of the developing optic nerve. These features presumably enhance the counts of axonal profiles in the optic nerve, already transiently elevated by an overproduction of ganglion cells (Perry et al., 1983; Rakic and Riley, 1983; Braekevelt et al., 1986; Williams et al., 1986; Robinson et al., 1987; Kirby et al., 1988). It is even possible that filopodia of growth cones were included in counts of axonal profiles for the chick optic nerve (Rager and Rager, 1978).

Regeneration

In the final section of this chapter, we turn to an experimental model used to such advantage by Gaze throughout his career: the regenerating optic nerve of the frog. Here we describe the morphology of individual regenerating optic axons at stages when axons penetrate the crush site and grow toward the brain (Figure 12.3A-G). Optic nerves were crushed unilaterally in the green and golden tree frog *Litoria (Hyla) moorei*. This procedure, which severs all axons but leaves the nerve sheath and the blood supply to the eye intact, leads to a rapid and consistent sequence of regeneration (Humphrey and Beazley, 1982, 1983, 1985).

Within 1–3 days of optic nerve crush, axons retract somewhat from the crush site, form retraction bulbs, and then develop growth cones (Figure 12.3A,B). The growth cones are round or elongate and sometimes have one or two filopodia of

---

bands of Fontana; these structures are thought to allow passive stretching and avoid mechanical damage during motor activity (Haninec, 1986). Axons were labeled with HRP in 2% dimethyl sulfoxide *in vitro* (Dunlop et al., 1990; Fraley et al., 1991; Dunlop et al., in press). Our technique of applying HRP to severed axons in restricted regions of the retinal nerve fiber layer labels a small percentage of axons and allows the morphology of individual axons to be assessed. Longitudinal vibratome sections of optic nerves were reacted with cobalt-enhanced diaminobenzidine. Scale bars = 10 μm (bar on **H** applies to **A-I**).

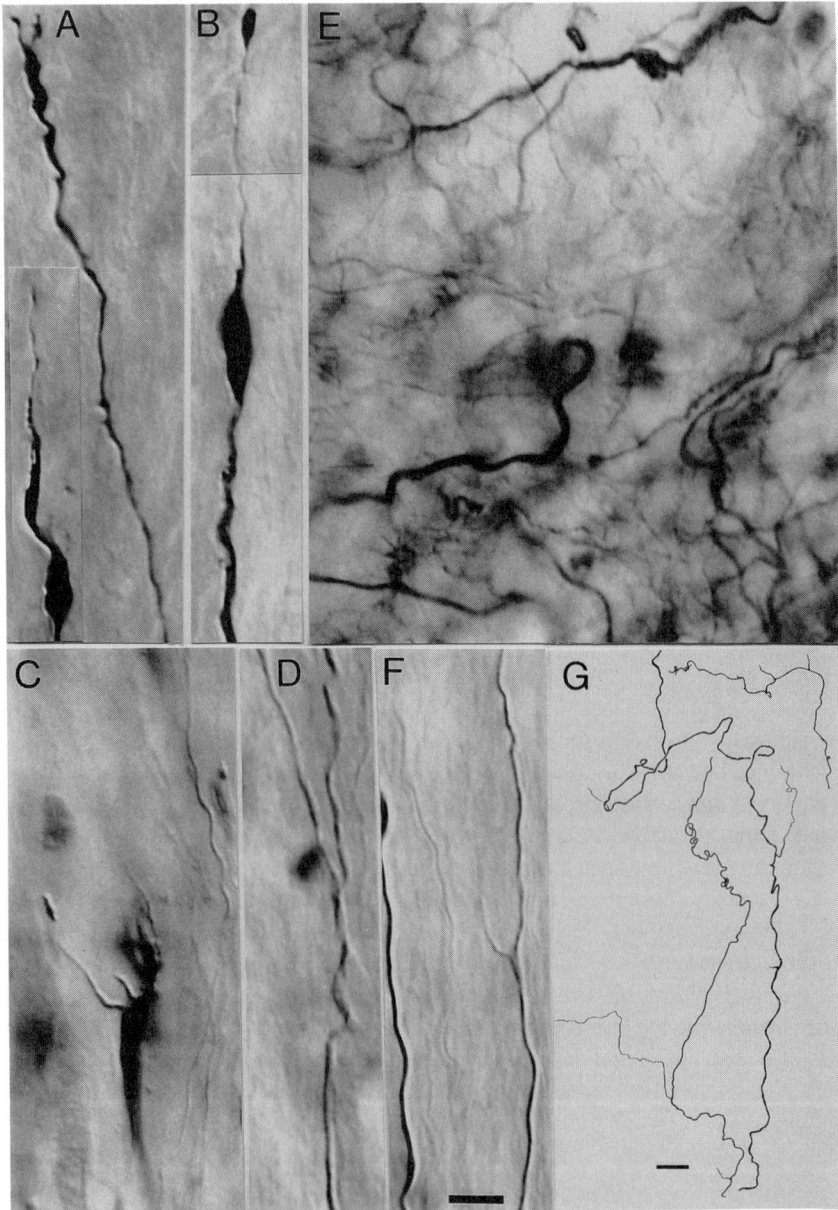

FIGURE 12.3. Regenerating axons in the frog optic nerve. The eye is to the bottom and the brain to the top of the page. **A,B,D,F** are in the region of the nerve between the eye and the crush site; **C,E,G** are within the crush site. Photographed or drawn with a camera lucida using Nomarski optics at 1,500×. **A** and **B** are montages of photographs at more than one plane of focus. Procedures are as in Figure 12.2, except that axons were labeled with a 1:1 mixture of HRP and wheat-germ agglutinin. Scale bars = 10 μm (bar on **F** applies to **A-F**).

varying length that extend in front of the body of the growth cone. Once within the crush site, growth cones either have a simple morphology or possess a few long, fingerlike filopodia that radiate away from the body of the growth cone (Figure 12.3C). This radiating orientation is reminiscent of filopodia on growth cones in the developing optic chiasm of the chick (Dunlop et al., 1992) and mouse (Godemont et al., 1990).

In the region proximal to the crush site, a minority of axons branch to form two axons of similar caliber to the parent axon (Figure 12.3D). Some axons turn and track toward the eye without approaching the crush site. Within 1 week, many regenerating axons enter the crush site. In this region axon trajectories become highly aberrant (Figure 12.3E). Axonal branches and disoriented axons are still present within the crush site at 4 weeks postcrush. We suggest that retinal ganglion cells with misdirected axons may be among the 40% or so of the ganglion cell population that does not survive regeneration (Humphrey and Beazley, 1985; Stelzner and Strauss, 1986).

Once beyond the crush site, axons tend to congregate at a subpial location before progressing toward the brain. Their trajectories at this point are less aberrant, in that backtracking, looping, or meandering axons are less prominent.

Axonal spins are another characteristic of regenerating axons (Figure 12.3F,G). These structures are unbranched, thinner than the parent axon, and often 5–10 μm long, sometimes extending for up to 50 μm. As with the developing optic axons of chick and mammals we have described, spines in the frog are most prevalent on those regenerating axons with aberrant trajectories. The favored location of spines is, as in the developing system, that part of an axon which has curved to change direction (Figure 12.3G).

Our studies illustrate the complex morphology of regenerating axons. Although both spines and collateral branches on regenerating frog axons are described here for the first time, their existence had been surmised from counts of axonal profiles. These are elevated severalfold between the crush site and the brain (Stelzner and Strauss, 1986), despite the concurrent loss of ganglion cells (Humphrey and Beazley, 1985; Stelzner and Strauss, 1986).

The misdirection of axons at the crush site was not unexpected, in that a tangled mass of axons is seen in the crush site using silver-stained preparations (Sperry, 1963; Gaze, 1970). However, we were surprised by the number of axons backtracking toward the eye without approaching the crush site. Presumably such axons, alluded to previously by Stelzner et al., (1986), contribute to the doubling in numbers of axonal profiles in the portion of the optic nerve between the eye and the crush site (Stelzner and Strauss, 1986).

Our next task is to examine the morphology of regenerating axons in the frog as they advance into the brain. We wish to determine whether features seen within the nerve, such as growth cone filopida, axonal spines, collateral branches, and misdirected axons, are maintained en route to and within the visual centers. In this way, we will increase our understanding of the factors that allow regenerating axons to navigate successfully and establish appropriate connections within the brain.

Acknowledgments

S.A.D. and L.D.B. are, respectively, Research Fellow and Principal Research Fellow, National Health and Medical Research Council (NH & MRC). We thank W.M. Ross for expert technical assistance and Dr. P.W. Sheard for comments on the manuscript. This work was supported by NH & MRC.

REFERENCES

Beazley LD, Dunlop SA, Harman AM, Coleman L-A (1989): Development of ganglion cells density gradients in the retinal ganglion cell layer of amphibians and marsupials: Two solutions to one problem. In: *Development of the Vertebrate Retina*, Finlay BL, Sengelaub DR eds. New York: Plenum Press.

Bovolenta P, Mason C (1987): Growth cone morphology varies with position in the developing mouse visual pathway from retina to first targets. *J Neurosci* 7: 1447–1460

Boycott BB, Wassle H (1974): The morphological types of ganglion cells of the domestic cat retina. *J Physiol (Lond)* 249: 397–419

Braekevelt CR, Beazley LD, Dunlop SA, Darby JE (1986): Numbers of axons in the optic nerve and of retinal ganglion cells during development in the maruspial *Setonix brachyurus*. *Dev Brain Res* 25: 117–125

Crespo D, O'Leary DDM, Cowan M (1985): Changes in the number of optic nerve fibres during late prenatal and postnatal development in the albino rat. *Dev Brain Res* 19: 129–134

Dann JF, Buhl EH, Peichl L (1987): Dendritic maturation in cat retinal ganglion cells: A Lucifer Yellow study. *Neurosic Lett* 80: 21–26

Dann JF, Buhl EH, Peich L (1988): Postnatal dendritic maturation of alpha and beta ganglion cells in cat retina. *J Neurosci* 8: 1485–1499

Dunlop SA (1990): Early development of retinal ganglion cell dendrites in the marsupial *Setonix brachyurus*, quokka. *J Comp Neurol* 293: 425–447

Dunlop SA, Beazley LD (1987): Cell death in the developing retinal ganglion cell layer of the wallaby *Setonix brachyurus*. *J Comp Neurol* 264: 14–23

Dunlop SA, Ross WM, Beazley LD (1990): Retinal projections and axonal morphology in the developing visual system of the quokka. *Neurosci Abst* 16: 147.12

Dunlop SA, Ross WM, Fraley SM (1992): Growth cone axon morphology in the developing chick optic nerve. *Proc Aust Neurosci Soc* 3: 94

Fraley SM, Ross WM, Dunlop SA (1991): Axonal morphology in the developing chick optic nerve. *Neurosci Abst* 17: 269.12

Fujisawa HN, Tani N, Watanabe K, Ibata Y (1982): Branching of regenerating retinal axons and preferential selection of appropriate branches for specific neuronal connections in the newt. *Dev Biol* 90: 43–57

Gaze RM (1970: *The Formation of Nerve Connections*. London: Academic Press

Godemont P, Salaun J, Mason CA (1990): Retinal axon pathfinding in the optic chiasm: Divergence of crossed and uncrossed fibres. *Neuron* 5: 173–186

Goffinet AM (1984): Events governing organization of postmigratory neurons: Studies on brain development in normal and reeler mice. *Brain res Rev* 7: 261–296

Haninec P (1986): Undulating course of nerve fibres and Bands of Fontana in peripheral nerves of the rat. *Anat Embryol* 174: 407–411

Harris WA, Holt CE, Bohnhoeffer F (1987): Retinal axons with and without their somata, growing to and arborizing in the tectum of *Xenopus* embryos: A time-lapse video study of single fibers *in vivo*. *Development* 101: 123–133

Holt C (1988): A single-cell analysis of the early stages of retinal ganglion cell differentiation in *Xenopus:* From soma to axon tip. *J Neurosci* 9: 3123–4145

Horsburgh G, Sefton A (1986): The early development of the optic nerve and chiasm in the embryonic rat. *J Comp Neurol* 243: 547–560

Humphrey MF, Beazley LD (1982): An electrophysiological study of early retinotectal projection patterns during optic nerve regeneration in *Hyla moorei*. *Brain Res* 239: 595–602

Humphrey MF, Beazley LD (1983): An electrophysiological study of early retinotectal patterns during regeneration following optic nerve crush inside the cranium in *Hyla moorei*. *Brain Res* 269: 153–158

Humphrey MF, Beazley LD (1985): Retinal ganglion cell death during optic nerve regeneration in the frog *Hyla moorei*. *J Comp Neurol* 236: 382–402

Kirby MA, Wilson PD, Fischer TM (1988): Development of the optic nerve of the opossum *(Didelphis virginiana)*. *Dev Brain Res* 44: 59–67

Lau KC, So K-F, Tay D (1990): Effects of visual or light deprivation on the morphology, and the elimination of the transient features during development, of Type I retinal ganglion cells in hamsters. *J Comp Neurol* 300: 583–592

Leventhal AG, Rodieck RW, Dreher B (1985): Central projections of cat retinal ganglion cells. *J Comp Neurol* 237: 216–226

Leventhal AG, Schall JD (1983): Structural basis of orientation selectivity of cat retinal ganglion cells. *J Comp Neurol* 220: 465–475

Levick WR, Thibos LN (1982): Analysis of orientation bias in cat retina. *J Physiol (Lond)* 329: 243–261

Maffei L, Perry VH (1988): The axon initial segment as a possible determinant of retinal ganglion cell dendritic geometry. *Dev Brain Res* 41: 185–194

Maslim J, Stone J (1986): Synaptogenesis in the retina of the cat. *Brain Res* 373: 35–48

Maslim J, Webster M, Stone J (1986): Stages in the structural differentiation of retinal ganglion cells. *J Comp Neurol* 254: 382–402

Mastronarde DN (1899): Correlated firing of retinal ganglion cells. *Trends Neurosci*. 12: 75–80

Morest DK (1970): The pattern of neurogenesis in the retina of the rat. *Z Anat Entwickl Gesch* 131: 45–67

Murray M (1982): A quantitative study of regenerative sprouting by optic axons in goldfish. *J Comp Neurol* 209: 352–362

Ng AKY, Stone J (1982): The optic nerve of the cat: Appearance and loss of axons during normal development. *Brain Res* 5: 263–271

Peichl L, Wassle H (1981): Morphological identification of on- and off-centre brisk transient (Y) cells in the cat retina. *Proc Roy Soc Lond B* 212: 139–156

Perry VH (1982): The ganglion cell layer of the mammalian retina. In: *Progress in Retinal Research*, Chader N, Chader G (eds). Oxford: Pergamon Press

Perry VH, Henderson Z, Linden R (1983): Postnatal changes in retinal ganglion cell and axon populations in the pigmented rat. *J Comp Neurol* 219: 356–368

Perry VH, Walker M (1980): Morphology of cells in the ganglion cell layer during development of the rat retina. *Proc Roy Soc Lond B* 208: 433–445

Rager G, Rager U (1978): Systems matching by degeneration. I. A quantitative electron-microscopic study of the generation and degeneration of retinal ganglion cells in the chicken. *Exp Brain Res* 33: 65–78

Rakic P, Riley KP (1983): Over-production and elimination of retinal axons in the fetal rhesus monkey *Science* 219: 1441–1444

Ramoa AS, Campbell G, Shatz CJ (1987): Transient morphological features of identified ganglion cells in living fetal and neonatal retina. *Science* 237: 522–525

Ramoa AS, Campbell G, Shatz CJ (1988): Dendritic growth and remodelling of cat retinal ganglion cells during fetal and postnatal development. *J Neurosci* 8: 4245–4261

Reh TA, Constantine-Paton M (1985): Growth cone-target interaction in the frog retino-tectal pathway. *J Neurosci Res* 13: 89–100

Robinson SR, Horsburgh GM, McCall MJ, Dreher B (1987): Changes in the numbers of retinal ganglion cells and optic nerve axons in the developing albino rabbit. *Dev Brain Res* 35: 161–174

Rowe MH, Dreher B (1982): Functional morphology of beta cells in the area centralis of the cat's retina: A model for the evolution of central retinal specializations. *Brain Behav Evol* 21: 1–23

Sagaguchi DS, Murphey RK (1985): Map formation in the developing *Xenopus* retinotectal system: An examination of ganglion cell terminal arborizations. *J Neurosci* 5: 3228–3245

Saito H-A (1983): Morphology of physiologically identified X-, Y- and W-type retinal ganglion cells of the cat. *J Comp Neurol* 221: 279–288

Schall JD, Leventhal AG (1987): Relationships between ganglion cell dendritic structure and retinal topography in the cat. *J Comp Neurol* 257: 149–159

Schall JD, Perry VH, Leventhal AG (1986): Retinal ganglion cell dendritic fields in old-world monkeys are oriented radially. *Brain Res* 368: 18–23

Schmidt J, Turcotte JC, Buzzard M, Tieman DG (1988): Staining of regenerated optic arbors in goldfish tectum: Progressive changes in immature arbors and a comparison of mature regenerated arbors with normal arbors. *J Comp Neurol* 269: 565–591

Sengelaub DR, Finlay BL (1982): Cell death in the mammalian visual system during normal development. I. Retinal ganglion cells. *J Comp Neurol* 204: 311–317

Silver J, Rutishauser U (1984): Guidance of optic axons *in vivo* by a preformed adhesive pathway on neuroepithelial endfeet. *Dev Biol* 106: 485–499

Sperry RW (1963): Chemoaffinity in the orderly growth of nerve fiber patterns and connections. *Proc Natl Acad Sci USA* 50: 703–709

Sretevan D, Shatz CJ (1984): Prenatal development of individual retinogeniculate axons during the period of segregation. *Nature* 308: 845–848

Stelzner DJ, Bohn RC, Strauss JA (1986): Regeneration of the frog optic nerve—comparison with development. *Neurochem Pathol* 5: 255–288

Stelzner DJ, Strauss JA (1986): A quantitative analysis of frog optic nerve regeneration: Is retrograde ganglion cell death or collateral axonal loss related to selective reinnervation? *J Comp Neurol* 245: 83–106

Stone J, Clarke R (1980): Correlation between soma size and dendritic morphology in cat retinal ganglion cells: Evidence for further variation in the gamma class size. *J Comp Neurol* 192: 211–217

Stone J, Maslim J, Rapaport D (1984): The development of the topographical organisation of the cat's retina. In: *Development of Visual Pathways in Mammals,* Stone J, Dreher B, Rapaport D, eds. New York: Alan R Liss

Van der Loos H (1965): The improperly oriented pyramidal cell in the cerebellar cortex and its possible bearing on problems of neuronal growth and cell orientation. *Bull Johns Hopk Hosp* 117: 228–250

Walsh C, Polley EH (1985): The topography of ganglion cell production in the cat's retina. *J Neurosci* 5: 741–750

Walsh C, Polley EH, Hickey TL, Guillery RW (1983): Generation of cat retinal ganglion cells in relation to central pathways. *Nature* 302: 611–614

Wassle H, Peichl L, Boycott BB (1981): Dendritic territories of cat retinal ganglion cells. *Nature* 292: 344–345

Williams RW, Bastiani M, Lia B, Chalupa L (1986): Growth cones, dying axons and developmental fluctuations in the fiber population of the cat's optic nerve. *J Comp Neurol* 246: 32–69

Wong ROL (1990): Differential growth and remodeling of ganglion cell dendrites in the postnatal rabbit retina. *J Comp Neurol* 294: 109–132

CHAPTER 13

Development of the Anuran Retina: Past and Present

CHARLES STRAZNICKY

HISTORICAL OVERVIEW

I was fortunate in spending a year in 1969 in Mike Gaze's laboratory in the Department of Physiology, Edinburgh. This period coincided with the time when the research into the factors controlling the formation and reformation of the retinotectal projection in fish and amphibia was at its height. I was at that time involved in studying the formation of connections between the axons of ventral horn motor neurons and limb muscles in chick embryos (Straznicky, 1963, 1965, but see also 1983). I was utterly frustrated by the lack of appropriate techniques to assay motor fiber outgrowth and the establishment of connections between axons and developing wing muscles. I therefore did not need much persuasion to change my topic of research from neuromuscular specificity studies to the more exciting developmental studies on the retinotectal connections. Indeed, I am greatly indebted to Mike who gave me enormous encouragement during my Edinburgh stay and thereafter a number of times when we had a very useful collaboration in his laboratory at Mill Hill, London. In a way I still regret that in the early 1980s I changed my interest from the study of retinotectal connectivity to retinal microanatomical research. Thus I am no longer involved in the recent advances of this field so well represented by a number of younger colleagues (Fujisawa, 1984, 1987; Fujisawa et al., 1982; Harris, 1986, 1989; Stuermer, 1984).

The prime concern of the late 1960s was to confirm, reject, or modify the neuronal specificity hypothesis formulated so succinctly by Roger Sperry (1951, 1963). Sperry's proposition assumed a rather rigid genetic specification of retinal ganglion cells based on chemical codes that corresponded to their position in the retina. Furthermore, he postulated the formation in early development of a matching specification of corresponding tectal points in concordance with the specification of the retina. Thus, during development or in regeneration, the likely mechanism for the formation or the reformation of the point-to-point retinotectal projection involved optic axons actively seeking out their tectal counterparts with matching cytochemical specification. The early results of the compound eye paradigm (eyes made by the fusion of two nasal or temporal halves of the right and left eye cups of *Xenopus laevis* embryos) have clearly shown that rigid chemospec-

Formation and Regeneration of Nerve Connections
Sansar C. Sharma and James W. Fawcett, Editors
© 1993 Birkhäuser Boston

ificity may not be the only answer to the possible rules governing the formation and reformation of neuronal connections. The whole concept of the neuronal specificity hypothesis was elegantly evaluated in the light of the results of compound eye experiments by Mike in his book (Gaze, 1970).

In 1969 Mike had the foresight to realize that the mechanisms of the formation of retinotectal projection cannot be unraveled unless we understand the mode of retinal and tectal development from early embryonic stages to adult. Surprisingly, after 25 years of study, starting with Sperry's (1943, 1944) illuminating series of experiments on frog optic nerve regeneration combined with eye rotation, and continuing with Mike Gaze's optic fiber regeneration studies in adult *Xenopus laevis*, utilizing extracellular recording from optic terminals in the tectum (Gaze, 1959) and compound eye studies (Gaze et al., 1963, 1965), only Joe Hollyfield took the trouble to look at retinal cell generation in *Rana pipiens* in 1968. He clearly established that from late larval stages onwards virtually all mitoses in the retina take place at the ciliary margin. Mike suggested that we take the matter further from Joe's initial observation. Using ^{3}H-thymidine autoradiography on a retinal developmental series in *Xenopus laevis*, we have demonstrated that cell addition at the ciliary margin progresses throughout the lifespan of the animal (Straznicky and Gaze, 1971). To our delight, the pattern of tectal cell generation was quite different from that of the retina. In contrast to the ringlike cell addition and consequently a center-to-periphery retinal growth, the tectum grows linearly in a rostrolateral to caudomedial direction (Straznicky and Gaze, 1972). These two findings clearly indicated that the pattern of retinotectal projection had to be changing over the whole life of the animal, with the projection of the retinal center, and of course the entire retinal field, moving continuously on the tectum from rostrolateral to caudomedial. The concept of shifting retinotectal connections has been amply supported by the result of subsequent experiments (Gaze et al., 1974, 1979; Lázár, 1973; Scott and Lázár, 1976; Jacobson, 1977; Reh and Constantine-Paton, 1984).

INTRODUCTION

The development and maturation of the adult retina include a number of morphogenetic steps. During early development, as a result of intensive cell division (1) and postmitotic cell migration (2) the characteristic laminar structure of the retina is laid down. These early steps are followed by cell differentiation (3), the acquisition of morphological, cytochemical, and functional identities of retinal neurons. Cell differentiation is accompanied by naturally occurring cell death (4) and followed later by dendritic maturation (5) that is assumed to be based on cell–cell interactions. The last step of retinal development includes retinal areal growth (6), to a large extent due to growth and stretching of the already formed elements of the retina. The protracted development of the retina in fish and amphibia allows an especially clear resolution of the mode of neuron generation at the ciliary margin, not only during early development but at any chosen postmetamorphic stage. In this essay some of these topics (1, 3, 5, and 6) will be addressed in more detail than others, reflecting our present research interest in anuran retinal development.

The anuran retina has been shown to grow by the addition of newly formed cells at the ciliary margin (Hollyfield, 1968; Straznicky and Gaze, 1971). Consequently, cells formed earliest occupy the retinal center and are surrounded by more peripheral cells that are generated later. The latest-formed cells are located at or adjacent to the ciliary margin. After the initial observations in the retina of *Xenopus laevis* (Straznicky and Gaze, 1971), the pattern of center-to-periphery retinal growth has been confirmed in fish (Johns, 1977), in the chick (Kahn, 1974), and in mammalian species (Hinds and Hinds, 1974). The most detailed descriptions of retinal development are available in *Xenopus laevis*; these results show a well-defined spatiotemporal pattern of cell generation at the ciliary margin. During early development more cells are laid down in the dorsal retina (Grant et al., 1980), followed from midlarval stages onwards by a greater cell generation in the ventral retina (Beach and Jacobson, 1979). The frequency of cell mitoses at the ciliary margin decreases after metamorphosis and actually stops along the dorsal pole of the retina in juvenile animals (Dunlop and Beazley, 1984). In contrast, cell addition at the ventral ciliary margin continues in a crescentic fashion that eventually changes the position of the optic nerve head.

Another important aspect of retinal development concerns the differentiation of its component neurons. Within each phenotype, morphologically and cytochemically distinct subtypes have been identified (Kolb et al., 1981). Conservative estimates may include at least 30 or 40 different amacrine cell types of the adult retina in mammals (Masland, 1988). It is very difficult to perceive how this diversity is achieved during retinal development. Although the factors leading from the undifferentiated state to mature cell types are not yet fully elucidated, it is clear that a genetically controlled cell lineage and the interaction of already differentiated cells with each other and with other still uncommitted cells may be some of the determinants (Wetts et al., 1989; Negishi et al., 1990).

THE ORGANIZATION OF THE ADULT ANURAN RETINA

The organization of the adult retina, including the structure and function of its component cells and their retinal distribution, is best described in the cat and rabbit (Wässle and Boycott, 1991). In most of the vertebrate species so far studied, retinal neurons are nonuniformly distributed across the retina. In terrestrial frogs and toads, the highest cell density in the ganglion cell layer (GCL) has been found within the center of a streak across the nasotemporal meridian of the retina called the visual streak (Bousfield and Pessoa, 1980; Dunlop and Beazley, 1984; Nguyen and Straznicky, 1989). A fovea centralis-like specialization, typical of the retina of some mammalian, bird, and lizard species, is not present in anura. A nonuniform cell distribution has also been found in the inner nuclear layer (INL) and photoreceptor layer (PRL) of the *Bufo marinus* retina (Zhu et al., 1990; Zhang and Straznicky, 1991) matching the cell distribution of the GCL. The organization of the *Xenopus* retina is quite different from that of terrestrial frogs and toads in that it lacks a visual streak, and apart from a slight increase of cell densities toward the ciliary margin, cell distribution is more or less uniform.

Xenopus *Retina*

In adult *Xenopus laevis* the retinal surface area is about 22 to 26 mm^2, with a total number of 60,000 (Dunlop and Beazley, 1984) to 90,000 (Straznicky and Hiscock, 1984; Jenkins and Straznicky, 1986) neurons in the GCL, according to the size of the animals and the methods of cell counts used (Table 13.1). These estimations include about 16% to 25% displaced amacrine cells (DAs), Ganglion cell (GC) and DA densities increase slightly toward the retinal periphery, more markedly along a crescent of the ventral ciliary margin. In fully adult animals the lowest density of cells is located in the geometric center of the retina, ventral to the optic nerve head, and a slightly elevated density is found in the central temporal retina (Graydon and Giorgi, 1984). In general, the cell density is higher in the temporoventral than in the nasodorsal retina, the former corresponding to the binocular visual field. Following optic nerve section and after a few weeks'

TABLE 13.1. Changing cell numbers and densities of the inner nuclear (INL) and ganglion cell (GCL) layers in the retina of *Xenopus laevis* and *Bufo marinus* during development

Age/size	Retina in mm^2	Density in GCL/mm^{-2}	in INL/mm$^3 \times 10^{-5}$	Cell Numbers GCL	INL
Xenopus laevis[1]					
St52	1.2	150	—	17,000	—
St58	2.1	125	—	25,000	—
M*	3.5	100	82	35,000	375,000
3MAM†					
25 mm	5.7	95	—	53,000	—
6MAM					
45 mm	10.2	70	60	72,000	550,000
Adult					
85 mm	24.0	37	43	90,000	840,000
Bufo marinus[2]					
M					
9 mm	3.3	165	96	52,000	750,000
3MAM					
15 mm	13.0	160	87	225,000	3,400,000
6MAM					
40 mm	35.5	115	69	440,000	9,600,000
Adult					
100 mm	140.0	68	55	950,000	18,500,000

*M: Metamorphosis
†3MAM: 3 months after metamorphosis
[1]Straznicky and Hiscock (1984); Jenkins and Straznicky (1986); Zhu et al. (1990)
[2]Nguyen and Straznicky (1989); Zhu et al. (1990)

survival, GCs and DAs can easily be distinguished from one another. The former cells show signs of retrograde degeneration, while the latter remain intact. On that basis the numbers and retinal distribution of DAs can be determined (Nguyen and Straznicky, 1989; Chng and Straznicky, 1990). Alternatively, counts of optic axons in the optic nerve and counts of cells in the GLC can be compared and the difference will indicate the numbers of DAs (Wilson, 1971; Dunlop and Beazley, 1984). However, this approach appears to be somewhat less reliable than revealing healthy DAs in the affected retina. According to our estimation, the distribution of DAs in the *Xenopus* retina is similar to the distribution of all neurons in the GCL.

Bufo Marinus *Retina*

The retinal surface area in fully grown adult animals varies between 120 and 164 mm^2 (Table 13.1). Consequently, cell numbers in the GCL are high, approaching 1,000,000 (Nguyen and Straznicky, 1989). Cells in all three layers of the retina are nonuniformly distributed. The visual streak of the GCL is well marked, extending nasotemporally across the retina approximately 2 mm dorsal to the optic nerve head. There is a steep density slope of 6 to 1 (GCs and DAs combined) from the central temporal part of the visual streak to the dorsal and ventral poles of the retina. Hence, the peak cell density of $16,000/mm^2$ decreases to less than $3,000/mm^2$. The density slope of GCs alone is about 7 to 1 (Figure 13.1A), in contrast to the much shallower density slope of 1.6 to 1 of DAs (Figure 13.1B). Because of this differential density gradient, the incidence of DAs within the cell population of the GCL increases peripherally. The number of DAs is 20% of all neurons in the GCL, or about 210,000 cells across the retina. The remaining

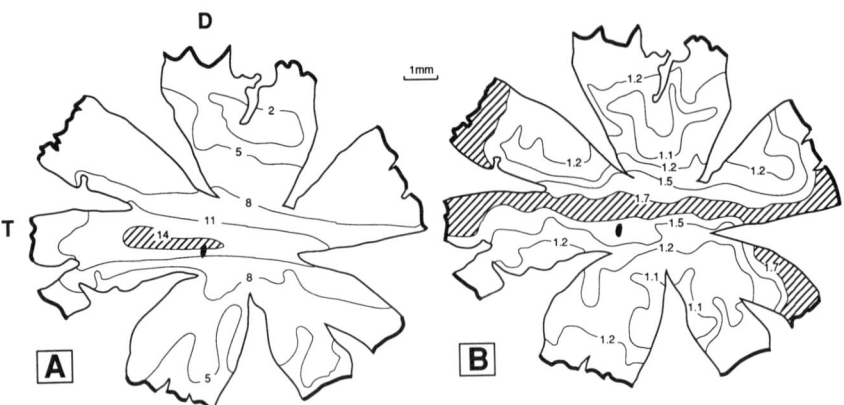

FIGURE 13.1. Isodensity map of the retinal distribution of GCs (**A**) and DAs (**B**) in the GCL of the adult *Bufo marinus* retina. Numbers indicates cells × $1,000/mm^2$. D = dorsal, T = temporal poles of the retina. Filled area corresponds to the optic nerve head. Hatched areas correspond to the position of highest cell densities. Note the decreasing cell density toward the ciliary margin.

800,000 neurons are GCs, the highest number so far recorded in an anuran species. Similar to the organization of the GCL, cells in the INL and PRL are also nonuniformly distributed, although the density slopes are less steep than in the GCL (Figure 13.2). The highest cell densities in the three layers overlap, forming an area within the visual streak that corresponds to the site of acute vision of the retina in *Bufo marinus* (Jagger, 1988).

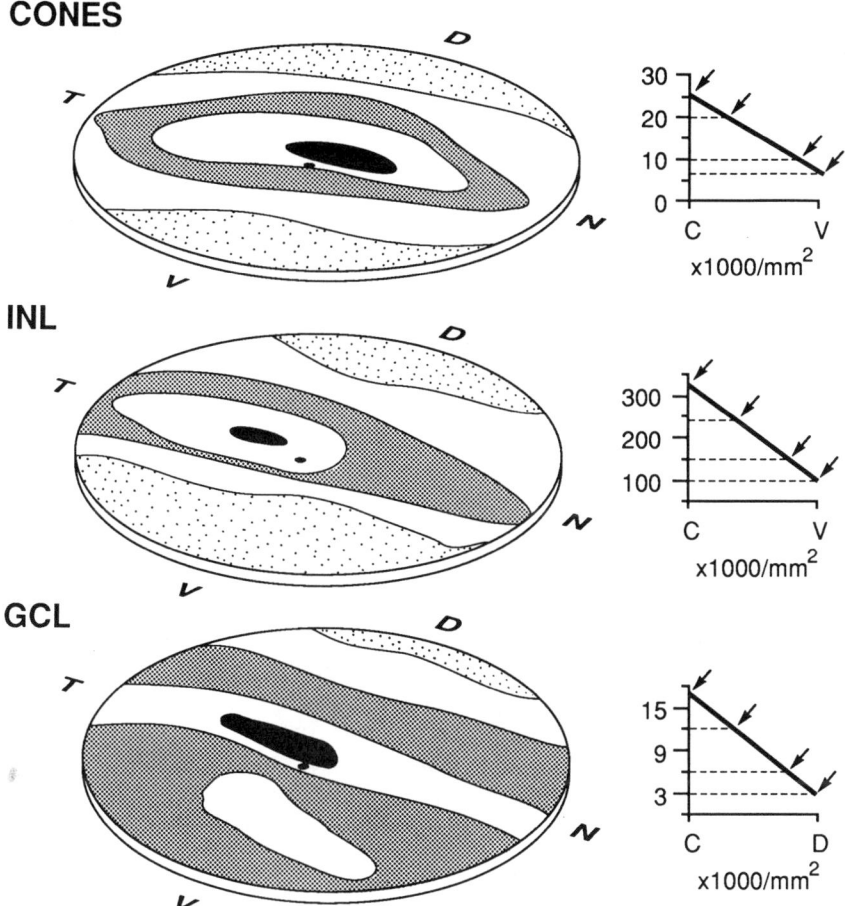

FIGURE 13.2. Diagrammatic representation of neuron distributions in the PRL, INL, and GCL of the *Bufo* retina. Values of isodensity lines for each layer are also given. Note that the cell densities and the gradation of the density slope from central to dorsal and ventral are different for cones and neurons of the INL and GCL. Reprinted with permission of Springer–Verlag from Zhang Y, Straznicky C (1991): The morphology and distribution of photoreceptors in the retina of *Bufo marinus. Anat Embryol* 183: 97–104.

RETINAL HISTOGENESIS AND AREAL GROWTH

The timetable of generation of different cell types of the cat retina has been determined over a period of 45 days, from E20 to birth (Zimmerman et al., 1988). First, GCs are generated, followed by axonless horizontal cells, amacrine cells (ACs), cones, bipolar cells, rods, and axon-bearing horizontal cells. In contrast, cell production continues throughout life and is confined to the narrow ciliary margin of the anuran retina; consequently, the component neurons of the retina are generated simultaneously (Hollyfield, 1971; Straznicky and Gaze, 1971). It is likely that further morphological differentiation of particular phenotypes of ACs and GCs is associated with the production of enzymes for neurotransmitter synthesis (Schnitzer and Rusoff, 1984; Mitrofanis et al., 1989). The initiation of cell differentiation is assumed to result from genetic determination; however, the molecular mechanism underlying it is not yet known.

Retinal expansion in anura occurs partly by cell addition and partly by areal growth of the already formed elements. Up to metamorphosis, most of the retinal growth is due to cell addition. As the cell generation slows down after metamorphosis, areal expansion becomes the main source of retinal growth. It has been shown in mammals (Mastronarde et al., 1984) and birds (Straznicky and Chehade, 1987) that retinal areal expansion is differential: more expansion occurs in the peripheral than in the central retina. This has been shown to be the case in *Xenopus* (Straznicky and Hiscock, 1984) and in *Bufo* (Zhu et al., 1990).

The Pattern of Cell Generation

Cell generation from late embryonic stages onwards (stages 36–38, Nieuwkoop and Faber, 1975) is restricted to the ciliary margin. Occasional mitoses are seen in the INL and PRL of the retinal fundus in pre- and postmetamorphic frogs (Hollyfield, 1971). The bulk of these newly generated cells contribute to the formation of the INL, carrying the largest numbers of neurons of the three cellular layers of the neural retina (Zhu et al., 1990). The pattern of cell generation changes considerably along the ciliary margin and over the period of active cell production. This actually contributes to the shaping of the adult-type retinal cell distributions in the PRL, INL, and GCL of the retina. This trend has been well documented in the frog *Limnodynastes dorsalis* (Coleman et al., 1984) and in the toad *Bufo marinus* (Zhu et al., 1990) during the formation of the visual streak. In both species, significantly more cells are generated at the nasal and temporal poles of the ciliary margin than at its dorsal and ventral poles.

In *Xenopus* there is a greater cell generation along the ventral crescent of the ciliary margin. The most apparent result of this differential cell generation is seen in the continuous displacement of the optic nerve head further dorsal in the retina. After metamorphosis, the location of the highest cell production is at the nasal pole of the retina, followed by a higher rate of cell generation at the temporal pole; consequently, the optic nerve head moves not only in the vertical but also in the horizontal direction across the retina (Straznicky and Hiscock, 1984). Besides its larger area, cell densities in the GCL are consistently higher ventrally than dorsally (Dunlop and Beazley, 1984; Straznicky and Hiscock, 1984; Jenkins and Straznicky, 1986). The overgrown ventral retinal halves of the right and left eyes have an important bearing on the formation of the binocular visual field in *Xenopus*. Since the eyes look up dorsally and forwards, there is at least a 70% overlap of the two monocular visual fields, and the overlap corresponds to the fields reflected onto the temporoventral retina.

The Development of the Regional Differences of the Retina

One of the characteristic features of retinal organization is the nonuniformity of cell distribution of the GCL in the form of the visual streak. Initially, during development, cells in the three cellular layers of the retina are about uniformly distributed, from which the adult pattern gradually emerges. Retinal developmental studies in various mammalian and nonmammalian species have indicated a number of possible mechanisms for the emergence of the adult-type nonuniform cell distribution. It has been postulated that a higher rate of cell production in the retinal center, migration of cells towards the retinal center, greater cell death in the peripheral retina, and a possible differential stretching of the retinal surface area may each contribute to the emergence of the adult nonuniform cell distribution in mammals (Stone et al., 1982, 1984).

Similar to other species, regional differences of the adult anuran retina evolve from a uniform cell distribution which is present in tadpoles. It takes a relatively long period of time from the metamorphic climax to young adulthood when the final, fully matured stage is reached. This period of retinal development coincides with the rapid increase of cell numbers in the GCL (Nguyen and Straznicky, 1989) and INL (Zhu et al., 1990), as well as a considerable increase of retinal surface area. Since the cell generation at the ciliary margin does not keep abreast with the rapid increase of retinal surface area, there is an overall decline of cell densities in both the GCL and the INL. Changing cell numbers, changing cell densities, and increasing retinal area in *Xenopus laevis* and *Bufo marinus* during development are summarized in Table 13.1.

In fully grown terrestrial frogs and toads, a well-defined horizontal anisotropy in the GCL of the retina in the form of the nasotemporally extending visual streak is present (Bousfield and Pessoa, 1980; Dunlop and Beazley, 1981; Nguyen and Straznicky, 1989). Tadpole and premetamorphic retinae do not have this horizontal anisoptropy. Instead, cell densities in the GCL increase from the center to the periphery of the retina, due to very high levels of cell production at the ciliary margin in tadpoles. The high cell density ridge of the GCL appears first in the temporal retina at the beginning of the metamorphic climax. Over the next few weeks it extends across the temporonasal axis of the retina, and the density difference between the evolving visual streak and the dorsal and ventral peripheral retinal poles becomes progressively greater (Nguyen and Straznicky, 1989).

Regional differences in cell production of the ciliary margin appear to be controlled by an intraretinal mechanism. It has been shown that a 180° rotation of the eye cup during early development results in a corresponding change of the alignment of the visual streak in postmetamorphic frogs (Dunlop and Beazley, 1985). The development of compound double ventral and double dorsal eyes—eyes made by the fusion of two ventral or dorsal halves of the eye cup—corresponds to the mitotic pattern of the half-retina (Straznicky and Tay, 1977). Consequently, double ventral eyes grow bigger than normal, in contrast to the smaller than normal size of double dorsal eyes. These observations indicate that the mitotic clock of the ciliary margin is controlled intrinsically by the retinal tissue and not by the surrounding extraretinal tissue.

Retinal Areal Growth

The retinal surface area grows continually throughout life. This areal growth, on the one hand, reduces the cell densities of the GCL and the INL, and on the other, contributes to the formation of the nonuniform distribution of retinal neurons. The retinal areal increase from metamorphosis to adulthood is different in various anuran species, but it is considerable in each case. An areal increase of four to fivefold has been reported in *Limnodynastes* (Coleman et al., 1984), 10-fold in *Xenopus* (Jenkins and Straznicky, 1986), 25-fold in *Heleioporus* (Dunlop and Beazley, 1981), and a massive 46-fold in *Bufo marinus* (Nguyen and Straznicky, 1989).

Our previous observations in *Bufo* have shown that the retinal area growth is differential; the peripheral retina expands more than the central retina. This proposition is based on the observation that cell densities decrease more markedly in peripheral than in central retinal regions (Nguyen and Straznicky, 1989). Areal enlargement involves radial and orthogonal expansions. The latter is more substantial, due to the displacement of cells generated at the ciliary margin toward the equator of the eye. Cells are passively displaced from the ciliary margin by waves of cells generated later. Since the diameter of the narrow ciliary ring is much less than that of the broader equatorial diameter, the neighborhood distance between cells increases considerably (Figure 13.3). As the retina grows partly by cell addition at the ciliary margin and partly by passive stretching, cells will be farther apart from each other toward the equator of the eye than at the ciliary margin.

Neuron Death in the Retina During Development

Neuron death is a major feature of retinal development in the chick (Rager and Rager, 1978; Straznicky and Chehade, 1987) and in mammals (Sengelaub and Finley, 1982; Dunlop and Beazley, 1987). Estimates showed that up to 40% of the originally generated GCs do not reach the fully differentiated state and they subsequently die. The evolving nonuniform GC distribution is partly due to differential cell death, with more cells dying peripherally than centrally.

In the anuran retina, cell numbers, which continue to increase throughout life, overshadow a possible cell loss when comparing numbers of live cells between

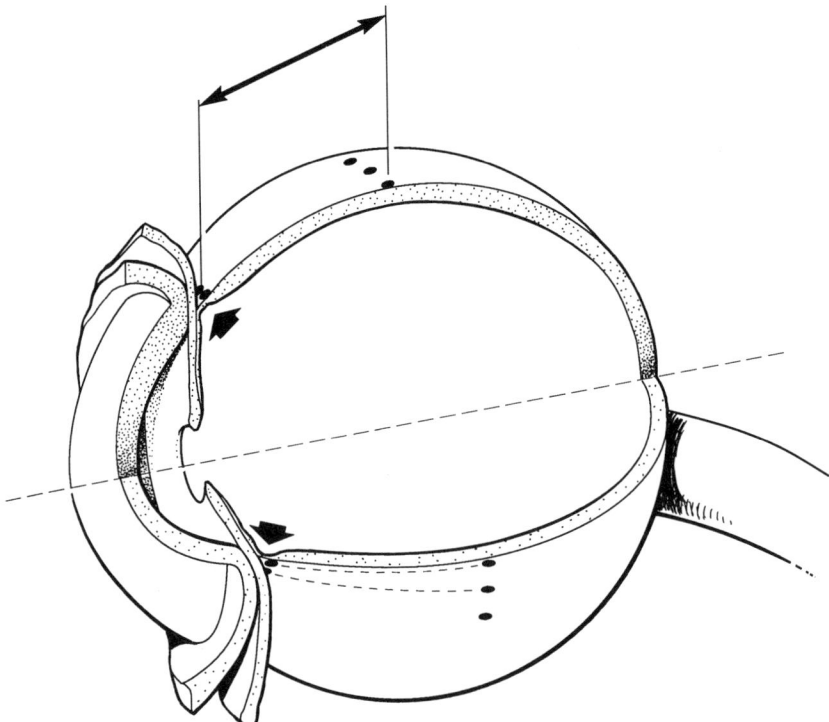

FIGURE 13.3. Schematic representation of retinal area expansion as seen by the relative displacement of cells along the orthogonal retinal axis during development. Notice that cells generated at the ciliary margin are passively pushed toward the equator of the eye (double-headed arrow) by cells generated subsequently. Since the diameter of the ciliary ring is much smaller than that of the equatorial diameter of the retina, neighborhood distances of adjacent cells increase.

two developmental stages. Indeed, all previous reports agree that if there is a neuron loss it is very minor in magnitude, in comparison with other vertebrate groups. Apart from early embryonic stages (Glücksmann, 1940), dying, pyknotic cells were not seen in the three cellular layers of the developing anuran retina unless the optic nerve had been cut (Scalia et al., 1985; Humphrey and Beazley, 1985; Jenkins and Straznicky, 1986; Nguyen and Straznicky, 1989).

However, the results of previous ^{3}H-thymidine autoradiographic studies have documented the loss of neurons in the GCL of the retinal center over long periods of time. These observations have been based on either counting the number of labeled cells at different times (*Rana pipiens*: Cigarroa and Constantine-Paton, 1980) or determining the number of unlabeled cells of the GCL within a ring of cells generated and autoradiographically labeled at stage 53 (*Xenopus laevis*: Jenkins and Straznicky, 1986). The estimated cell loss among neurons of the GCL of the retinal center was in the vicinity of 20% or 4,000 cells over a period of 7 months of survival. The pyknotic cycle of cells undergoing naturally occurring cell death has been documented to be short, about 3–4 hr in dorsal root ganglia (Pannese, 1976). The actual duration of the pyknotic cycle of retinal neurons is not known, but it can be assumed to be within the same time range of a few hours. From these data one can calculate daily cell loss of the central retina to be about 20 cells, but due to the short pyknotic cycle, at any one time, only a few pyknotic neurons may be present whose detection is very difficult.

The low level and regional occurrence of cell loss in the GCL are taken to indicate that it does not have a significant impact on the shaping of the adult-type nonuniform cell distribution in the three cellular layers of the retina.

The Emergence of Retinal Neuron Types

The uncommitted progenitor cells at the neurogenic zone of the ciliary margin undergo a series of steps that lead from an undifferentiated state to full maturity. These steps include the specification of particular phenotype, dendritic development, acquisition of cytochemical marker molecules, and functional maturation. Recent cell lineage studies have shown that the diverse cell types of the vertebrate retina arise from a single progenitor (Turner and Cepko, 1987). Utilizing intracellular injection of fluorescent dextran into uncommitted cells of the ciliary margin of *Xenopus laevis*, clonal analyses of the descendant cells confirmed that the progeny included all neural, glial, and pigmented epithelial cells (Wetts and Fraser, 1988; Wetts et al., 1989). Apart from the retinal pigment epithelial cells, which are determined earliest, all neuron types arrive after a number of subsequent mitoses of the precursor cell. These observations support the view that specification in the cell lineage of the neural retina occurs relatively late and that the precursor cells are pluripotent. However, the steps that progressively restrict the developmental potential to particular phenotypes and their large numbers of subtypes are yet to be established.

Although all retinal neuron types are represented in each part of the retina, their ratios may vary significantly according to retinal location. For example, the relative frequency of DAs in the GCL increases from 15% in the central to 30% in the peripheral retina of *Bufo marinus* (Nguyen and Straznicky, 1989). Since cell death is not a major factor in regulating cell numbers, genetic and local environmental signals are likely to be involved in controlling ratios of retinal neurons. Selective elimination of dopamine-containing ACs of the frog retina with 6-hydroxydopamine injection into the eye has resulted in a transiently increased production of these cells at the ciliary margin (Reh and Tully, 1986). Similarly, kainic acid-induced depletion of ACs has transiently up-regulated the production of these cells (Reh, 1987). The increase of the number of newly produced cells at the ciliary margin following neurotoxin treatment may be the result of either changing ratios between neuron types, i.e., more cells differentiating into one type, or a transient increase of mitotic activity of the ciliary margin. ^{3}H-Thymidine studies of the same authors suggest that the up-regulation of the production of replacement cells is by way of increased mitotic activity. Our recent observations agree with this conclusion. Early larval removal of the optic nerve brings about a complete degeneration of GCs, leaving behind only DAs. The number of DAs in the optic nerve-lesioned retina has been found to be comparable with the number of DAs on the intact side (Chng and Straznicky, 1990). Thus the elimination of one cell type, in this case GCs, did not result in the depletion of the number of cells of another type that arose from the same clonal line. Our observations and those of Reh and co-workers appear to suggest that the regulatory influence was local. An initial up-regulation of neurons that had been previously depleted was followed by a slight rebound, and then a normal rate of cell production in the next annulus of cells to be generated. It is also apparent from the results of lesion experiments that transformation from one to another committed cell type does not occur.

Of the various neuron types of the anuran retina, GCs appear to be the first to acquire specification followed by characteristic dendritic growth (Sakaguchi et al., 1984). Intracellular injections of Lucifer Yellow into these cells enabled the detection of early dendritic development that leads to the formation of various GC types. The early appearance of these cell types is a clear indication that the distinguishable morphological features are determined soon after their exclusion from the mitotic cycle. Immunohistochemical studies on ACs of the fish retina have shown that dendritic development is closely followed by the expression of catecholamine, indolamine, and neuropeptide marker molecules (Negishi et al., 1990). It is likely that codes which determine morphological and cytochemical features of the cell are coupled. After the acquisition of the particular phenotype, further dendritic growth occurs to the full maturation of the retinal tissue. Dendritic growth has been shown to be an active process, not just the passive stretch of the already established dendritic tree (Wong, 1990).

DEVELOPMENT OF IDENTIFIED AMACRINE CELLS

ACs can be subdivided into a number of distinct types on the basis of their dendritic morphology and/or cytochemical markers. In the retinae of *Bufo marinus* and *Xenopus laevis* we have described substance P-, neuropeptide Y-, serotonin (5HT)-, and dopamine-containing ACs (Hiscock and Straznicky, 1989a, 1989b, 1990; Zhu and Straznicky, 1990a, 1990b) (Table 13.2). Each of these types has a well-defined dendritic morphology and branching pattern in the inner plexiform layer of the retina. Like other neurons of the anuran retina, these cells are generated sequentially at the ciliary margin throughout life. After their exclusion from the mitotic cycle, they undergo rapid differentiation that can be followed by the application of immunohistochemistry either on whole-mount preparations or on retinal sections. We are particularly interested in determining the chronology of the appearance of these neurochemical markers and the accompanying dendritic development of these cells from early larval stages to adulthood.

TABLE 13.2. Numbers and distributions of immunohistochemically identified orthotopic and displaced amacrine cells in the retina of adult *Xenopus laevis* (X) and *Bufo marinus* (B)

	Numbers orthotopic/displaced	Densities in mm^2			Dendritic field size in mm^2
		Highest	Lowest	Average	
Substance P[1]					
B	2,300/25	—	—	11	0.65 ± 0.12
X	600/5–10	—	—	25	0.28 ± 0.08
NPY[2]					
B	3,400/20	—	—	32	0.53 ± 0.27
X	1,000/10	—	—	38	—
Dopamine[3]					
B	5,650/50–130	70	30	53	0.07 ± 0.01 (central) 0.05 ± 0.01 (peripheral)
X	735/30	—	—	31	0.16 ± 0.02
Serotonin[4]					
B	25,000/120	—	—	198	—
GABA[5]					
B	2,000,000/70,000	8,000	20,000	14,600	—

[1]Hiscock and Straznicky (1989a)
[2]Hiscock and Straznicky (1989b)
[3]Zhu and Straznicky (1990a)
[4]Zhu and Straznicky (1990b)
[5]Gábriel et al. (1992)

THE ACQUISITION OF IMMUNOCYTOCHEMICAL CHARACTERS

Previous observations in the retina of developing *Xenopus laevis* have shown a very early appearance of the ability to take up, manufacture, accumulate, and/or express cytochemical marker molecules. Soon after the cessation of mitotic activity in the central part of the eye cup (at stages 33–36), postmitotic cells are able to accumulate gamma-aminobutyric acid (GABA) (Hollyfield et al., 1979), glycine (Rayborn et al., 1981), dopamine (Sarthy et al., 1981), and serotonin (Frederick et al., 1989). Our recent studies have identified serotonin immunoreactivity in ACs as early as stage 40 (Zhu and Straznicky, 1992), substance P and neuropeptide Y (NPY) in cells at stages 40–45 (Hiscock and Straznicky, 1990), and tyrosine hydroxylase immunoreactivity for dopamine in cells at stage 42 (Zhu and Straznicky, 1991). At any stage during development, numerous immunoreactive cells can be observed adjacent to the ciliary margin. First the immunoreactivity is localized to the somata of newly generated cells, followed, within a short period of a few days, by the appearance of dendrite-bound immunoreactivity (Hiscock and Straznicky, 1990; Zhu and Straznicky, 1991 and 1992). It is not certain that the timetable of dendritic maturation as seen on immunohistochemical preparations corresponds strictly with steps of dendritic development. There may be a slight time delay between the dendritic development and the appearance of immunoreactivity along the dendritic arbor of these cells, due to the presence of a low concentration of marker molecules that cannot be revealed by immunohistochemistry. Nevertheless, these observations are consistent with the view that the expression of these marker molecules occurs shortly after the exclusion of ACs from the mitotic cycle in conjunction with the onset of dendritic development (Frederick et al., 1989). Furthermore, it is quite likely that the cytochemical character and the associated distinct morphological features of various AC types are determined very early, even though these cells derive from a common cell lineage (Wetts et al., 1989).

Dendritic Development

The numbers of immunohistochemically identified neurons increase very substantially from larval stages to adulthood. The numerical increase of these cells in the retina is accompanied by the development of their dendritic processes and the growth of their dendritic fields. Our analyses have shown that morphological maturation of NPY-, substance P-, and dopamine-containing ACs occurs rapidly over a short period of time (Hiscock and Straznicky, 1990; Zhu and Straznicky, 1991), followed thereafter by a further significant increase of dendritic field sizes. The increase of dendritic field sizes could be correlated with the rate of retinal areal expansion and with the decreasing cell densities (Figure 13.4). The dendritic arbors of dopamine-containing cells in the postmetamorphic and adult retinae are scaled versions of one another. This indicates that the dendritic growth of these cells during retinal expansion is the result of increasing dendritic mass of an arbor whose basic geometry remains unchanged.

In the *Bufo* retina the dendritic fields of dopamine-containing ACs were significantly larger in the dorsal and ventral retina than in the visual streak, where the former areas correspond to lower and the latter to higher dopamine cell densities. These observations support the notion that the adult dendritic field size is determined by interactions of neighboring cells. The overall expansion of the retinal surface area and the concomitant decrease of cell densities in all retinal layers contribute to the reduction of dendritic competition for arborization space (Perry and Linden, 1982). Consequently, the dendritic fields of immunoreactive ACs continue to grow until the final adult size is reached.

Changing Retinal Distribution

The distribution of neurons in the three cellular layers of the retina has been shown to change during development. This trend can be seen both in ACs of orthotopic position in the INL and in ACs of displaced position in the GCL. The immunohistochemical labeling of groups of ACs and the identification of DAs enabled us to follow the changing patterns of their retinal distribution throughout life.

Development-related changes are more marked in the retina *Bufo marinus* than of *Xenopus laevis*, due to the evolving visual streak of the retina in the former. Of the immunohistochemically characterized ACs, those that contain dopamine or

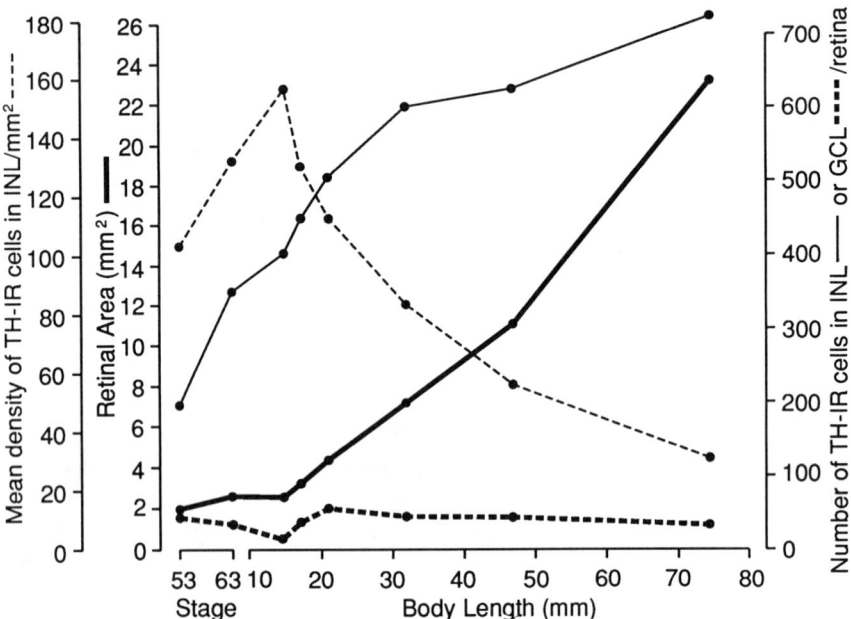

FIGURE 13.4. Histogram showing changing numbers and densities of dopamine-containing ACs during development in *Xenopus laevis*. Reprinted with permission of Springer-Verlag from Zhu B-S, Straznicky C (1991): Morphology and retinal distribution of tyrosine hydroxylase-like immunoreactive amacrine cells in the retina of developing *Xenopus laevis*. *Anat Embryol* 184: 33–45.

serotonin are nonuniformly distributed in the adult retina (Zhu and Straznicky, 1990a, 1990b), in contrast to the uniform distribution of substance P- and NPY-containing cells in both *Bufo* and *Xenopus* (Hiscock and Straznicky, 1989a, 1989b).

Immunohistochemically Characterized Amacrine Cells

DOPAMINE-CONTAINING AMACRINE CELLS

The number of these cells in the *Bufo* retina increases from about 600 at metamorphosis to about 6,000 in the adult, and in *Xenopus* from 250 to 735 (Table 13.2). In both species a few dopamine-containing cells are displaced into the GCL. In adult *Bufo* we have found a density gradient of 3.4 to 1 from the visual streak to the dorsal and ventral retinal poles. This observed nonuniform cell distribution develops after metamorphosis simultaneously with the evolving visual streak of the GCL (Nguyen and Straznicky, 1989) and INL (Zhu et al., 1990). In contrast to the increasing number of dopamine-containing ACs in the INL, the number of DAs remains at a relatively constant level (Zhu and Straznicky, 1990a). It is possible that some of these cells express immunoreactivity transiently, and hence their number does not increase in the GCL (Figure 13.5). Substance P immunore-activity present during larval life in the somata of GCs of the *Rana pipiens* retina has been shown to disappear by the time of the metamorphic climax (Kuljis and Karten, 1986). Alternatively, some of the DAs that express tyrosine hydroxylase immunoreactivity for dopamine may perish during development, as has been shown to be the case in the cat retina (Mitrofanis et al., 1989). We favor the second alternative, since during development a limited cell loss has been previously reported in the central retina (Cigarroa and Constantine-Paton, 1980; Jenkins and Straznicky, 1986; Gaze and Grant, 1992).

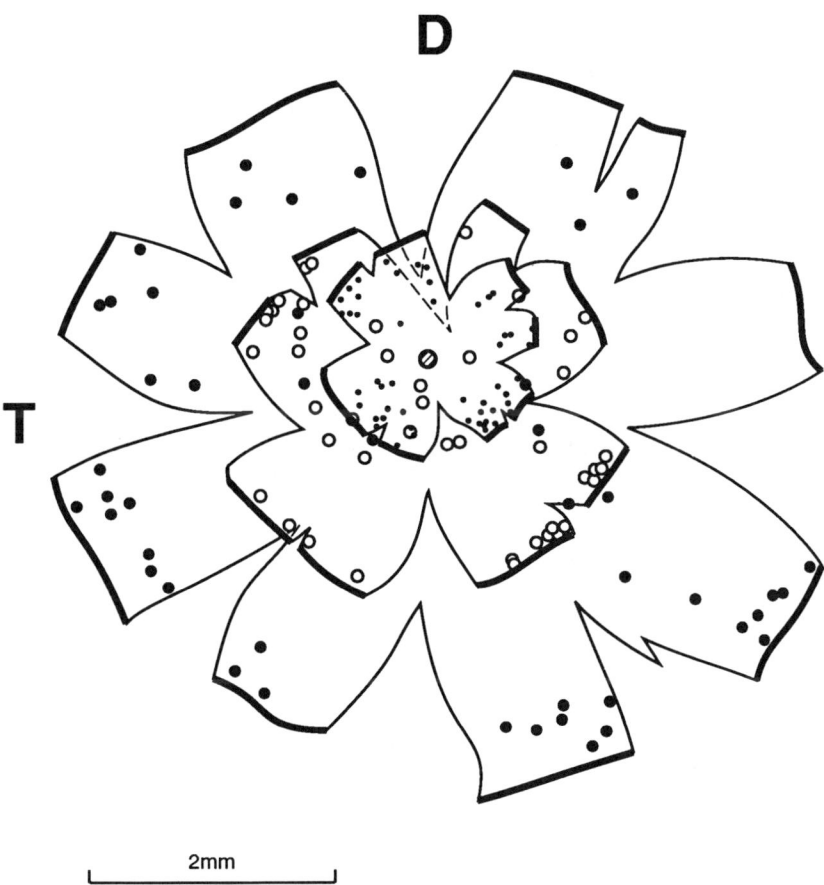

FIGURE 13.5. Diagram showing the numbers and distributions of dopamine-containing DAs in stage 53 tadpole (•), 15-mm postmetamorphic (○), and 75-mm adult (●) *Xenopus laevis*. Note that many of the immunoreactive cells generated before stage 53 or 15-mm size cannot be traced in the adult retina.

SEROTONIN-CONTAINING AMACRINE CELLS

These cells appear at stage 40 in *Xenopus*, and by stage 46 two populations of large, heavily stained and small, lightly stained cells can be distinguished. The number of large ACs increases from 53 cells per retina at stage 42 to 800 in adult, while the density decreases from $176/mm^2$ to $34/mm^2$ over the same time (Zhu and Straznicky, 1992). The retinal distribution of large, serotonin-containing ACs in adult *Xenopus* has a peak in the central temporoventral retina, observed recently by Schütte and Witkovsky (1990). In contrast, in *Bufo* there is only a slight center-to-periphery density slope of serotonin-containing cells (Zhu and Straznicky, 1990b). It appears that the nonuniform distribution of serotonin-containing cells also evolves during development with a similar time scale to that observed in dopamine-containing ACs.

It is interesting to note that some of the immunoreactive ACs are uniformly distributed across the anuran retina. This may indicate that substance P and NPY cells are generated at an even rate, both during development and along various parts of the ciliary margin (Hiscock and Straznicky, 1990). This is apparently not the case with dopamine- and serotonin-containing ACs. The evolving nonuniform distribution of these cells in the adult retina implies that the rate of generation changes during development and that more cells are generated at the nasal and temporal parts of the ciliary margin than at its dorsal and ventral parts. The particular mechanism(s) involved in this delicate spatiotemporal control of generation of various AC types requires further elucidation.

DISPLACED AMACRINE CELLS

The GCL of the retina is composed of two cells types, GCs and DAs. Following optic nerve section, GCs undergo retrograde degeneration, while DAs will remain intact and, therefore, are easily distinguishable from GCs (Table 13.3). We have

TABLE 13.3. Changing numbers and distributions of displaced amacrine cells (DAs) during development in the retina of *Bufo marinus*[1,2]

Age/size	DA numbers	Percentage of DAs in the GCL		
		Average	Central	Dorsal/Ventral
M				
10 mm	11,000	18.0	—	—
4MAM				
25 mm	68,000	18.5	17.7	25.8
7MAM				
50 mm	126,000	20.2	18.8	26.0
Young adult				
80 mm	179,000	21.4	15.2	28.1
Adult				
120 mm	211,000	22.0	14.9	30.6

[1]Chng and Straznicky (1990)
[2]Chng and Straznicky (1991)

shown in *Bufo marinus* (Chng and Straznicky, 1990, 1992) that the uniform distribution of DAs at metamorphosis develops into a nonuniform distribution in the adult with a shallow density slope of 1.6 to 1 from center to periphery. Since the density of GCs decreases more toward the periphery than that of the ACs, the incidence of DAs within the GCL steadily increases from the retinal center to the periphery. The possible control of production of these cells at the ciliary margin is intriguing and raises a number of questions. Since the central retina in *Bufo* is formed before and around metamorphosis and the more peripheral retina is formed after metamorphosis (Nguyen and Straznicky, 1989), this implies that more of the newly generated cells differentiate into the GC line than into the AC line before metamorphosis. Conversely, after metamorphosis, although cell production slows down at the ciliary margin, more cells follow the AC line. The regional difference in the ratios of GCs and DAs may be achieved by either shifting more uncommitted cells of the ciliary margin toward the AC/DA line, or more ACs may become displaced into the peripheral part of the GCL where the GC density is low.

We have argued previously that the displacement of neurons into inappropriate layers of the retina, GCs into the INL and ACs into the GCL, may be due to population pressure (Tóth and Straznicky, 1989). At times of high cell production at the ciliary margin it is likely that more cells fail to migrate to their destination, thereby increasing the numbers of displaced ganglion cells and amacrine cells. It is also possible that decreasing GC production and the resulting lower cell densities in the GCL may facilitate the migration of ACs into the GCL. The present available evidence does not allow us to choose between the two alternatives: increased AC and decreased GC production or increased migration of ACs into the GCL. In either case the GC/DA ratio will change. It appears likely that local environmental cues and interactions of cells of similar or different types determine the actual ratio and numbers of the two cell types of the GCL, whatever the underlying mechanism(s) may be (Reh and Tully, 1986; Reh, 1987).

SUMMARY AND CONCLUSIONS

The studies outlined in this chapter have described morphogenetic steps that are involved in anuran retinal development. Cell generation is confined to a narrow ring of germinal neuroepithelial cells at the ciliary margin, and all retinal neurons and pigment cells derive from this source. Recent experimental evidence strongly supports the view of a common cell lineage for all retinal neurons and non-neural cells. Furthermore, the germinal epithelial cells of the ciliary margin are determined (1) to differentiate into the various neuron types, (2) to yield the required numbers of various neuron types, and (3) to regulate regional differences of retinal cell distribution.

The regional organization of the adult retina is inexorably related to the spatiotemporal sequence of cell generation at the ciliary margin. Higher rates of cell generation at or after metamorphosis at the ciliary margin of the nasal and temporal poles of the retina contribute to the formation of the visual streak and to the nonuniform retinal distribution of other neuron types.

Morphological differentiation of retinal neurons is accompanied by cytochemical specification. High-affinity uptake of GABA, glycine, dopamine, and serotonin by ACs precedes synthesis of these marker molecules. All transmitter candidates present in ACs of the adult Xenopus retina can be identified by stage 45. The cytochemical character of ACs is determined very early during development, well before their morphological uniqueness appears. Further dendritic growth of these cells is associated with retinal areal expansion, and the rate-limiting factor is competition for arborization space.

Although we are beginning to understand the diversified morphogenetic steps of retinal development, the underlying mechanism, in particular the nature of genetic control and the role of cellular and environmental interactions, remains to be elucidated.

Acknowledgments

The author gratefully acknowledges Mr. Denis Jones' assistance with the computer graphics and valuable comments on the manuscript by, Drs. Robert Gábriel, Jenny Hiscock, Judy Wye-Dvorak, and Baosong Zhu. The work was supported by the National Health and Medical Research Council of Australia, the Australian Research Council, and Flinders University Research Budget grants.

REFERENCES

Beach DH, Jacobson M (1979): Patterns of cell proliferation in the retina of the clawed frog during development. *J Comp Neurol* 183: 603–614

Bousfield JD, Pessoa VF (1980): Changes in ganglion cell density during post-metamorphic development in a neotropical tree frog *Hyla raniceps*. *Vision Res* 20: 501–510

Chng SK, Straznicky C (1990): The generation of displaced amacrine cells in the retina of *Bufo marinus*. *Proc Austr Physiol Pharmacol Soc* 21: p113

Chng SK, Straznicky C (1992): The generation and changing retinal distribution of displaced amacrine cells in *Bufo marinus* from metamorphosis to adult. *Anat Embryol* 186: 175–181

Cigarroa RG, Constantine-Paton M (1980): Centripetal cell loss in the retina of *Rana pipens*. *Soc Neurosci Abstr* 6: 296

Coleman L-A, Dunlop SA, Beazley LD (1984): Patterns of cell division during visual streak formation in the frog *Lumnodynastes dorsalis*. *J Embryol Exp Morphol* 83: 119–135

Dunlop SA, Beazley LD (1981): Changing retinal ganglion cell distribution in the frog *Heleioporus eyrei*. *J Comp Neurol* 202: 221–236

Dunlop SA, Beazley LD (1984): A morphometric study of the retinal ganglion cell layer and the optic nerve from metamorphosis in *Xenopus laevis*. *Vision Res* 24: 417–427

Dunlop SA, Beazley LD (1985): Cell distributions in the retinal ganglion cell layer of adult Leptodactylid frogs and premetamorphic eye rotation. *J Embryol Exp Morphol* 89: 159–173

Dunlop SA, Beazley LD (1987): Cell death in the developing retinal ganglion cell layer of the wallaby *Setonix brachiurus*. *J Comp Neurol* 264: 14–23

Frederick JM, Rayborn ME, Hollyfield JG (1989): Serotoninergic neurons in the retina of *Xenopus laevis:* Selective staining, identification, development and content. *J Comp Neurol* 281: 516–531

Fujisawa H (1984): Pathways of retinotectal projection in developing *Xenopus* tadpoles revealed by selective labeling of retinal axons with horseradish peroxidase (HRP). *Dev Growth Differ* 26: 545–553

Fujisawa H (1987): Mode of growth of retinal axons within the tectum of *Xenopus* tadpoles, and implications in the ordered neuronal connection between the retina and the tectum. *J Comp Neurol* 260: 127–139

Fujisawa H, Tani N, Watanabe K, Ibata Y (1982): Branching of regenerating retinal axons and preferential selection of appropriate branches for specific neuronal connection in the newt. *Dev Biol* 90: 43–57

Gábriel R, Straznicky C, Wye-Dvorak J (1992): GABA-like immunoreactive neurons in the retina of *Bufo marinus:* Evidence for the presence of GABA-containing ganglion cells *Brain Res* 571: 175–179

Gaze RM (1959): Regeneration of the optic nerve in *Xenopus laevis. Q J Exp Physiol* 44: 290–308

Gaze RM (1970): *The Formation of Nerve Connections.* London and New York: Academic Press

Gaze RM, Grant P (1992) Spatio-temporal patterns of retinal ganglion cell death during *Xenopus* development. *Comp Neurol* 315: 264–274

Gaze RM, Jacobson M, Székely G (1963): The retinotectal projection in *Xenopus* with compound eyes. *J Physiol (Lond)* 165: 484–499

Gaze RM, Jacobson M, Székely G (1965): On the formation of connexions by compound eyes in *Xenopus. J Physiol (Lond)* 176: 409–417

Gaze RM, Keating MJ, Chung SH (1974): The evolution of the retinotectal map during development in *Xenopus. Proc R Soc Lond Ser B* 185: 301–330

Gaze RM, Keating MJ, Ostberg A, Chung SH (1979): The relationship between retinal and tectal growth in larval *Xenopus:* Implications for the development of the retino-tectal projection. *J Embryol Exp Morphol* 53: 103–143

Glücksmann A (1940): Development and differentiation of the tadpole eye. *Br J Ophthalmol* 25: 154–178

Grant P, Rubin E, Cima P (1980): Ontogeny of the retina and optic nerve in *Xenopus laevis.* I: Stages in the early development of the retina. *J Comp Neurol* 189: 593–613

Graydon ML, Giorgi PP (1984): Topography of the retinal ganglion cell layer of *Xenopus.* J Anat 139: 145–157

Harris WA (1986): Homing behavior of axons in the embryonic vertebrate brain. *Nature (Lond)* 320: 266–269

Harris WA (1989): Local positional cues in the neuroepithelium guide retinal axons in embryonic *Xenopus* brain. *Nature (Lond)* 339: 218–221

Hinds JW, Hinds PL (1974): Early ganglion cell differentiation in the mouse retina: An electron microscopic analysis utilizing serial sections. *Dev Biol* 37: 381–416

Hiscock J, Straznicky C (1989a): Morphological characterization of substance P-like immunoreactive amacrine cells in the anuran retina. *Vision Res* 29: 293–301

Hiscock J, Straznicky C (1989b): Neuropeptide Y-like immunoreactive amacrine cells in the retina of *Bufo marinus. Brain Res* 494: 55–64

Hiscock J, Straznicky C (1990): Neuropeptide Y- and substance P-like immunoreactive amacrine cells in the retina of the developing *Xenopus laevis. Dev Brain Res* 54: 105–113

Hollyfield JG (1968): Differential addition of cells to the retina in *Rana pipens* tadpoles. *Dev Biol* 18: 163–179

Hollyfield JG (1971): Differential growth of the neural retina in *Xenopus laevis* larvae. *Dev Biol* 24: 264–286

Hollyfield JG, Rayborn ME, Sarthy PV, Lam DMK (1979): The emergence, localization and maturation of neurotransmitter systems during development of the retina in *Xenopus laevis*. I. γ-aminobutyric acid. *J Comp Neurol* 188: 587–598

Humphrey MF, Beazley LD (1985): Retinal ganglion cell death during optic nerve regeneration in the frog. *Hyla moorei*. *J Comp Neurol* 236: 383–402

Jacobson M (1977): Mapping the developing retino-tectal projection in frog tadpoles by a double label autoradiographic technique. *Brain Res* 127: 55–67

Jagger WS (1988): Optical quality of the eye of the cane toad *Bufo Marinus*. *Vision Res* 28: 105–114

Jenkins S, Straznicky C (1986): Naturally occurring and induced ganglion cell death: A retinal whole-mount autoradigraphic study in *Xenopus*. *Anat Embryol* 174: 59–66

Johns PR (1977): Growth of the adult goldfish eye. III. Source of the new retinal cells. *J Comp Neurol* 176: 343–358

Kahn AJ (1974): An autoradiographic analysis of the time of appearance of neurons in the developing chick neural retina. *Dev Biol* 38: 30–40

Kolb H, Nelson R, Mariani A (1981): Amacrine cells, bipolar cells and ganglion cells of the cat retina: A Golgi study. *Vision Res* 21: 1081–1114

Kuljis RO, Karten HJ (1986): Substance P-containing ganglion cells become progressively less detectable during retinotectal development in the frog *Rana pipens*. *Proc Natl Acad Sci USA* 83: 5736–5740

Lázár G (1973): The development of the optic tectum in *Xenopus laevis*: A Golgi study *J Anat* 116: 347–355

Masland RH (1988): Amacrine cells. *Trends Neurosci* 11: 405–410

Mastronarde DN, Thibeault MA, Dubin WM (1984): Non-uniform postnatal growth of the cat retina. *J Comp Neurol* 228: 598–608

Mitrofanis J, Maslim J, Stone J (1989): Ontogeny of catecholaminergic and cholinergic cell distribution in the cat's retina. *J Comp Neurol* 289: 228–246

Negishi K, Teranishi T, Karkhanis A, Stell WK (1990): Emergence and development of immunoreactive cells in telostean retinas during the perinatal period. *Dev Brain Res* 55: 127–137

Nguyen VS, Straznicky C (1989): The development and the topographic organization of the retinal ganglion cell layer in *Bufo marinus*. *Exp Brain Res* 75: 345–353

Nieuwkoop PD, Faber J (1975): *Normal Table of Xenopus laevis (Daudin)*. Amsterdam: North-Holland

Pannese E (1976): An electron microscopic study of cell degeneration in chick embryo spinal ganglia. *Neuropathol Appl Neurobiol* 2: 247–267

Perry VH, Linden R (1982): Evidence for dendritic competition in the developing retina. *Nature (Lond)* 297: 683–685

Rager GH, Rager U (1978): Systems-matching by degeneration. I. A quantitative electronmicroscopic study of the generation and degeneration of retinal ganglion cells in chicken. *Exp Brain Res* 33: 55–78

Rayborn ME, Sarthy PV, Lam DMK, Hollyfield JG (1981): Emergence, localization and maturation of transmitter systems during development of the retina in *Xenopus laevis*: II. Glycine. *J Comp Neurol* 195: 585–593

Reh TA (1987): Cell-specific regulation of neuronal production in larval frog retina. *J Neurosci* 7: 3317–3324

Reh TA, Constantine-Paton M (1984): Retinal ganglion cell terminals change their projection sites during larval development of *Rana pipens*. *J Neurosci* 4: 442–457

Reh TA, Tully T (1986): Regulation of tyrosine hydroxylase containing amacrine cell number in larval frog retina. *Dev Biol* 114: 463–469

Sakaguchi DS, Murphey RK, Hunt RK, Tompkins R (1984): The development of retinal ganglion cells in a tetraploid strain of *Xenopus laevis*: A morphological study utilizing intracellular dye injection. *J Comp Neurol* 224: 231–251

Sarthy PV, Rayborn ME, DMK, Hollyfield JG (1981): The emergence, localization and maturation of neurotransmitter systems during development of the retina in *Xenopus laevis*: III. Dopamine. *J Comp Neurol* 195: 595–602

Scalia F, Arango V, Singman EL (1985): Loss and displacement of ganglion cells after optic nerve regeneration in adult *Rana pipens*. *Brain Res* 344: 267–280

Schnitzer J, Rusoff AC (1984): Horizontal cells of the mouse retina contain glutamic acid decarboxylase-like immunoreactivity during early development stages. *J Neurosci* 4: 2948–2955

Schütte M, Witkovsky P (1990): Serotonin-like immunoreactivity in the retina of the clawed frog *Xenopus laevis*. *J Neuorocytol* 19: 504–518

Scott TM, Lázár G (1976): An investigation into the hypothesis of shifting neuronal relationships during development. *J Anat* 12: 485–496

Sengelaub DR, Finley BL (1982): Cell death in the mammalian visual system during normal development. I. Retinal ganglion cells. *J Comp Neurol* 204: 311–317

Sperry RW (1943): Effect of 180° degree rotation of the retina on visuomotor co-ordination. *J Exp Zool* 92: 263–279

Sperry RW (1944): Optic nerve regeneration with return of vision in anurans. *J Neurophysiol* 7: 57–70

Sperry RW (1951): Mechanisms of neural maturation. In *Handbook of Experimental Psychology*, Stevens SS, ed. New York: Wiley

Sperry RW (1963): Chemoaffinity in the orderly growth of nerve fiber patterns and connections. *Proc Natl Acad Sci USA* 50: 703–710

Stone J, Maslim J, Rapaport DH (1984): The development of the topographical organisation of the cat's retina. In *Development of Visual Pathways in Mammals*, Stone J, Dreher B, Rapaport DH, eds. New York: Alan R. Liss

Stone J, Rapaport DH, Williams RW, Chalupa LM (1982): Uniformity of cell distribution in the ganglion cell layer of prenatal cat retina: Implications for mechanisms of retinal development. *Dev Brain Res* 2: 231–242

Straznicky K (1963): Function of heterotopic spinal cord segments investigated in the chick. *Acta Biol Hung* 14: 145–155

Straznicky K (1965): The development of the innervation and the musculature of wings innervated by thoracic nerves. *Acta Biol Hung* 20: 185–192

Straznicky C (1983): The patterns of innervation and the movements of ectopic hindlimbs supplied by brachial spinal cord segments in the chick. *Anat Embryol* 167: 247–262

Straznicky K, Gaze RM (1971): The growth of the retina in *Xenopus laevis*: An autoradiographic study. *J Embryol Exp Morphol* 26: 67–79

Straznicky K, Gaze RM (1972): The development of the tectum in *Xenopus*: An autoradiographic study. *J Embryol Exp Morphol* 28: 87–115

Straznicky C, Chehade M (1987): The formation of the area centralis of the retinal ganglion cell layer in the chick. *Development* 101: 869–876

Straznicky C, Hiscock J (1984): Postmetamorphic retinal growth in *Xenopus*. *Anat Embryol* 169: 103–109

Straznicky K, Tay D (1977): Retinal growth in double dorsal and double ventral eyes in *Xenopus*. *J Embryol Exp Morphol* 40: 175–185

Stuermer CAO (1984): Rule for retinotectal arborization in the goldfish optic tectum: A whole-mount study. *J Comp Neurol* 229: 214–232

Tay D, Hiscock J, Straznicky C (1982): Temporo-nasal asymmetry in the accretion of retinal ganglion cells in late larval and postmetamorphic *Xenopus*. *Anat Embryol* 164: 75–83

Tóth P, Straznicky C (1989): The distribution and dendritic morphology of displaced ganglion cells of the anuran retina. *Visual Neurosci* 3: 551–561

Turner DL, Cepko CL (1987): A common progenitor for neurons and glia persist in rat retina late in development. *Nature (Lond) 328: 131–136*

Wässle H, Boycott BB (1991): Functional architecture of the mammalian retina. *Physiol Rev* 71: 447–480

Wetts R, Fraser SE (1988): Multipotent precursors can give rise to all major cell types of the frog retina. *Science* 239: 1142–1145

Wetts R, Serbedzija GN, Fraser SE (1989): Cell lineage analysis reveals multipotent precursors in the ciliary margin of the frog retina. *Dev Biol* 136: 254–263

Wilson MA (1971): Optic nerve fibre counts and retinal ganglion cell counts during development of *Xenopus laevis*. *Q J Exp Physiol* 56: 83–91

Wong ROL (1990): Differential growth and remodelling of ganglion cell dendrites in the postnatal rabbit retina. *J Comp Neurol* 294: 109–132

Zhang Y, Straznicky C (1991): The morphology and distribution of photoreceptors in the retina of *Bufo marinus*. Anat Embryol 183: 97–104

Zhu B-S, Straznicky C (1990a): Dendritic morphology and retinal distribution of tyrosine hydroxylase-like immunoreactive amacrine cells in *Bufo marinus*. *Anat Embryol* 181: 365–371

Zhu B-S, Straznicky C (1990b): Morphology and distribution of serotonin-like immunoreactive amacrine cells in the retina of *Bufo marinus*. *Visual Neurosci* 5: 371–378

Zhu B-S, Straznicky C (1991): Morphology and retinal distribution of tyrosine hydroxylase-like immunoreactive amacrine cells in the retina of developing *Xenopus laevis*. Anat Embryol 184: 33–45

Zhu, B-S, Straznicky C (1992) Large serotonin-like immunoreactive amacrine cells in the retina of developing *Xenopus laevis*. *Dev Brain Res* 69: 000–000

Zhu B-S, Hiscock J, Straznicky C (1990): The changing distribution of neurons in the inner nuclear layer from metamorphosis to adult: A morphometric analysis of the anuran retina. *Anat Embryol* 181: 585–594

Zimmerman RP, Polley EH, Fortney RL (1988): Cell birthdays and rate of differentiation of ganglion and horizontal cells of the developing cat's retina. *J Comp Neurol* 274: 77–90

CHAPTER 14

Activity-Driven Mechanisms for Sharpening Retinotopic Projections: Correlated Activity, NMDA Receptors, Calcium Entry, and Beyond

JOHN T. SCHMIDT

INTRODUCTION

The precise organization of visual projections develops by selective stabilization of appropriate synapses from a more diffuse set of initial connections, a process that requires normal patterns of activity during the sensitive period. This chapter covers research on the establishment of retinotopic precision in the direct retinal map on tectum of fish and frog. This research grew out of two separate lines of inquiry: early demonstrations of plasticity in both developing and adult retinotectal projection, and studies of the effects of experience on development of visual cortex in mammals. Of course, Mike Gaze was the leader in demonstrating plasticity within the retinotectal projection, which initially had been thought to be rigidly specified (Sperry, 1963).

By the early 1970s, two experiments from Gaze's lab had demonstrated dramatic plasticity. First, there was a startling capability to reorganize the mature projection following surgical ablations. When half of the tectum was removed in adult goldfish, the retinal fibers reorganized a full map on the remaining half-tectum in a process termed "compression" (Gaze and Sharma, 1970). Likewise, after removal of half the retina, the remaining half expanded its projection over the full tectal surface (for a review see Schmidt, 1982). Even though each retinal fiber moved to terminate at a new tectal site, the overall retinotopic precision of the compressed or expanded maps was often nearly as good as in the normal map.

Shortly afterward, Gaze et al. (1974) reported that the developing retinotectal projection undergoes a continuous synaptic reorganization necessitated by the differing growth patterns of retina and tectum. The retina adds cells in a closed annulus, but tectum adds cells in an open crescent. Consequently, the oldest retinal cells at the center initially connect with the oldest tectal cells at the rostral pole, but later shift their terminals to connect with the midtectal region to maintain a retinotopic map on tectum. The surgically induced plasticity could then be seen as a manifestation of this natural plasticity. In addition, it became unlikely that the mechanism of map formation involved a search by each retinal fiber for precisely marked tectal target neurons, since such markers would have to be

Formation and Regeneration of Nerve Connections
Sansar C. Sharma and James W. Fawcett, Editors
© 1993 Birkhäuser Boston

changed continuously during growth and rapidly after the surgical ablation. In both cases, one major question was, What is the mechanism by which the projection can maintain precise retinotopography if individual fibers are not seeking precisely marked tectal locations?

At this point, a suggested mechanism came from work on the cat visual cortex. Results from studies of monocular deprivation and selective stripe exposure (Wiesel, 1982; Hirsch and Tieman, 1987) suggested the possibility that activity might be involved in retinotopic sharpening. In fact several labs had begun to inject tetrodotoxin (TTX), a sodium channel blocker, into the eye to remove visual activity entirely. Such studies eventually showed that activity (whether visually driven or spontaneous) was important (1) in the segregation of the two eyes' inputs into ocular dominance patches at the cortical level (Stryker and Harris, 1986; Stryker and Strickland, 1984) and into eye-specific laminae in the lateral geniculate nucleus (LGN) (Shatz, 1990), and (2) in the segregation of retinal input terminals by receptive field type in LGN (Dubin et al., 1986; Hahm et al., 1991). The use of intraocular TTX injections to block activity in the retinotectal system, begun about the same time, launched a series of investigations in many labs on the role of activity both in map formation (reviewed below) and in the segregation of eye-specific patches when two eyes are allowed to innervate a single tectum (Meyer, 1982; Boss and Schmidt, 1984; Reh and Constantine-Paton, 1985). This latter phenomenon serves as a model system for the related process of ocular dominance patch formation in visual cortex.

Many of the experiments reviewed herein employ the regeneration of the optic nerve fibers following nerve crush, rather than initial development, for reasons of convenience. Regeneration of the projection recapitulates the major features seen during development (reproducible axonal outgrowth and synaptogenesis after arrival back at the tectum) within a large, adult animal that is more easily manipulated. Recent studies have shown that the activity-driven processes uncovered in regeneration are also active during development, and these will be discussed together.

ACTIVITY-DRIVEN RETINOTOPIC SHARPENING OF THE REGENERATED PROJECTION

Several mechanisms are involved in map formation, including activity mechanisms, competition (illustrated by the "compression" results above), and chemoaffinity between optic fibers and different regions of tectum (Sperry, 1963; Walter et al., 1987). Chemoaffinity, however, is regional in character and sufficient only to organize a very rough map. Thus, axons from temporal retina grow into rostral tectum, whereas those from nasal retina grow around the edges to enter caudal tectum. In the other dimension, dorsal retinal fibers grow into ventrolateral tectum, whereas those from ventral retina grow into dorsomedial tectum. However, chemoaffinity information is not precise enough to tell the axons exactly where to form their arbors, and a second mechanism, an activity-driven retinotopic sharpening, returns the map to the mature level of organization. Since this process has recently been reviewed in detail (Schmidt, 1992), only the main points will be considered here.

Anatomical Studies of HRP-Stained Optic Axons and Arbors

Normal optic axons in goldfish fall into three classes by caliber (Stuermer and Easter, 1984; Schmidt et al., 1988). Fine axons give rise to small, superficial arbors in tectum; medium-caliber axons give rise to somewhat larger axons that terminate in intermediate to deeper portions of the retinal terminal lamina; and coarse-caliber axons give rise to the largest arbors, which terminate at superficial to intermediate depths in tectum (Figure 14.1A). The depth at which each arbor terminates is tightly maintained, so that the arbors are essentially two-dimensional. Each ganglion cell generally gives rise to a single axon, although a collateral may be given off to the thalamic or hypothalamic visual areas.

Optic axons regenerating from the site of the crush each form several collaterals within the nerve and tract (Murray, 1982), but the subsequent elimination of these extra collaterals is probably not affected by activity cues (Hartlieb and Stuermer, 1987; Cook and Becker, 1988).

At the level of the optic tract, staining of individual fibers with horseradish peroxidase (HRP) showed that axons arriving back at the tectum at 2 to 3 weeks postcrush make many exploratory branches within the plane of the map (Schmidt et al., 1988; Stuermer, 1988a). Although the average normal arbor is about 200 μm across, these early arbors average almost 1,200 μm across, a substantial fraction of the tectal length of 3 to 4 mm (Figure 14.1B). At 5 to 6 weeks postcrush, the arbors are much smaller and simpler, suggesting that many branches are subsequently eliminated. The arbors return toward a normal appearance, both in extent and in number of branches (Figure 14.1C). In addition, the depths of termination that are characteristic of each fiber group are also roughly reestablished.

One of the cues used to determine which of the initial exploratory branches get eliminated is spike activity in the pathway, for regenerating arbors that were silenced by intraocular TTX or synchronized in their firing by strobe illumination were roughly two times larger than normal in extent and also failed to coalesce into a single cluster of branches (Schmidt and Buzzard, 1990). Examples are shown in Figure 14.2. The depths at which they terminated were not affected by the manipulations of activity, nor were the number of branches and branch endings per arbor. In addition, Hayes and Meyer (1989) found that intraocular TTX did not affect the numbers of synapses formed by the regenerating fibers. Thus, there was a relatively selective effect on *where and over how much of the tectum* those branches were deployed. This lack of sharpening following strobe exposure was also seen with retrograde HRP labeling (Cook and Rankin, 1986), and is also reflected in the map that is recorded electrophysiologically from the tectum (see below).

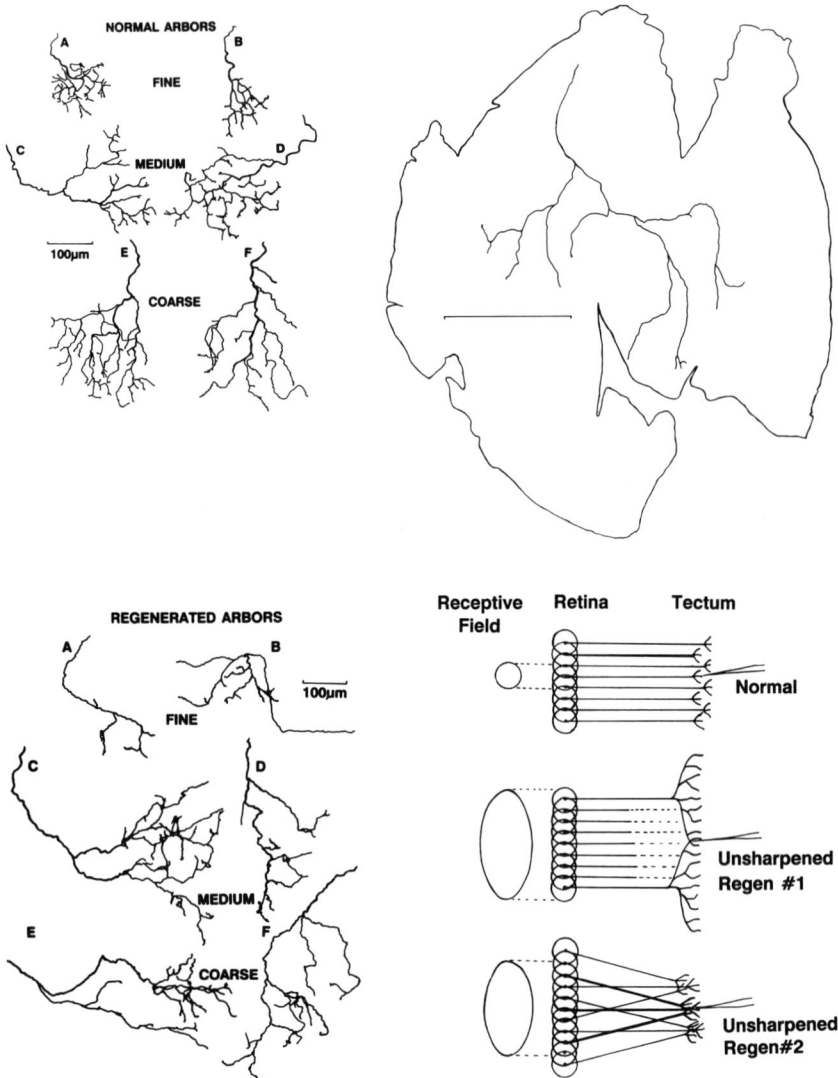

FIGURE 14.1. Optic arbors stained by HRP injection into the optic tract and drawn in camera lucida. **A:** Examples of normal arbors arising from fine, medium, and coarse caliber axons. **B:** An example of a regenerating arbor with many exploratory branches at 3 weeks postcrush. The outline shows the edges of the tectum, which had to be split in two places to make it lie flat on the slide. The scale bar is 1 mm. **C:** Examples of regenerated arbors from fine, medium, and coarse axons at 6 to 8 weeks postcrush. Note that most exploratory branches are gone and that the branches form distinct clusters. **D:** Possible ways that enlarged multiunit receptive fields are recorded in unsharpened projections. On the left are the ganglion cells and receptive fields, and on the right are their arbors in tectum. The recording electrode will record ganglion cells from a wide area of retina either if the arbors remain large and overlapping (center), or if they are misdirected but of normal size (bottom). Most of the lack of sharpening can be accounted for by the first possibility (see Figure 14.2). Parts **A, C,** and **D** reprinted from Schmidt and Buzzard, 1990.

STROBE REGENERATED ARBORS

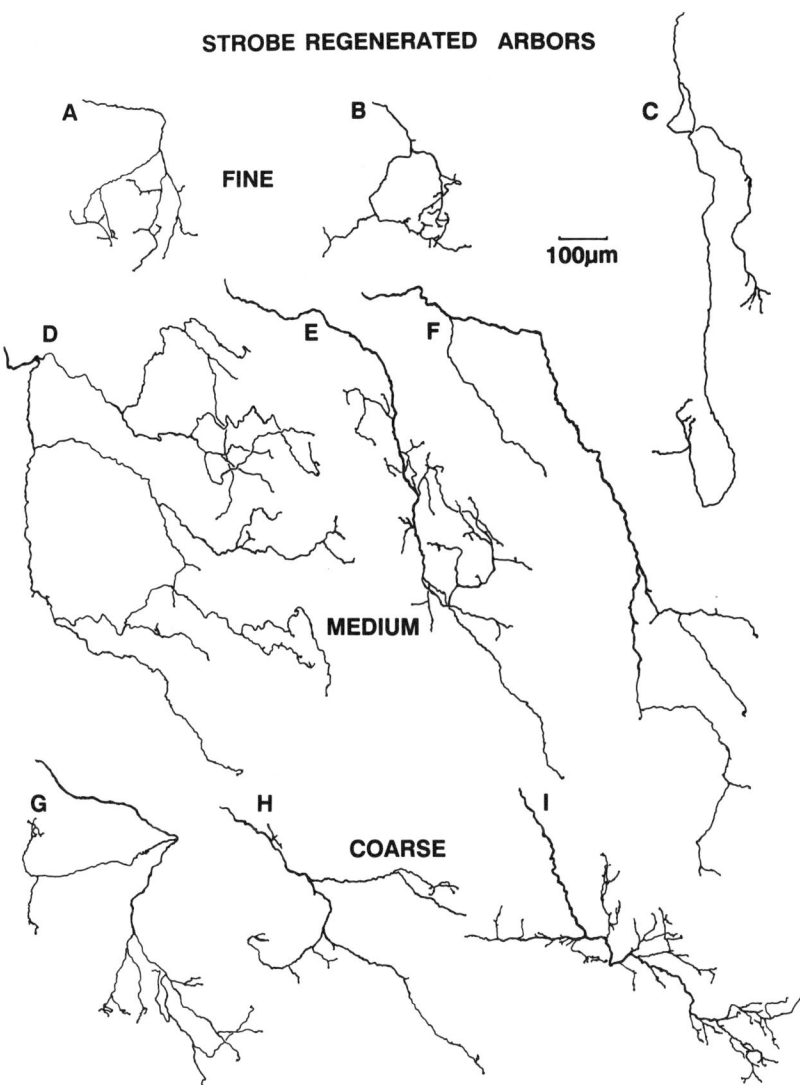

FIGURE 14.2. Examples of arbors regenerated under strobe illumination at 6 to 8 weeks postcrush. Three examples are given of arbors from fine, medium, and coarse axons. Note the scattered branches and the failure to focus branches in one area. (Reprinted from Schmidt and Buzzard, 1990.)

Manipulating activity (monocular deprivation) also exerts effects on developing retinogeniculate arbors in the kitten (Sur et al., 1982): X-cell arbors become wider in the plane of the map, while Y-cell arbors appear shrunken. Much earlier, during prenatal development, blocking all activity with intracranial infusion of TTX caused the arbors both to become enlarged in the plane of the retinotopic map and to grow across the normal laminar boundaries (Shatz, 1990). Thus, these effects are similar to those in goldfish tectum.

Electrophysiological Recordings of the Projection

Extracellular unit recordings from superficial tectal neuropil are mostly from the arbors of the retinal ganglion cells (Schmidt and Edwards, 1983; Adamson et al., 1984). Although several arbors are usually recorded simultaneously, the individual unit receptive fields normally overlap greatly and the inclusive area from which they can be driven is called the multiunit receptive field. In normal goldfish tectum this multiunit receptive field is approximately 11°, which is almost exactly the same as the average receptive field for single ganglion cells. Thus, in normal goldfish or frog the retinotectal projection is very precise (Schmidt and Edwards, 1983; Adamson et al., 1984).

The regenerating projection is far less precise. The fibers first arrive at the tectum at around 14 days after nerve crush (Schmidt et al., 1983; 1988), but orderly maps can first be recorded only at 35 to 40 days. This is about a week after the exploratory branches are withdrawn and about when normal-sized arbors are first seen anatomically. Before that point, low-amplitude multiunit activity seems to be driven from wide areas of the retina, but the multiunit receptive fields are difficult to define because the units are of very low amplitude and fatigue extremely rapidly (Schmidt and Edwards, 1983). In frog, however, recordings can be made during the entire process of sharpening, which is easily followed (Humphrey and Beazley, 1982; Adamson et al., 1984).

Even though these recordings are largely presynaptic, several studies have also shown that effective synaptic connections are made in nonretinotopic areas of tectum. In an electron microscope study, Kageyama and Meyer (1988) reported that nasal retinal fibers appropriate for caudal tectum made synapses while passing through inappropriate rostral tectum. In frog (Adamson et al., 1984), recordings from cells postsynaptic to the retinotectal synapses in the intertectal relay showed that their receptive fields were enlarged to the same extent as the multiunit receptive fields, as would be expected if effective synaptic transmission passed on the diffuse map from the contralateral tectum. Lastly, Matsumoto et al. (1987) used field potentials to examine the spread of monosynaptic activation in goldfish tectum created by retinal point stimulation. In the early stages of regeneration, synaptic activation was spread over very large areas (prior to the time of appearance of a sharp map), but became confined and more focused during maturation of the projection. Together, these studies leave no doubt that regenerating retinal fibers make effective synaptic connections in inappropriate regions of tectum.

In goldfish, if the projection is silenced with intraocular TTX, or alternatively synchronized by strobe illumination in a white featureless environment (Schmidt and Edwards, 1983; Schmidt and Eisele, 1985; Eisele and Schmidt, 1988) from 14 to 35 days after nerve crush, then the diffuse map is retained [see similar effect in Figure 14.3, right, a map unsharpened by *N*-methyl-D-aspartate (NMDA) block-

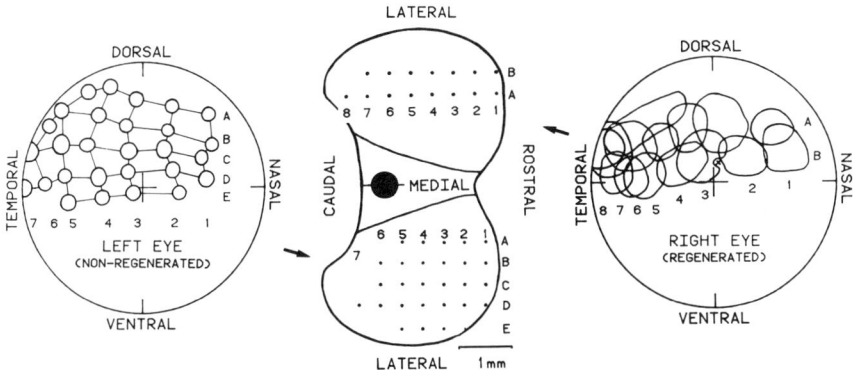

FIGURE 14.3. Maps of two retinotectal projections (one regenerating and one not regenerating) from a fish infused with AP7 from 21 to 35 days postcrush. The pump infused an average of 3.1 μl/day of 500 μM AP7, producing an estimated concentration of 10–20 μM after dilution in the fluids in and around the brain. The large circles represent the visual fields of the left and right eyes, and in the center is a drawing of the left and right tectal surfaces with the positions of electrode penetrations marked and numbered. Corresponding designations are given for the outlines of the respective multiunit receptive fields in the visual fields. On the left is the orderly map of the nonregenerating projection from the left eye to the right tectum, which was not affected by the block of NMDA receptors. On the opposite tectum of this fish, the regenerating projection from the right eye (right) was prevented from sharpening by the block of NMDA receptors (average multiunit receptive field size of 27.4° vs. 10.8°). The visual field of the left eye was reversed so that the array would directly correspond to that on the tectum. The blackened circle at the midline shows the entry point of the cannula through the tectal commissure to deliver the AP7. Reprinted with permission of Oxford University Press from Schmidt JT (1990): Long term potentiation and activity dependent retinotopic sharpening in the regenerating retinotectal projection of goldfish: Common sensitive period and sensitivity to NMDA blockers. *J Neurosci* 10: 233–246.

ers]. The units recorded from each point are driven from retinal areas about 30° or more across. The centers of these areas are approximately in the right place; it is their extent that reflects the uncorrected errors in targeting of the arbor branches. Both the separation of single units by spike height and the recording of single ganglion cells from retina showed that the ganglion cells retained normal receptive fields after TTX or strobe treatment; the enlarged multiunit receptive fields were therefore due to abnormal convergence onto each tectal point of ganglion cell arbors from wide retinal areas (Schmidt and Edwards, 1983; Schmidt and Eisele, 1985).

The effects of strobe, which causes the firing of all ganglion cells, point out the importance of the normal pattern of activity that must contain information instructive for the process of sharpening. Normally, activity is correlated between neighboring, but not distant, ganglion cells of the same type (ON, OFF, or

ON-OFF); this is especially true in the light, but also occurs in the dark due to retinal circuitry (for review see Mastronarde, 1989).

Comparison to Sharpening During Development

It is interesting that the development of the retinotectal projection in birds and mammals has now been shown to proceed through a similar diffuse stage in which exploratory branches and collaterals cover a wide area of tectum (Schneider et al., 1981; Sachs et al., 1986; Nakamura and O'Leary, 1989; Simon and O'Leary, 1990). Later such branches are not seen, and are presumably withdrawn, although the death of the parent ganglion cell may also play a role in eliminating errors (O'Leary et al., 1986).

The development of the projection in fish and frog is complicated because the eye and tectum are so small at the time of initial innervation, and the vast majority of the cells of each structure are added later. Thus the initial arbors are necessarily much smaller than in the adult. Nevertheless, if the size of the arbors is expressed as a percent of the total tectal area, they are then seen to decrease in size considerably during subsequent maturation. The relative tectal coverage may be a more relevant figure, since the same extent of visual field (in degrees) is represented on the tectum at all stages. The percent tectal coverage dropped at least ten-fold from about 2% to less than 0.2% in goldfish (Stuermer and Raymond, 1989), about sevenfold in zebrafish (Stuermer, 1988b), and from more than 25% to about 5% in the first few weeks in larval *Xenopus* (Sakaguchi and Murphey, 1985). Only in late larval *Xenopus* were arbors reported to be transiently larger in absolute terms than in the adult (Fujisawa, 1987).

The effects of manipulating activity during development have been studied in several species. In zebrafish, the initial ingrowth into tectum is not affected by blocking activity with TTX (Stuermer, 1990). However, the arbors shift dynamically across the tectum and maintain order during the subsequent growth, and the manipulation of activity during this stage in goldfish produces effects similar to those seen during regeneration. When newly hatched goldfish were reared for more than a year under alternate 12-h periods of darkness and strobe illumination, the retinotectal map failed to sharpen. Multiunit receptive fields averaged 25°, about the same as in strobe-regenerates, and HRP-stained arbors in these fish were abnormal in ways similar to those of the strobe-regenerates (Schmidt and Buzzard, unpublished results). In the developing *Rana* tectum, blocking NMDA receptors caused a disruption of retinotopic organization assessed by retrogarde HRP transport (Cline and Constantine-Paton, 1989). The NMDA receptors are thought to mediate the detection of coincident synaptic activation (see below). In *Xenopus*, however, dark rearing had no effect on the size of multiunit receptive fields of event detectors (class III, Keating et al., 1986). These results may be due to the facts that the size of individual receptive fields of class III ganglion cells is normally already very large (21°) and that the map from the smaller (3–6°) class II bug detectors was inexplicably not examined. Alternatively, they may be due to the fact that spontaneous activity occurring in the dark might cause some limited degree of sharpening. Thus, there is substantial support for a role of activity-driven sharpening of diffuse initial projections in development as well as in regeneration.

A HEBBIAN MODEL FOR SHARPENING BASED ON NMDA RECEPTORS

A model for retinotopic sharpening is based upon four premises: (1) correlated firing of neighboring ganglion cells, (2) retinal arbors that are initially large and highly overlapping, (3) resultant summation of retinal excitatory postsynaptic potentials (EPSPs) in postsynaptic tectal cells, and (4) Hebbian strengthening of those synapses participating in the largest synaptic depolarizations (Hebb, 1949). The number of ganglion cells reporting on each retinal point is large. With about 200,000 ganglion cells, an average receptive field diameter of 11°, and a 183° visual field, I calculate that about 700 ganglion cell receptive fields overlap at each point, producing large populations with correlated firing. On each postsynaptic cell, inputs from neighboring ganglion cells would fire with the highest degree of correlation, experience the greatest summation of their EPSPs, and be stabilized by the Hebbian mechanism. Arbors from some more distant ganglion cells would also overlap on tectal cells in this region. However, they would not have correlated firing and would not be stabilized, but would instead be withdrawn. On other postsynaptic cells, a different set of inputs would have the greatest summation and thus capture exclusive control. Eventually each tectal cell would be captured by neighboring ganglion cells from the appropriate retinal area to form a continuous retinotopic map. The mechanism whereby that state is reached could involve either (1) a stimulated growth of new synaptic contacts from appropriate fibers in the immediate region of their stabilized synapses or (2) an iterative process of random reinsertion followed by further stabilization. From this model it is easy to see that if no cells had activity, no synapses would be differentially stabilized. Conversely, if all ganglion cells fired in synchrony due to the strobe, the cue for finding neighboring cells would be removed and all synapses would be stabilized equally regardless of their origin. In either case, no synapse would be stabilized more than any other.

If innervation from a second eye is introduced, the fibers within each eye should have highly correlated activity, whereas those from different eyes should not. This would, therefore, drive segregation into eye-specific patches by the same Hebbian mechanism, since the fibers from each eye would cooperatively stabilize each others' synapses, driving out the other eye's synapses and consolidating their hold on their areas. In this case, it is obvious that synapse elimination is also associated with synapse stabilization, since large gaps are left in the once continuous projections. Thus, the mechanism that normally sharpens the map can also produce great discontinuities.

For this model to work, there must be first a mechanism in the postsynaptic neuron for detecting synchrony of synaptic activation, and second a mechanism whereby the growth of presynaptic terminals is influenced by this postsynaptic detection system. The NMDA receptor seems to subserve the first mechanism, and arachidonic acid (AA) is a possible candidate for the second.

NMDA RECEPTORS AS TRIGGERS OF HEBBIAN STABILIZATION

Retinotectal transmission appears to be mediated primarily by glutamate (or some similar excitatory amino acid or small peptide) acting on several types of postsynaptic receptor, and most of the EPSP amplitude seems to be mediated by non-NMDA receptors (Langdon and Freeman, 1986; Schmidt, 1990, 1991a; Cline, 1991). These receptors, often referred to as quisqualate (α-amino-3-hydroxy-5-methyl-isoxazole-4-propionic acid; AMPA) and kainate receptors, operate in a conventional fashion upon binding the neurotransmitter, and they open channels that admit primarily Na^+ and K^+ ions. The NMDA receptors contribute much less to EPSP amplitude (5–10%: Schmidt, 1990, 1991a), but they appear to play a tremendously important role in controlling the developmental plasticity and maintenance of synaptic connections.

NMDA receptors, which have been implicated in triggering long-term potentiation (LTP) in the CA1 region of hippocampus (Collingridge and Bliss, 1987) and several areas of cortex (Tsumoto et al., 1990), are well suited to the task of signaling whether the synapse has participated in a substantial synaptic depolarization. Although activated by neurotransmitter, the channel requires a degree of depolarization in order to conduct fully. The depolarization removes a magnesium block from the open channel (Nowak et al., 1984) and allows substantial calcium entry (MacDermott et al., 1986). Thus, NMDA receptors both have the appropriate trigger properties and generate a second messenger (calcium) that can signal Hebbian stabilization.

If NMDA receptors play such a role in the stabilization of retinotectal synapses, then (1) the retinotopic sharpening should be blocked by NMDA receptor blockers, and (2) there might also be an enhanced capacity for LTP during the period of sharpening, which should also be blocked by NMDA receptor blockers.

NMDA Receptor Blockers Prevent Sharpening

Since sharpening occurs over several weeks, NMDA receptors had to be blocked continuously for a long period. For this purpose, osmotic minipumps, attached to the fish's head, were used to infuse the NMDA receptor blockers AP5 (2-amino-5-phosphonovalerate) or AP7 (2-amino-7-phosphonoheptanoate) into the tectal ventricle (Schmidt, 1990). Infusions, begun at 21 days postcrush when field potentials were first detectable, delivered an average of 3.5 µl/day over the next 12 to 24 days. The 500 µM solution in the pumps was diluted about 25–50× within the fluid spaces in and around the brain, giving an estimated 10 µM AP5 in tectum. These concentrations should not have reduced retinotectal transmission to any substantial degree, and fish injected intracranially with larger amounts of AP5 or AP7 showed no performance deficit in a tectally mediated behavioral assay, the optomotor response to a rotating drum (Schmidt, 1990).

In eight control fish infused with Ringer's solution alone or solutions of inactive substances, the projections sharpened normally. The projections of seven fish infused with NMDA receptor blockers all failed to sharpen. The results (Figure 14.3) were both qualitatively and quantitatively comparable to those of strobe exposure for the same period during regeneration. The gross organization of the map was correct, but multiunit receptive fields were much enlarged, averaging 28°. These enlarged values in the blocked projections must reflect uncorrected errors in targeting of optic arbors, since single units in these projections had normal receptive fields of about 11°. In contrast, all of the projections receiving control infusions [either with Ringer's solution alone, or with the chemically similar but inactive AP6 (2-amino-6-phosphonohexanoate), or with AP5 at too low a concentration] had multiunit receptive fields averaging 11.6°, a value that was indistinguishable from that for uninfused regenerated projections.

Figure 14.3 also shows the lack of effect on the intact projection. Since the cannula delivered the AP7 at the midline, the intact projection must have received the same concentration of the blocker as the regenerating one, but it was not rendered unsharpened as a result. This indicates a selective interference with the activity-driven synapse stabilization and elimination that goes on in the regenerating projection. A similar lack of effect on intact projections was also found in strobe and TTX experiments (Schmidt and Eisele, 1985).

Enhanced Capacity for LTP During Sharpening

In studying the return of field potentials during regeneration, I noticed that, although the amplitude after maximal optic nerve stimulation was very small during the early phase, it tended to increase markedly after a series of maximal stimuli were administered to it, suggesting that the projection was undergoing LTP (Figure 14.4). In systematic studies, the maximum potentiation during the time of sharpening averaged 400% and was stable overnight. This potentiation is reviewed in more detail elsewhere (Schmidt, 1991a).

Although the frequencies used had to be kept very low because the immature projection fatigues rapidly, such low-frequency stimulation is also able to elicit LTP in developing cortex (Tsumoto et al., 1990) and in hippocampus if sufficient depolarization is assured by blocking inhibition with picrotoxin, by pairing with intracellular injection of depolarizing current, or by pairing with trains to another input (Collingridge and Bliss, 1987). The exact reason why the tectal cells might be sufficiently depolarized or disinhibited during regeneration is not known, but tectal cells were reported to become more excitable and fire in bursts until reinnervated (Brink and Meyer, 1986).

LTP was greatest early in regeneration, when the fibers were forming synapses at the fastest rate and sharpening was taking place (Schmidt, 1990). Potentiations between 20 and 40 days averaged 380%, while those from 40 to 70 days averaged 80–90%. Since a sharp map can first be recorded at 35–40 days, the match to the period of sharpening (Eisele and Schmidt, 1988) is particularly good. This time period also corresponds to the period of greatest transport and accumulation at the retinal terminals of GAP 43 (Benowitz and Schmidt, 1987), a protein that is

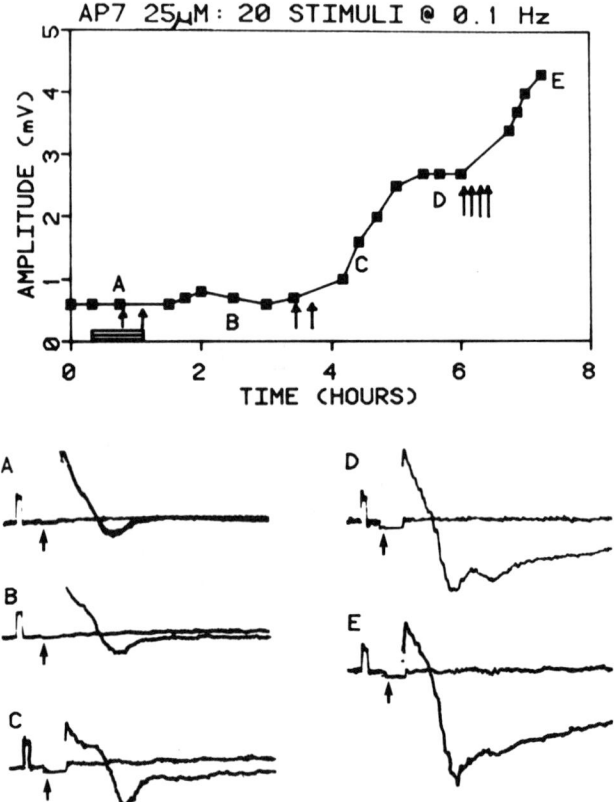

FIGURE 14.4. Block of LTP in a regenerating retinotectal projection by the NMDA receptor antagonist AP7. **Top:** Amplitude of the responses is plotted vs. time. **Bottom:** Sample responses from times marked by letters in the plot above. In the plot, the arrows show the time of the potentiating trains of supramaximal stimuli (20 stimuli at 0.1 Hz). The first time that these trains were applied, the tectal NMDA receptors were blocked by topical application of 25 μM AP7 (striped bar), and no consistent gain was noted. Two hours after washout of AP7, the same trains resulted in a large potentiation. Test stimuli were given at 15- to 30-min intervals to avoid causing potentiation. Traces are 100 msec long and the calibration pulse is 1 mV × 2 msec. Negative is downward in all traces. The stimulus artifact was electronically suppressed. Reprinted with permission of Oxford University Press from Schmidt JT (1990): Long term potentiation and activity dependent retinotopic sharpening in the regenerating retinotectal projection of goldfish: Common sensitive period and sensitivity to NMDA blockers. *J Neurosci* 10: 233–246.

phosphorylated by C kinase during LTP in hippocampus (Benowitz and Routtenberg, 1987). The level of LTP after 40 days is still much higher than that found in the mature projection (5%) and remains high for at least a year. Exactly what causes the enhanced capacity for LTP during the period of sharpening is not known, although increased excitability or disinhibition of the tectal cells are likely possibilities, along with the reexpression of GAP43. The field potentials showed no evidence of a greater role for NMDA receptors in transmission (Figure 14.4), as was suggested by Tsumoto et al. (1990) for developing visual cortex.

NMDA Receptor Blockers Also Block LTP

In goldfish tectum, LTP normally elicited by a train of stimuli was prevented by topical application of the NMDA receptor blockers AP5 or AP7 (but not by the closely related AP6; see Figure 14.4). Later, after the blocker was washed out, the same trains of stimuli elicited robust LTP, averaging 104% in the same projections (Schmidt, 1990).

Two factors link the LTP and the activity-driven retinotopic sharpening. First, the greatest capacity for LTP occurs precisely during the time when the sharpening is taking place. Second, both are blocked by low concentrations of NMDA receptor blockers. Although the exact relationship between the two phenomena is not completely defined, the intriguing possibility now presents itself that LTP, an increase in the gain of individual synapses, may constitute the first step in the stabilization process, one that leads to the later anatomical rearrangements.

These results directly parallel those from developing kitten visual cortex. During the sensitive period for development of ocular dominance, Komatsu et al. (1988) found a significant capacity for LTP that was not present in the adult. In addition, a capacity for LTP is also present in developing rat visual cortex, and it can be blocked by NMDA receptor blockers (Tsumoto et al., 1990).

NMDA Receptor Blockers Prevent the Formation of Eye-Specific Patches

Using the three-eyed tadpole, Cline et al. (1987, 1990) have tested the role of NMDA receptors in the segregation of visual afferents into eye-specific stripes. Slow-release plastic Elvax pellets impregnated with NMDA receptor blockers were placed onto the tectal surface, and several weeks later the fibers from the ectopic third eye were labeled with HRP. After several weeks of treatment with AP5, the stripes in the area around the plastic implant had merged, leaving a continuous distribution of labeled fibers. The NMDA channel blocker MK-801 caused desegregation of the stripes, but only if it was applied with NMDA to activate the receptors. NMDA alone caused the stripes to become sharper than in controls, with many fewer fibers crossing the open areas between the stripes. The NMDA and AP5 also seemed to have effects on individual arbor morphology, causing them to be less highly branched. Thus, there is very strong evidence that segregation into eye-specific stripes depends upon NMDA receptor activation, as would be predicted from the Hebbian model above.

In cat visual cortex, infusion of AP5 prevents the shift in ocular dominance caused by monocular lid suture (Kleinschmidt et al., 1987; Bear et al., 1990). In control areas, lid suture caused the open eye to dominate responses of cortical cells at the expense of the closed eye, which lost effectiveness, whereas in the area infused with AP5, the closed eye was still effective in driving cells. Likewise, Gu et al. (1989) showed that AP5 infusion prevents the shift back to the reopened eye in cases of reverse suture. These experiments are somewhat different from those in frog tectum, in that an electrophysiological assay was used to study whether each eye gained or lost in the proportion of cells that it dominated. Anatomical studies, on the other hand, followed segregation of inputs without gain or loss of territory. Nevertheless, both support the general idea that the interactions between the two eyes' inputs depend on NMDA receptors (for more discussion, see Schmidt, 1991a).

Although visual afferents from the two eyes usually segregate because they carry differing patterns of information, in special cases some afferents carry the same information and end up connected to the same cells, such as in mediating binocular vision. In fact, most neurons in the extragranular layers of normal cat visual cortex remain largely binocular. In addition, binocular convergence is set up in frog tectum by activity-driven cues and mediated by NMDA receptors (Chapter 5, this volume).

POSTSYNAPTIC TRIGGER AND PRESYNAPTIC EFFECTS

Calcium entry through the NMDA receptor channel serves as a signal that the synapse participated in a successful depolarization of the neuron, but this postsynaptic signal must somehow direct the presynaptic changes seen in branch retraction and eye-specific segregation. Therefore, one must postulate some type of retrograde signaling mechanism so that the presynaptic terminal can react appropriately. This retrograde signal might take one of several forms (Schmidt, 1992): (1) a soluble signal released by the postsynaptic element, (2) an addition to, or modification of, adhesion molecules mediating a stronger synaptic attachment, or (3) the polymerization of ependymin in the synaptic cleft due to calcium depletion as calcium enters the postsynaptic cell. LTP is also thought to require a retrograde signal, since part of the mechanism may be increased presynaptic release of glutamate (Bekkers and Stevens, 1990; Madison et al., 1991).

One candidate for a soluble retrograde signal is AA, since (1) NMDA receptor stimulation activates phospholipase A_2, the enzyme producing AA (Dumuis et al., 1988; Dorman et al., 1988; Sanfeliu et al., 1990); (2) LTP induction is prevented by either nordihydro-guaiaretic acid or bromophenacyl bromide, inhibitors of AA production (Williams and Bliss, 1988; Okada et al., 1989; Massicotte et al., 1990); (3) application of AA with mild stimulation increases synaptic gain, mimicking LTP (Williams et al., 1989; Drapeau et al., 1990); and (4) AA stimulates increased release of neurotransmitters (Freeman et al., 1990), possibly via the activation of the γ form of C-kinase (Shearman et al., 1989; Shapira et al., 1987) or the synergistic activation (with diacyl glycerol) of other forms of C kinase (Keyser and Alger, 1991).

Kinases are very important in the induction of LTP; both the C kinase and the calcium/calmodulin kinase II are apparently required (Madison et al., 1991). Induction can be blocked by intracellular injection into the postsynaptic cell of specific blockers of either kinase. Expression of previously induced LTP, on the other hand, is not dependent on either kinase in the postsynaptic cell (Madison et al., 1991), but may be dependent on C kinase in the presynaptic terminal, where it is known to phosphorylate GAP-43 and increase transmitter release by increasing Ca^{2+} entry (Norden et al., 1991).

If LTP is indeed an important step in the stabilization of synaptic connections, then synapse stabilization might also be expected to depend on the functioning of these kinases. Infusion of agents that block these kinases (such as sphingosine and H7) or that indiscriminately activate and down-regulate the C kinase (phorbol esters such as tetradecanoyllphorbol acetate, TPA) should block the activity-driven stabilization.

In preliminary experiments (Schmidt, 1991b), I have tested the effects of the phorbol ester TPA in several fish during the period of sharpening and obtained results supporting a role for C kinase (unpublished data). Whether the TPA was repeatedly injected into the regenerating eye or infused into the tectum with minipumps, the projection failed to undergo activity-driven retinotopic sharpening. The resulting projections were much like those seen with NMDA blockers with enlarged multiunit receptive fields. Control injections into the opposite eye had no effect, thus establishing the eye as a second site of action and emphasizing the importance of *presynaptic* C kinase in synapse stabilization. In the eye or the tectum, the TPA probably both stimulates the phosphorylation of proteins such as GAP 43 that are transported to tectum, and down-regulates the amount of C kinase present. We are in the process of measuring to detect such decreases. Of course, other (postsynaptic) effects could be produced by intracranial TPA, but apparently only the presynaptic effects are necessary to prevent sharpening. We have found that intracranial AA injections also prevent sharpening, as predicted by the hypothesis that this amphiphilic agent may serve as the natural retrograde agent.

Our results appear to be quite different from those reported for maintenance of the eye-specific stripes. Cline and Constantine-Paton (1990) reported that chronic release of such agents from Elvax pellets did not prevent the maintenance of eye-specific stripes in three-eyed frogs, although the arbors became much smaller and had fewer branches. They concluded that protein kinase activity is not required for selective stabilization of coactive inputs in this case.

The differences between these two studies may be instructive. First, retinotopic sharpening, because it is such an active process, may be much more sensitive to blocking kinases than is the maintenance of preexisting stripes. This would parallel a difference in the effects of kinase blockers on LTP induction and maintenance. Second, the fact that the kinase blockers caused the arbors to become smaller suggests that they might have interfered with growth, thus preventing arbor branches from entering the interstripe zones. This possibility is consistent with our results from a culture assay showing that kinase blockers (such as H7 at 200 μM and sphingosine at 100 μM) arrest outgrowth of neurites, collapse growth cones, and cause their retraction. This causes a complication in the interpretation of their lack of effect on stripes. TPA (1 μM), which activates and down-regulates C kinase, also slows neurite outgrowth, but AA (100 μM) has no effect on their growth rate. Thus, some kinase function (particularly C kinase) is probably required for growth as well as for stabilization. When TPA or AA is present, C kinase is active, allowing exploratory growth but perhaps not differential synaptic stabilization. These results are preliminary and further experiments are underway. In addition, a culture system for studying retinotectal synaptogenesis is being developed so that the effects on these different components of the sharpening process can be studied separately, using both electrophysiological recording and direct visual observation.

Acknowledgments

This work was supported by NIH grant EY03736 to J.S.

REFERENCES

Adamson J, Burke J, Grobstein P (1984): Recovery of the ipsilateral oculotectal projection following nerve crush in the frog: Evidence that retinal afferents make synapses at abnormal tectal locations. *J Neurosci* 4: 2635–2649

Bear MF, Kleinschmidt A, Gu Q, Singer W (1990): Disruption of experience-dependent synaptic modifications in striate cortex by infusion of an NMDA receptor antagonist. *J Neurosci* 10: 909–925

Bekkers JM, Stevens CF (1990): Presynaptic mechanism for long-term potentiation in the hippocampus. *Nature* 346: 724–728

Benowitz LI, Routtenberg A (1987): A membrane phosphoprotein associated with neural development, axonal regeneration, phospholipid metabolism, and synaptic plasticity. *Trends Neurosci* 10: 527–532

Benowitz LI, Schmidt JT (1987): Activity dependent sharpening of the regenerating retinotectal projection in goldfish: Relationship to the expression of growth associated protein. *Brain Res* 417: 118–112

Boss VC, Schmidt JT (1984): Activity and the formation of ocular dominance patches in dually innervated tectum of goldfish. *J Neurosci* 4: 2891–2905

Brink DL, Meyer RL (1986): Locally correlated spontaneous activity in the goldfish optic tectum after nerve crush. *Neurosci Abstr* 12: 438

Cine HT (1991): Activity-dependent plasticity in the visual system of frogs and fish. *Trends Neurosci* 14: 104–111

Cline HT, Constantine-Paton M (1989): NMDA receptor antagonists disrupt the retinotectal topographic map. *Neuron* 3: 413–426

Cline HT, Constantine-Paton M (1990): The differential influence of protein kinase inhibitors on retinal arbor morphology and eye-specific stripes in the frog retinotectal system. *Neuron* 4: 899–908

Cline HT, Debski EA, Constantine-Paton M (1987): N-Methyl-D-aspartate receptor antagonist desegregates eye specific stripes. *Proc Natl Acad Sci USA* 84: 4342–4345

Cline HT, Debski EA, Constantine-Paton M (1990): NMDA receptor agonists and antagonists alter retinal ganglion cell arbor structure in the developing frog retinotectal projection. *J Neurosci* 10: 1197–1215

Collingridge GL, Bliss TVP (1987): NMDA receptors—their role in long term potentiation. *Trends Neurosci* 10: 288–293

Cook JE, Becker DL (1988): Retinotopic refinement of the regenerating goldfish optic tract is not linked to activity-dependent refinement of the retinotectal map. *Development* 104: 321–329

Cook JE, Rankin ECC (1986): Impaired refinement of the regenerated retinotectal projection of the goldfish in stroboscopic light: A quantitative WGA-HRP study. *Exp Brain Res* 63: 421–430

Dorman RV, Schwartz MA, Terrian DM (1988): Depolarization induced [^{3}H]arachidonic acid accumulation: Effects of external Ca^{++} and phospholipase inhibitors. *Brain Res Bull* 21: 445–450

Drapeau C, Pellerin L, Wolfe LS, Avoli M (1990): Long-term changes of synaptic transmission induced by arachidonic acid in the CA1 subfield of the rat hippocampus. *Neurosci Lett* 115: 286–292

Dubin MW, Stark LA, Archer SM (1986): A role for action potential activity in the development of neuronal connections in the kitten retinogeniculate pathway. *J Neurosci* 6: 1021–1036

Dumuis A, Sebben M, Haynes L, Pin J-P, Bockaert J (1988): NMDA receptors activate the arachidonic acid cascade system in striatal neurons. *Nature* 336: 68–70

Eisele LE, Schmidt JT (1988): Activity sharpens the regenerating retinotectal projection in goldfish: Sensitive period for strobe illumination and lack of effect on synaptogenesis and on ganglion cell receptive field properties. *J Neurobiol* 19: 395–411

Freeman EJ, Terrian DM, Dorman RV (1990): Presynaptic facilitation of glutamate release from isolated hippocampal mossy fiber endings by arachidonic acid. *Neurochem Res* 15: 743–750

Fujisawa H (1987): Mode of growth of retinal axons within the tectum of *Xenopus* tadpoles, and implications in the ordered neuronal connection between the retina and the tectum. *J Comp Neurol* 260: 127–139

Gaze RM, Sharma SC (1970): Axial differences in the reinnervation of the goldfish optic tectum by regenerating optic nerve fibres. *Exp Brain Res* 10: 171–181

Gaze RM, Keating MJ, Chung SH (1974): The evolution of the retinotectal map during development in *Xenopus*. *Proc R Soc Lond B* 185: 301–330

Gu Q, Bear M, Singer W (1989): Blockade of NMDA receptors prevents ocularity changes in kitten visual cortex after reversed monocular deprivation. *Dev Brain Res* 47: 281–288

Hahm J-O, Langdon RB, Sur M (1991): Disruption of retinogeniculate afferent segregation by antagonists to NMDA receptors. *Nature* 351: 568–570

Hartlieb E, Stuermer CAO (1987): Preferential loss of collaterals from goldfish retinal axons in the optic tract is delayed by tetrodotoxin. *Neurosci Lett* 79: 1–5

Hayes WP, Meyer RL (1989): Impulse blockade by intraocular tetrodotoxin during optic regeneration in goldfish: HRP-EM evidence that the formation of normal numbers of optic synapses and the elimination of exuberant optic fibers is activity independent. *J Neurosci* 9: 1414–1423

Hebb DO (1949): *The Organization of Behavior*. New York: John Wiley and Sons

Hirsch HVB, Tieman SB (1987): Perceptual development and experience-dependent changes in cat visual cortex. In: *Sensitive Periods in Development: Interdisciplinary Perspectives*, Bernstein M, ed. Hillsdale, NJ: Lawrence Erlbaum Assoc

Humphrey MF, Beazley LD (1982): An electrophysiological study of early retinotectal projection patterns during optic nerve regeneration in *Hyla moorei*. *Brain Res* 239: 595–602

Kageyama GH, Meyer RL (1988): Regenerating optic axons form transient topographically inappropriate synapses in goldfish tectum: A WGA-HRP study. *Neurosci Abstr* 14: 674

Keating MJ, Grant S, Dawes EA, Nanchanal K (1986): Visual deprivation of the retinotectal projection in *Xenopus laevis*. *J Embryol Exp Morphol* 91: 101–115

Keyser DO, Alger BE (1991): Synergistic activation of protein kinase C by arachidonic acid and diacyl glycerol in hippocampal neurons. *Neurosci Abstr* 17: 16

Kleinschmidt A, Bear MF, Singer W (1987): Blockade of "NMDA" receptors disrupts experience-dependent plasticity of kitten striate cortex. *Science* 238: 355–357

Komatsu Y, Fujii K, Maeda J, Sakaguchi H, Toyoma K (1988): Long term potentiation of synaptic transmission in kitten visual cortex. *J Neurophysiol* 59: 124–141

Langdon RB, Freeman JA (1986): Antagonists of glutaminergic neurotransmission block retinotectal transmission in goldfish. *Brain Res* 398: 169–174

MacDermott AB, Mayer ML, Westbrook GL, Smith SJ, Barker JL (1986): NMDA-receptor activation increases cytoplasmic calcium concentration in cultured spinal cord neurones. *Nature* 321: 519–522

Madison DV, Malenka RC, Nicoll RA (1991): Mechanisms underlying long-term potentiation of synaptic transmission. *Ann Rev Neurosci* 14: 379–397

Massicotte G, Oliver MW, Lynch G, Baudry M (1990): Effect of bromophenacyl bromide, a phospholipase A_2 inhibitor, on the induction and maintenance of LTP in hippocampal slices. *Brain Res* 537: 49–53

Mastronarde DN (1989): Correlated firing of retinal ganglion cells. *Trends Neurosci* 12: 75–80

Matsumoto N, Kometani M, Nagano K (1987): Regenerating retinal fibers of the goldfish make temporary and unspecific but functional synapses before forming the final retinotopic projection. *Neuroscience* 22: 1103–1110

Meyer RL (1982): Tetrodotoxin blocks the formation of ocular dominance columns in goldfish. *Science* 218: 589–591

Murray M (1982): A quantitative study of regenerative sprouting by optic axons in goldfish. *J Comp Neurol* 209: 352–362

Nakamura H, O'Leary DDM (1989): Inaccuracies in initial growth and arborization of chick retinotectal axons followed by course corrections and axon remodelling to develop topographic order. *J Neurosci* 9: 3776–3795

Norden JJ, Lettes A, Costello B, Lin LH, Wouters B, Bock S, Freeman JA (1991): Possible role of GAP-43 in calcium regulation/neurotransmitter release. *Ann NY Acad Sci* 627: 75–93

Nowak L, Bregestovski P, Ascher P, Herbert A, Prochianz A (1984): Magnesium gates glutamate activated channels in mouse central neurons. *Nature* 307: 462–465

Okada D, Yamagishi S, Sugiyama H (1989): Differential effects of phospholipase inhibitors in long-term potentiation in the rat hippocampal mossy fiber synapses and schaffer/commissural synapses. *Neurosci Lett* 100: 141–146

O'Leary DDM, Fawcett JW, Cowan WM (1986): Topographic targeting errors in the retinocollicular projection and their elimination by selective ganglion cell death. *J Neurosci* 6: 3692–3705

Reh TA, Constantine-Paton M (1985): Eye-specific segregation requires neural activity in three-eyed *Rana pipiens*. *J Neurosci* 5: 1132–1143

Sachs GM, Jacobson M, Caviness VS (1986): Postnatal changes in arborization patterns of murine retinocollicular axons. *J Comp Neurol* 246: 395–408

Sakaguchi DS, Murphey RK (1985): Map formation in the developing *Xenopus* retinotectal system: An examination of ganglion cell terminal arborizations. *J Neurosci* 5: 3229–3245

Sanfeliu C, Hunt A, Patel AJ (1990): Exposure to N-methyl-D-aspartate increased release of arachidonic acid in primary cultures of rat hippocampal neurones and not in astrocytes. *Brain Res* 526: 241–248

Schmidt JT (1982): The formation of retinotectal projections. *Trends Neurosci* 5: 111–116

Schmidt JT (1990): Long term potentiation and activity dependent retinotopic sharpening in the regenerating retinotectal projection of goldfish: Common sensitive period and sensitivity to NMDA blockers. *J Neurosci* 10: 233–246

Schmidt JT (1991a): Long-term potentiation during the activity-dependent sharpening of the retinotopic map in goldfish. *Ann NY Acad Sci 627: 10–25*

Schmidt JT (1991b): Kinase manipulations disrupt activity-driven retinotopic sharpening in goldfish tectum. *Neurosci Abstr* 17:215

Schmidt JT (1992): The roles of activity, competition and continued growth in the formation and stabilization of retinotectal connections in fish and frog. In: *Advances in Neural and Behavioral Development, Vol. 4,* Casagrande VA, ed. Norwood, NJ: Ableman

Schmidt JT, Buzzard M (1990): Activity-driven sharpening of the regenerating retinotectal projection: Effects of blocking or synchronizing activity on the morphology of individual optic regenerating arbors. *J Neurobiol* 21: 900–917

Schmidt JT, Edwards DL (1983): Activity sharpens the map during the regeneration of the retinotectal projection in goldfish. *Brain Res* 269: 29–39

Schmidt JT, Eisele LE (1985): Stroboscopic illumination and dark rearing block the sharpening of the regenerated retinotectal map in goldfish. *Neuroscience* 14: 535–546

Schmidt JT, Edwards DL, Steurmer CAO (1983): The reestablishment of synaptic transmission by regenerating optic axons in goldfish: Time course and effects of blocking activity by intraocular injection of tetrodotoxin. *Brain Res* 269: 15–27

Schmidt JT, Turcotte JC, Buzzard M, Tieman DG (1988): Staining of regenerated optic arbors in goldfish tectum: Progressive changes in immature arbors and a comparison of mature regenerated arbors with normal arbors. *J Comp Neurol* 269: 565–591

Schneider GE, Rava L, Sachs GM, Jhaveri S (1981): Widespread branching of retinotectal axons: Transient in normal development and anomalous in adults with neonatal lesions. *Neurosci Abstr* 7: 732

Shapira R, Silberberg SD, Ginsberg S, Rahaminoff R (1987): Activation of protein kinase C augments evoked transmitter release. *Nature* 325: 58–60

Shatz CJ (1990): Competitive interactions between retinal ganglion cells during prenatal development. *J Neurobiol* 21: 197–211

Shearman MS, Naor Z, Sekiguchi A, Kishimoto A, Nishizuka (1989): Selective activation of the gamma subspecies of protein kinase C from bovine cerebellum by arachidonic acid and its lipoxygenase metabolites. *FEBS Lett.* 243: 177–182

Simon DK, O'Leary DDM (1990): Limited topographic specificity in the targeting and branching of mammalian retinal axons. *Dev Biol* 137: 125–134

Sperry RW (1963): Chemoaffinity in the orderly growth of nerve fiber patterns and connections. *Proc Natl Acad Sci USA* 50: 703–709

Stryker MP, Harris WA (1986): Binocular impulse blockade prevents the formation of ocular dominance columns in cat visual cortex. *J Neurosci* 6: 2117–2133

Stryker MP, Strickland SL (1984): Physiological segregation of ocular dominance columns depends upon the pattern of afferent electrical activity. *Invest Ophthalmol Vis Sci (Suppl)* 25: 278

Stuermer CAO (1988a): Trajectories of regenerating retinal axons in the goldfish tectum: II. Exploratory branches and growth cones on axons at early regeneration stages. *J Comp Neurol* 267: 69–91

Stuermer CAO (1988b): Retinotopic organization of the developing retinotectal projection in the zebrafish embryo. *J Neurosci* 8: 4513–4530

Stuermer CAO (1990): Development of the retinotectal projection in zebrafish embryos under TTX-induced neural-impulse blockade. *J Neurosci* 10: 3615–3626

Stuermer CAO, Easter SS (1984): A comparison of the normal and regenerated retinotectal pathways of goldfish. *J Comp Neurol* 223: 57–76

Stuermer CAO, Raymond PA (1989): Developing retinotectal projection in larval goldfish. *J Comp Neurol* 281: 630–640

Sur M, Humphrey AL, Sherman SM (1982): Monocular deprivation affects X- and Y-cell retinogeniculate terminations in cats. *Nature* 300: 183–185

Tsumoto T, Kimura F, Nishigori A (1990): A role for NMDA receptors and Ca^{++} influx in synaptic plasticity in the developing visual cortex. In: *Excitatory Amino Acids and Neuronal Plasticity*, Ben-Ari Y, ed. New York: Plenum Press

Walter J, Henke-Fahle S, Bonhoeffer F (1987): Avoidance of posterior tectal membranes by temporal retinal axons. *Development* 101: 909–913

Wiesel TN (1982): Postnatal development of the visual cortex and the influence of environment. *Nature* 299: 583–591

Williams JH, Bliss TVP (1988): Induction but not maintenance of calcium-induced long term potentiation in dentate gyrus and area CA1 of the hippocampal slice is blocked by nordihydroguaiaretic acid. *Neurosci Lett* 88: 81–85

Williams JH, Errington ML, Lynch MA, Bliss TVP (1989): Arachidonic acid induces a long-term activity-dependent enhancement of synaptic transmission in hippocampus. *Nature* 341: 739–742

CHAPTER 15

Concerning the Logic of Growth and Form in Living Systems

JOHN CRONLY-DILLON

INTRODUCTION

This article was written in part as a tribute to Professor Michael Gaze, who perhaps more than most of his contemporaries, has thought and delved deeply into the question of pattern formation in the developing and regenerating nervous system. Among the contributions he has made that have had a major impact on our thinking of how ordered arrays of nerve fibers are assembled and maintained in the nervous system are the notions of "sliding connections" and fiber–fiber interactions (see Gaze et al., 1974; Meyer, 1979; Stuermer, 1991) where the position occupied by a growing nerve fiber in a retinotectal array depends not only on its relationship with its immediate neighbors, but also on what may be happening some distance away: a form of behavior that clearly has holistic overtones. Part of the aim of this essay is to explore some of the limitations of a reductionist approach to problems of pattern formation in biological systems.

The essence of reductionism, whether in the physical sciences or in biology, is that it deals with systems in which certain characteristics of the building blocks remain invariant throughout the range of operation of the system. This article questions the uncritical application of this notion in biology and points to a number of new developments in mathematics such as fractals and far-from-equilibrium thermodynamics that may provide scientists with a new set of mathematical tools and physical paradigms with which to formulate and represent the global features of organization and function of the nervous system.

MATHEMATICS AS THE LANGUAGE OF SCIENCE

The last 40 years have brought major changes to our understanding of the structure, development, and function of the nervous system. Despite these achievements, some problems, such as pattern formation in the developing embryo or the problem of conscious experience, continue to defy a proper understanding in conventional neurophysiological terms. As we approach the end of the millennium, it is perhaps appropriate not only that we reflect upon past

Formation and Regeneration of Nerve Connections
Sansar C. Sharma and James W. Fawcett, Editors
© 1993 Birkhäuser Boston

achievements, but also that we reevaluate some of the central concepts that lie at the core of current thinking in neuroscience. In this respect it may also be helpful to contrast developments in the neurosciences with those that occurred in the physical sciences, where following the publication in 1687 of Newton's. *Mathematical Principles of Natural Philosophy* our understanding and ability to represent the diverse aspects of the physical world were transformed by mathematics. Indeed, Newton's invention of the differential calculus (which generally treats "change" as smooth and regular) provided scientists with an appropriate vehicle with which to represent the dynamic behavior of a seemingly all-embracing body of natural phenomena. On the other hand, this has not happened to anything like the same degree in the biological sciences, and it is therefore appropriate that we ask why this is so.

Until recently the prevailing outlook of Western science was the fragmentation of reality into a number of building blocks. Linear differential equations allowed scientists to analyze a great body of natural phenomena where the change in dynamic behavior is smooth and linear, and where the behavior of the whole could be represented as a sum of the behavior of its parts. On the other hand, certain nonlinear phenomena such as turbulent flow in gases or liquids and the many-body problem, where the behavior of the "parts" was not easily separated from the "whole," defied a "general" solution in these terms. Today we recognize that some branches of nonlinear mathematics may provide a "language" that is appropriate for representing many systems that behave "globally." In this regard it is worth recalling that during the classical period, Greek mathematics dealt mainly with rational numbers, even though the Greeks were well aware of the existence of some irrational numbers such as pi and the square root of two. Since then we have discovered that most numbers along an interval are irrational. Indeed, rationals are very rare. Likewise there is an increasing suspicion among certain groups of contemporary scientists that while many phenomena in the natural world behave linearly, the vast majority may be nonlinear and that some influences which affect their dynamic behavior may be nonlocal.

SELF-ORGANIZATION IN COMPLEX SYSTEMS

Returning to the field of biology, we note that in his remarkable book *On Growth and Form*, d'Arcy Thompson (1917) set out to try to outline an explanation of biological phenomena in terms of physics and to systematize the great diversity of biological forms through the medium of mathematics. However, the extent to which he achieved this aim was limited by the state of physical knowledge and the mathematical tools available to him at that time. Even today, much of the physics and mathematics we employ to represent the bulk of biological phenomena dates from the nineteenth century. Until relatively recently most of the developments in the physical sciences focused on the behavior of linear systems and on systems close to equilibrium. Indeed, it is only in the last few years that there has been a resurgence of interest in the behavior and mathematics of complex nonlinear dynamic systems. Some of these new developments are particularly relevant in that they may provide biologists with an appropriate set of physical paradigms and levels of description with which to represent global behavior and self-organization in biological systems. A brief survey of some of these developments and how they may affect biological thinking is given below.

A number of distinguished physicists (Davies, 1987; Penrose, 1989b) have commented on the possible relevance of the holistic qualities of quantum mechanics to self-organization in biological systems. A particularly illuminating example of global behavior in a physical system that is relevant to biology is the way proteins fold to achieve a state of energy minimum. There are indeed a multitude of conformations whose energies are very nearly the same, and if the protein had to explore all the likely possibilities it would take a very long time before it attained the right configuration. Instead, the protein appears to sense the needed final form to achieve the energy minimum (the most stable state), and widely separated portions of the protein appear to move in unison to attain this state according to some (nonlocal) global schedule (Davies, 1987). Another example discussed recently by Penrose (1989a, 1989b) concerns the growth of quasi-crystals (crystals with fivefold symmetry). We are all familiar with tiling patterns that have fourfold, or hexagonal, symmetry, but until Penrose's relatively recent study on the mathematics of tiling it was thought impossible to tile the plane using pentagons. Equally, a crystal with fivefold symmetry was thought to be an impossibility until metallurgists encountered just such a crystal in an alloy of aluminum, lithium, and copper. The symmetry in such crystals differs from that in the more familiar variety in a characteristic way, and so such crystals have been called "quasi-crystals." Penrose has discussed the peculiar symmetrical properties of these structures and notes that a quasi-crystal, unlike classical crystal growth, does not grow by individual atoms attaching themselves at a continuously moving growth line. Instead, he suggests that we must consider an evolving linear superposition of many different arrangements of attaching atoms. He has argued that the problem of quasi-crystal growth may only be solved in some global fashion and that "it is not just one thing that happens; many alternative atomic arrangements must co-exist in complex linear superposition. . . . it must be a co-operation between a large number of atoms that is achieved quantum mechanically by trying different combined arrangements of atoms simultaneously in linear superposition." The problem of how proteins fold and the growth of quasi-crystals may equally apply to some aspects of the assembly of the cell cytoskeleton. Here too certain aspects of the process may be determined nonlocally (in a manner akin to that of protein folding and quasi-crystal growth).

Another field of research that has yielded dynamic effects where the behavior seems to require the intervention of global processes has emerged from the study of thermodynamical systems that are far from equilibrium. Interest in the self-organizing properties of such systems has been stimulated by the pioneering work of Professor Ilya Prigogine and his co-workers (Prigogine, 1980; Nicolis and Prigogine, 1977; see also Nicolis, 1990). He showed that such systems sometimes display collective behavior that may not always be comprehensible in terms of their constituent parts. Of particular interest is the observation that in some chemical systems whose state is far from thermodynamic equilibrium, new emergent properties may arise from collective behavior which appear to derive from some nonlocal influences. Moreover, their collective behavior is such as to suggest something akin to a change in phase. An interesting illustration of this effect is the phenomenon of Benard convection. If a horizontal layer of fluid is heated uniformly from below, a critical temperature is reached after which the liquid organizes itself into distinctive rolls or cells with a hexagonal structure. Thus an initially homogeneous state gives way to a spatial pattern with distinctive long-range (nonlocal) order. The convection cells, which Prigogine refers to as "dissipative structures," are a mechanism for heat transport. A new level of molecular order appears that basically corresponds to a giant fluctuation, stabilized by the exchange of energy with the outside world. What seems to emerge from this discussion is that under certain conditions, some systems that are far from equilibrium may self-assemble or organize themselves in some global fashion to generate higher-order "building blocks." Prigogine notes that far-from-equilibrium chemical systems that include catalytic mechanisms (such as those present in living cells) may lead to dissipative structures (see also Nicolis and Prigogine, 1977). These structures are very sensitive to global features such as the size and form of the system, the boundary conditions imposed on its surface, and so forth. Indeed, the occurrence of dissipative structures generally requires that the system's size exceed some critical value, a complex function of the parameters describing the reaction–diffusion process. Therefore we may say that such chemical instabilities involve long-range order through which the system acts as a whole. All these features influence in a decisive way the type of instabilities that lead to dissipative structures that are relatively stable and display a surprising degree of permanence. However, in some cases, macroscopic fluctuations may lead to new types of instabilities and lead to other dissipative structures and the generation of new stable forms.

INTERACTIVE BEHAVIOR IN THE FORMATION OF PATTERNED NEURAL NETWORKS

The pattern of connectivity within brains is bewilderingly complex. Some interconnections are structured in ordered fashion, while others are random. In perhaps the majority of instances, the output of a given neuron depends in a nonlinear way on the combined input it receives from its connected partners and so may depend in a complex way on what is happening elsewhere in the system. It is therefore no surprise that a system with such a high degree of nonlinearity and feedback should display self-organizing properties and evolve collective modes of behavior that lead to the establishment of holistic patterns in the activity of the brain. The lability of individual synapses is useful from the point of view of brain plasticity, but equally it presents us with something of a problem when it comes to explaining the physiological basis of stable memory. The notion of the Hebbian synapse attributes memory to a change in connectivity between, say, neurons A and B that are driven repeatedly to fire almost contiguously, the firing of A just preceding or in synergy with that of B. This repeated sequence is supposed to strengthen the functional connections between A and B by making synapses more efficient. As stated, the Hebbian model suggests the change that occurs in learning is localized to individual synapses, yet we now recognize that many of these are labile dynamic structures that seem to retract and reform at different loci on the neuron (Purves et al., 1987). Despite this lability, the effects of learning endure, which suggests that what is preserved during learning is a learning-induced change in excitability within a pattern of synapses, whose precise location on a neuron may shift with time. To preserve the overall distribution of synaptic contacts on the cell and their relative excitability requires a global process that preserves the form or spatial relationship of the contacts, despite the relocation of individual synapses. This task depends on the dynamic activity of the neuronal cytoskeleton and genome, which cooperate in a holistic fashion to determine the shape, structure, and function of the cell. Despite its limitations, the notion of the Hebbian synapse has been enormously successful in accounting for many of the functionally driven changes in connectivity that lead to the establishment of orderly synaptic relations in development.

In the frog the retina projects and is represented topographically over the surface of the tectum, and this topography is preserved throughout the course of development from tadpole to frog. Yet the growth of the tectum and retina is dissimilar (Gaze, 1970; Gaze et al., 1974; Stuermer, 1991). The retina grows concentrically by adding annuli of new cells to the retinal periphery, while the tectum grows by the addition of "wedges of cells" at the mediocaudal boundary of the tectum. Given this dissimilar growth, the only way to allow spatial vision to continue throughout development is for the retinotectal connections to shift in a coordinated manner which preserves the orderly representation of retinal topography in the projection onto the tectum. In the frog, part of this orderliness is determined genetically, but a great deal of the refinement, of the projection occurs as a result of visual experience (Cline and Constantine-Paton, 1991). In the cat and monkey, the dependence on visual experience for the development of a topographic representation in the visual cortex is even more striking. Here the development of an orderly representation of the visual field during the visual critical period is very largely dependent on normal visual experience (Blakemore and Molnar, 1990).

In both cases the application of the Hebbian model is able in large measure to account for the refinement of connectivity that leads to a topographically organized projection. Some early studies on the development of the ipsilateral binocular projection in the frog (Keating, 1977; Meyer, 1982) suggest that ganglion cells located in corresponding points in both eyes will tend to fire synergistically in response to the image of an object lying in their corresponding receptive fields (RFs). If the terminal arbors of these retinal ganglion cells (RGCs) overlap in the tectum, the synergy of firing of these terminals causes some underlying neurons to also fire. In accordance with Hebb's hypothesis, this will strengthen the connections which these terminals make with the active cell(s) (Keating, 1977).

In the case of the refinement of the frog retinotectal projection (or the development of the cortical visual field representation in cat and monkey), the explanation seems to depend on the fact that RGCs closely adjacent have overlapping RFs and will tend to fire synergistically in response to a stimulus located in a region of the visual field whose image lies in the overlapping area of their RFs. As in the previous example, terminals that are firing synergistically will tend to converge (by strengthening their connections) onto common or closely adjacent central neurons. Indeed, the behavior of some "artificial learning nets" currently being investigated by computer engineers is quite capable of accounting for how the topography of the retina may be reconstructed in the cortex from a random pattern of terminal connections. In Hopfield's (1983) model, for example, the various possible states of the network can be represented by a bumpy surface in space, with the current state corresponding to an imaginary ball rolling on the surface. The ball will tend to roll down into the valleys or basins of attraction seeking out local minima. Without going into detail, the favored valley states are those which strongly connected neurons tend to fire together. If we assume some minimal principle whereby fibers which "on average" have closely similar firing patterns will tend to stay together, then it is not difficult to show how, with repeated and varied pattern visual stimulation, the system will self-organize to recreate the topography of the retina. An additional feature of interest in the Hopfield model is that information is stored holistically; it is the collective pattern of activity (of synergistically firing groups of terminals) and their interactions throughout the net that represents the information needed to recreate the retinal topography. Equally, one may perhaps think of an ensemble of interactions which collectively organizes synergistically firing groups into a "dissipative structure" that recreates retinal topography, in a manner reminiscent of a convection cell. Thus, at the level both of the individual neuron and of a patterned assembly of neurons, what is preserved is the overall form and not necessarily the individual connections. In each case this seems to be achieved by a dynamic global process.

ARE BRAIN GROWTH AND BEHAVIOR GOVERNED BY A PRINCIPLE OF "LEAST ACTION"?

Some of the most intriguing and remarkable ideas to emerge from classical mathematical physics were the minimal principles of least time and least action that were developed in connection with studies of optics, electrodynamics, and mechanics. The ideas traces back to Greek antiquity when Hero of Alexandria noted that when a ray of light issuing from a point A reaches a point B after being reflected by a mirror, it invariably follows the shortest path. The statement is not strictly accurate, in that the path is not necessarily the shortest (although this is true for a plane mirror). It may also be the longest (as is the case if the mirror is concave). Thus the characteristic of an actual path is that it is an extremum: the path can be a minimum or a maximum. These ideas were further developed by Fermat for wave optics in the principle of least time. Fermat stated:

A ray of light proceeding from a point A and reaching a point B, after any number of reflections and refractions will always lie along the path for which the time of transit is a relative extremum (usually a minimum).

For the corpuscular theory of light Maupertius developed a corresponding (minimal) principle of least action which was extended to mechanics by Lagrange and Euler. Lagrange in fact formulated a more accurate statement of Maupertius's principle because he recognized that the action was not always a minimum but may alternatively be a maximum. However, in *Essai de Cosmologie* Mauperitus wrote:

Here then is this principle, so wise, so worthy of the Supreme Being: Whenever any change takes place in Nature, the amount of action expended in this change is always the smallest possible.

Others, such as Hamilton (principle of stationary action) and Gauss (principle of least constraint) later introduced laws that were more general and convenient than that of Maupertius.

All the minimal principles have a characteristic in common. All state that in certain classes of natural phenomena, the process of change is such that some appropriate physical magnitude will be an extremum (usually a minimum).

In the last analysis, the minimal principles illustrate equivalent ways of expressing the same dynamical equations. Indeed the dynamical equations of relativity may be given the form of Maupertius's and of Hamilton's principles. Einstein's gravitational equations, Maxwell's electromagnetic equations, and Schrodinger's wave equation may likewise be expressed by stationary or extremum principles. Equally, the behavior of Hopfield's neural net model (which mimics some of the features of brain organization) may likewise be represented as operating according to a minimal principle.

COMPLEXITY AND SELF-ORGANIZATION IN NEURAL NETWORKS

Similar considerations apply to the formation of an orderly retinotopic central representation from a randomly distributed array of visual terminals. The orderly retinotopic representation is one which on average minimizes the differences in temporal firing patterns between adjacent points over the distribution of terminals. The problem is, How do the terminals attain and settle into their topographical distribution so quickly? This situation has certain features which recall the problem of protein folding. These remarks do not underestimate the importance of cell specific markers and adhesion molecules in the determination of tissue form. However, no single theory to date is able to account for the remarkable plasticity displayed in the regeneration, of the visual pathway in fish and amphibia. Sharma (1975), for example, showed in one set of experiments that a severed optic nerve in goldfish will reconstitute a normal retinotopic projection (with normal polarity) even when the tectum is rotated 180°. Other experiments also showed that the goldfish optic nerve may form a normal projection on a nonvisual area (cerebellum) after bilateral tectal ablation. Such results are difficult to explain on a strictly chemospecific basis. In contrast, if only a few regenerating axons enter the tectum, they tend to connect haphazardly. Thus, to establish an orderly retinotopic projection requires (a) a critical level of complexity, i.e., it is necessary for a minimum number of regenerating fibers to interact to achieve self-organization, and (b) that these terminals fire in some coherent but nonuniform pattern (as is provided by normal visual experience). Under these conditions the central terminals and connections self-organize to match the retinal topography.

As with Benard convection, a certain level of complexity is required before self-organization will occur. In his book *Theory of Self-Reproducing Automata* John Von Neumann shows that self-reproduction can only occur when the device exceeds a certain threshold of complication. This would seem to indicate that a system may take on new properties when it exceeds a certain threshold level of complexity.

THE CELL AS A NONLINEAR DYNAMIC SYSTEM

Today the picture that emerges from biochemical and physiological investigations of the neuron is that of a dynamic entity whose form and function depend on a complex spatiotemporal ordering of the chemical reactions taking place within the cell interior. Indeed, it is now apparent that many enzymes in the cytosol are physically clustered pathway by pathway and attached to the cytoskeleton to permit a more efficient channeling of intermediates within the pathway. Thus, despite the continuous turnover of cellular components and minor perturbations to the chemical environment due to extraneous influences, a great many, perhaps even the majority, of chemical reactions taking place within the cell appear to be spatially and temporally organized in a manner that sustains a stable organization and function.

Modern cell biology has also revealed that autocatalytic oscillating reactions are a common feature of biological reactions and play an important role in many cellular functions (Hess and Markus, 1987; Crutchfield et al., 1986). Certain biochemical systems, such as glycolysis and peroxidase reactions, may under certain conditions display oscillations with ever-changing and unpredictable frequencies: behavior with the characteristics of (deterministic) chaos. Such systems may reveal many different varieties of oscillation, including periodic, quasi-periodic, chaotic, and multiple oscillatory states. In phase space, periodic changes are represented by closed orbits. For quasi-periodic oscillations the trajectories are not closed but fill a surface. Chaotic oscillations are represented by a domain that is self-similar at any scale of magnification (a fractal). The periodic orbits, the quasi-periodic surfaces, and the chaotic domains are examples of attractor states, such that the reaction system is attracted to these phase space regions when the process is initiated somewhere else.

Viewed as a dynamic system, the cell's interacting autocatalytic processes behave as an "attractor" and a "dissipative" structure that determines the cell's form and function. One of the interesting characteristics of an attractor is that for periodic orbits, it constrains the system to resist small perturbations, restoring the latter to its original stable state. Also, if the perturbation exceeds some critical value, then because some of the dynamic processes taking place within the cell may be inherently nonlinear, the system may in some instances be displaced to a totally different equilibrium state (Hess and Markus, 1987). This may be reflected in an altered morphological and functional organization for the cell. The notion that a neuron's morphological configuration is an "attractor" is consistent with observations made on the phenomenon of compensatory sprouting, in which cells tend to maintain the total number of their terminal arbors. If some of the branches are cut, others sprout to take their place until the original number of branches is restored (Schneider, 1981). Thus the system tends to resist small disturbances that displace it away from a stable attractor. The dynamic character of the living cell or neuron which enables it to maintain its form is revealed in the observation that many synaptic junctions are not permanent structures but tend to move around or be relocated from day to day at different sites on the neuron, yet the neuron retains its original functional characteristics. What is preserved is the global organization of the synaptic input and not necessarily individual contacts. Indeed, the cell's components, including those of plasma and synaptic membranes, are continually turning over. Recent experiments by Purves et al. (1987) showed that individual synapses that are established with a neuron are continually changing. Although the overall pattern may endure, individual synapses disappear and are either relocated or replaced by others, sometimes at a different location, yet the overall functional characteristics of the neuron continue to be preserved.

Similarly, with regard to growth regulation in neuronal networks, Rotshenker (1988) has argued that nerve cells are in a state of dynamic equilibrium in relation to axonal growth. Among others, he has drawn attention to the existence of transneuronal mechanisms which are signals for growth that are communicated between nerve cells in the central nervous system (CNS). The following provide a few illustrations of transneuronal effects:

1. A temporary interruption of axonal transport in a motor nerve of the frog (insufficient to block nerve activity or synaptic transmission) was followed by sprouting and synapse formation by the intact nerve innervating the homologous muscle on the opposite site of the body.
2. A growing neuron may die unless it is able to establish and maintain a certain number of terminal contacts. Incidentally, all of these effects involve an altered equilibrium state for the cell that is stabilized by a corresponding change in cytoskeletal organization.
3. Developing dopaminergic neuron of the olfactory bulb in the rat do not express tyrosine hydroxylase until after they have migrated and received synaptic contact.

BEHAVIOR OF NONLINEAR DYNAMIC SYSTEMS AND FRACTAL GEOMETRY

The recent interest in the behavior of nonlinear dynamic systems (deterministic chaos) and developments in mathematics, particularly fractal geometry, which are suited to represent such behavior, open up new perspectives for the representation and understanding of global behavior in biology. There is now accumulating evidence that many structures in biology have a fractal geometry which may have been generated by just the kind of nonlinear oscillatory behavior described above (see Meinhardt, 1991). Examples of biological structures in biology which display fractal characteristics include the peculiar geometry of tree ferns, the branching of alveoli in the lung, and the branching of neurites in the nervous system. Equally, the geometry of the cytoskeleton, which is a determinant of the pattern of neurite branching, may also be represented as a fractal structure (see Mandelbrot, 1983; Pietgen and Richter, 1986). Within the complex networks of the brain, some interconnections are structured in ordered fashion, while other are random. In perhaps the majority of instances the output of a given neuron depends in a nonlinear way on the combined input it receives from its connected partners, with which it may also share a metabolic interdependence.

A possibly useful way to picture a cell is as a nonlinear system of coupled chemical oscillators which organize themselves dynamically into a relatively stable configuration (in phase space) that determines the cell's form and function. In the nervous system, spontaneous activity may be one of several means through which the oscillating chemical systems of one neuron are coupled to those of neighboring cells. Hence the output of a neuron, as well as its assembly into a coherent neuronal circuit, may depend in a complex way on what is happening elsewhere in the system. It is therefore no surprise that a system with such a high degree of nonlinearity and feedback should display self-organizing properties and develop collective modes of behavior.

Commenting on the relationship between contemporary neurophysiology and mental processes, Sperry (1976) has expressed a view which is consistent with the notion that mental activity associated with the collective behavior of neurons in neuronal networks have a holistic quality that resembles that of a dissipative structure. Sperry talks of "the lower level entities (neurons, etc.) becoming caught up in a holistic pattern much as a water droplet is caught up in a whirlpool and constrained to act co-operatively to the overall organized activity" (see also Freeman, 1991).

CELL SYMBIOSIS AS A PRINCIPLE UNDERLYING BRAIN ORGANIZATION AND DEVELOPMENT

The demonstration that transneuronal effects may regulate cellular growth and morphology reveals the metabolic interdependence of the cellular components of nervous tissue, not only between neurons and glia, but also between neurons assembled in a neuronal network. Until relatively recently, much of the metabolic interdependence between the neuronal elements of nervous tissue seems to have been largely overshadowed by the attention given to neuronal networks as systems specialized for the integration and communication of electrical signals. Indeed, it is not uncommon to find modern neuroscience textbooks comparing neurons with computer microchips. While this may be a useful metaphor to illustrate the integrative properties of nerve cells, such a comparison tends to dissociate the functional attributes of nerve cells from the processes which regulate their growth, differentiation, and survival. Whatever the similarities, nerve circuits are fundamentally different from their computer analogues. Unlike microchips, neurons are living reactive elements, ready in most cases to adapt to local extraneous influences to survive. Within the CNS, the metabolic interaction of a neuron with other neurons and with glia may be important both for its survival and for the development and maintenance of its morphological and functional characteristics. Such interdependence may be likened to the metabolic interdependence between the members of a symbiotic community in which cells adjust their morphological and functional properties collectively to achieve a particular equilibrium state for the whole assembly. The extent to which this occurs may depend on the strength of the interdependent relationship. This may vary in degree, ranging from some where the metabolic coupling between cells is weak, so that they are virtually independent, to others where the coupling is strong and they must behave collectively to maintain a particular equilibrium state. Also the interdependence may be subject to modulation, for example, (by learning or as an adaptive response to extraneous influences.

In a seminal monograph on symbiosis and cell evolution published some years ago, Margulis (1981) proposed a symbiotic origin for eukaryotic cells. She suggested that the eukaryotic cell is homologous to a stable community of tightly integrated microbial symbionts. She has argued that many of the organelles found in eukaryotic cells, such as mitochondria, chloroplasts, microtubules, microtubule organizing centers, and associated structures such as: kinetsomes together with their own DNA, derived from previously free-living microbes that settled to a new "equilibrium" mode of "social interdependence and interaction" by adapting to a symbiotic existence within the invaded host. Margulis set out a convincing case in support of the view that microtubules within eukaryotic cells derived from an early ancestral spirochete. In the new symbiotic lifestyle which their contemporary descendants adopted within eukaryotic cells, we find the microtubule framework defining the positioning of the main organelles, endomembranes, and interactions with other cytoskeletal elements. Margulis also views cell differentiation in animals and plants as having developed from the population dynamics and interaction of interdependent members of former bacterial communities. As in any community that grows and alters, the numbers of community members and their ratios change through time (Margulis, 1990), perhaps as a result of an interplay between endogenous and extraneous signals. Equally, symbiosis is likely to have been the evolutionary forerunner of multicellularity in living organisms. Margulis notes that the protist ancestors of the animals and plants never solved the problem, on the single-cell level, of how to retain both their motility and their ability to divide by mitosis. The cells that were motile by undulipodia and were therefore forced to relinquish mitosis were ancestral to animals and plants. They adopted multicellularity as a solution because a given cell could not solve the problem of simultaneously retaining mitotic reproduction and undulipodial motility; consequently it kept in contact with the mitotic cell from which it had originated. The swimming, dividing collective accomplished what was impossible for either cell alone.

Biologists usually employ the term "symbiosis" to refer to the close association established between two or more genetically different organisms that have adopted a "communal" lifestyle where they are metabolically interdependent, and whose metabolic and other activities complement one another in the struggle for survival. Usually the term is used to refer to communities where the members are either attached to one another or one is enclosed in the other. The term is also used in sociobiology to refer to the intimate, relatively protracted, and dependent relationship of members of one species with those of another. In the following we extend the meaning of symbiosis to refer also to the metabolic and functional interdependence of cells in a multicellular organism that develops from an embryo and in which all the cells of the organism (except for the gametes) possess the same complement of genes.

The classical and perhaps best-known example of symbiosis is provided by the lichens (Fink, 1935). Lichens are a group of plants unique in being a combination of two different groups, algae and fungi, where the shape of the lichen is usually determined by the specific combination of the alga and the fungus. Symbiosis in lichens stems from the fact that while fungi absorb water from the atmosphere, algae manufacture food by photosynthesis. In addition, there exists an exchange of complex organic compounds between them. Such association has made the lichen extremely adaptive. Within the context of evolutionary development, it is assumed that the earliest lichens were a loose association of fungal hyphae with free-living algae. Moreover, it is believed that some lichens are still on this borderline, and the evolution from a looser to a more definite structure is continually in process. Multicellularity is thus seen as having arisen from the compartmentalization of perhaps mutually exclusive functions in different cells and tissues and the establishment or strengthening of some symbiotic relationships and the breakdown of others.

A well-studied and particularly interesting illustration that is pertinent to our understanding of the processes of self-organization in embryonic development is the life cycle of the "social ameba" *Dictyostelium discoideum* (Loomis, 1975). This slime mold exists in two functional states in its life cycle: a population of unicellular, vegetative amebae and a multicellular pseudoplasmodium that develops into a fruiting body. All growth, DNA synthesis, and cell division are restricted to the vegetative phase, while morphogenesis, pattern formation, and cell differentiation occur during the multicellular phase. The life cycle of this organism starts with a haploid spore which splits on germination to liberate a single-celled ameba which feeds on bacteria. The amebae grow and divide repeatedly by binary fission, yielding thousands of independent cells. When the food supply is exhausted (or possibly as a result of the accumulation of a cell division inhibitor in the medium), aggregation begins and the amebae move toward a central point. Mobility is directional. Pseudopodia orient toward the aggregating center, which is the source of the chemotactic signal (cAMP) secreted in pulses by an ameba which acts as a "founder" cell. In addition, there is relaying and amplification of cAMP signaling by cells all along the radial paths leading to the main center. Once formed, the aggregate elongates into a cigar-shaped slug which moves off by coordinated ameboid activity of cells of the lower surface. The slug possesses an anterior-posterior polarity and responds to environmental stimuli such as light and temperature. It may migrate for a period of several hours until it reaches a territory where food (bacteria) is abundant. There migration ceases and fruiting begins. Soon after the slug forms, it contains two cell populations: anterior and posterior prestalk and prespore cells, respectively. Genetically identical cells differentiate into two distinct populations, apparently on the basis of the positions they occupy in the multicellular community. Later the slug undergoes further differentiation and transformation to form the fruiting body. Two details are worth noting:

1. In response to starvation, the amebae express cAMP receptors on their surface and signaling is mediated by the secretion of cAMP. The second event which takes place on the surface of starved preaggregative amebae is the acquisition of cohesive properties through the expression of new genes, some of which encode a series of cell adhesion molecules involved in cell aggregation and pattern formation (for further details see Alberts et al., 1989).
2. Many of the cells' autocatalytic reactions that lead to self-organization in *D. discoideum* involve oscillation-producing feedback mechanisms, mediated in some cases by membrane-bound enzymes and membrane receptors capable of sensing chemical gradients of attractants. It is this which leads to the symmetry-breaking transitions that determine cell differentiation and pattern formation in this and other developing systems (see Nicolis and Prigogine, 1977; Meinhardt, 1991).

To sum up, the response to starvation gives rise to a new level of "social" organization that involves aggregation, differentiation, and coordinated behavior. Cell differentiation is a means of assigning specific functions to certain cells to allow the system to behave collectively in a manner that enables the "social amebae" to respond flexibly to a hostile environment.

BRAIN DEVELOPMENT AND PLASTICITY VIEWED AS PROBLEMS
IN SOCIOBIOLOGY

The dynamics of embryonic development in multicellular organisms have many parallels with the development of society, where individuals of the same species are "assigned" specific functions or roles to enable the society as a whole to maintain a particular "equilibrium state." If external conditions change dramatically, the society may reorganize and, if necessary, settle in a different equilibrium condition. Indeed, the vision of society provided by Durkheim and Wheeler is that of a superorganism that evolves to greater complexity through the complementary processes of differentiation and integration. As the society becomes increasingly large (and/or food resources become relatively scarce), the "community" becomes more compartmentalized. Its members become specialized into roles and castes and their roles become more precisely defined through superior communication.

In *The Insect Societies* (1971) Edward Wilson considers the physiology of caste determination and outlines a theory of caste ergonomics that determines the mixture of castes that will minimize the energy cost to achieve a particular goal (e.g., the production of a given number of queens with a given number of workers and a given level of access to food). A similar treatment may be applied to the process of differentiation and growth in the life cycle of the social ameba. Embryonic development and cell differentiation share many of the characteristics found in the development of society. These include the generation of morphological and functional diversity among its members, compartmentalization of function, and pattern formation (for a review of these characteristics in embryonic development, see Price, 1991).

The extent of interdependence among members of a social assembly may vary in degree. At one extreme, the interdependence may be so great that survival depends on collective behavior and the coordinated performance of complementary functions by members of the different groups that are assigned particular roles. In other cases the interdependence may not be so strong or critical. These differences allow a hierarchy of interdependent relationships to exist even within a single society. The comparison with sociobiology has implications for our understanding of neuroscience that are far from trivial. For our purposes, what is important in the example of development of the social ameba is that it reveals how a labile and changing pattern of metabolic interdependence can lead to adaptive changes in tissue organization in response to environmental constraints.

Similarly, cell and tissue differentiation in embryonic development may be viewed as an adaptation to local differences in physiological constraints. For example, a diffuse assembly of a heterogeneous population of neurons in a particular brain region may differentiate into a number of functionally distinct regions in response to local metabolic demands, including those brought about by nervous activity. Within each of these brain regions, the cellular components cooperate and reorganize to establish a new equilibrium organization and, in so doing, generate a new set of metabolically interdependent relationships in response to local physiological constraints. Viewed from this perspective, the brain appears as a tissue constituted of relatively stable interconnected islands of metabolically interdependent cellular communities, embedded in a diffuse, more weakly coupled, and in some cases modifiable network of brain elements. The hypothesis also has implications for our understanding of the nature of neural plasticity. Any major locally induced metabolic change that occurs within a few brain cells (e.g., as the result of nervous activity), may precipitate a reorganization throughout the whole metabolically interdependent community of which these cells are a member. The result may be a collective reaction in which the assembly assumes a new equilibrium configuration. Furthermore, the metabolic behavior of such interdependent aggregates may resemble that of a syncytium in which the metabolic coupling/decoupling between constituent elements may be modifiable. In this respect it is perhaps interesting to note that some degree of metabolic coupling and uncoupling of cells through gap junctions is regarded as an important regulator of cell and tissue differentiation in developing tissues (Warner, 1985).

It is indeed ironic that 100 years or so after Ramon y Cajal proposed the neuron doctrine, we should return to a notion in which certain nerve networks in some respects behave in a manner reminiscent of Von Gerlach's discredited reticular theory. Undoubtedly the hypothesis that neurons are the functional building blocks of the nervous system which communicate with one another through chemical synapses has been instrumental in bringing about the remarkable advances in our understanding of the functioning of the nervous system. However, we should note that despite the successes engendered by the neuron doctrine, some distinguished neuroscientists (e.g., Sperry, 1976) have expressed some disquiet at current attempts by neurophysiologists and computer scientists to describe mental processes solely in terms of discrete neurons signaling to each other in a complex switching assembly. This unease betrays a suspicion that some of our basic assumptions about nervous system organization may be too restrictive and that the traditional picture of the nervous system, centered around the neuron doctrine, as it currently is, needs to be supplemented by a new conceptual paradigm. Moreover, the paradigm will need to take into account the fact that growth regulation, cell differentiation, and signal processing are interdependent, if it is to provide a scheme that is more suited to explain global phenomena. In this respect, a portrayal of the brain as an interacting hierarchy of self-organizing metabolically interdependent cell communities would appear to have some of the characteristics of collective behavior that may be just what is needed to account for some of the global aspects of developmental dynamics and mental processes. Until recently the physical principles underlying the dynamics of such collective behavior posed a problem for classical physics, upon which most of our current notions of physiological mechanisms are based. However, new developments in mathematics and physics may overcome this difficulty and open up a new level of description of biological phenomena.

Earlier the neuron was described as comprising a complex of autocatalytic reactions that embody a dissipative structure. These reactions define the stable properties of the cell which (for the most part) allow us to identify the cell as a single physiological entity. We have outlined a picture of nervous tissue as comprising a system of metabolically interdependent cells that display a changing pattern of "symbiotic relationships" in response to environmental and physiological demands. Such interdependence may manifest various degrees of permanence in which the cytoskeleton plays a critical role in regulating the stability of the metabolic and functional relationship between cellular components. A system whose cellular components are strongly coupled metabolically to one another may, at a certain threshold of activity, behave collectively as a dissipative structure and so operate globally as a single, higher-order physiological entity. We are thus led to consider a structure such as the brain as comprising a continuously changing population of functional building blocks that vary in size, form, complexity, and degree of permanence, each of which is capable of operating at some time as a single physiological unit. Viewed from this perspective, a neuron may operate as a functional unit of the brain only when the brain is in its "ground state." At higher levels of metabolic and functional brain activity, individual neurons and neuron assemblies may be driven beyond a critical "bifurcation threshold" when individual neurons and even neuronal assemblies get caught up in aggregates of cells that operate collectively "much as a drop of water is caught up in a whirlpool" (Sperry, 1976). Whatever the precise details, there would seem to be a need to extend and perhaps amend our current interpretation of the neuron doctrine in a manner that provides for certain groups of cells to self-organize and operate in a global fashion, i.e., collectively. Only then may we hope to find a comprehensive explanation of the function and assembly of the elaborate fabric of the brain.

REFERENCES

Alberts B, Bray D, Lewis J, Raff M, Roberts K, Watson JD (1989): *Molecular Biology of the Cell*. New York: Garland

Blakemore C, Molnar Z (1990): Factors involved in the establishment of specific interconnections between thalamus and cerebral cortex. Cold Spring Harbor Meeting on The Brain, Abstracts

Cline HT, Constantine-Paton M (1991): Synaptic rearrangements in the developing and regenerating visual system. In: *Vision and Visual Dysfunction*, Cronly-Dillon J, ed. London: Macmillan

Crutchfield JP, Farmer JD, Packard H, Shaw R (1986): Chaos. *Sci Am* 255: (6)46–57

Davies P (1987): *The Cosmic Blueprint*. London: Heinemann

Fink B (1935): *The Lichen Flora of the United States*. Ann Arbor: University of Michigan Press

Freeman J (1991): The physiology of perception. *Sci Am* 264: (2)34–41

Gaze MJ (1970): *The Formation of Nerve connections*. New York: Academic Press

Gaze RM, Keating MJ, Chung SH (1974): The evolution of the retinotectal map during development in *Xenopus*. *Proc. R. Soc. Lond. B* 185: 301–330

Hess B, Markus M (1987): Order and chaos in biochemistry. *TIBS* 12: 45–50

Hopfield (1983) *Sci Am*

Loomis WF (1975): *Dictyostelium discoidem. A Developmental System*. New York: Academic Press

Keating MJ (1977): Evidence for plasticity of intertectal neuronal connections in adult *Xenopus*. *Phil. Trans. Roy. Soc. B* 278: 277–294

Mandelbrot BB (1983): *The Fractal Geometry of Nature*. New York: W. H. Freman

Margulis L (1981): *Symbiosis in Cell Evolution*. San Francisco: W. H. Freeman

Margulis L (1990): *Origins of Sex*. New Haven: Yale University Press

Meinhardt H (1991): Theories of morphogenetic fields. In: *Vision and Visual Dysfunction*, Cronly-Dillon J, ed. Lond: MacMillan

Meyer RL (1979): 'Extra' optic fibres exclude normal fibres from tectal regions in goldfish. *J Comp Neurol* 183: 883–902

Meyer RL (1982): Tetrodotoxin blocks the formation of ocular dominance columns in goldfish. *Science* 218: 589–591

Nicolis G, Prigogine I (1977): *Self Organization in Non-Equilibrium Systems*. New York: Wiley

Nicolis G (1990): Physics of far from equilibrium systems and self organization. In *The New Physics*, Davies P, ed. Cambridge: Cambridge University Press

Penrose R (1989a): Tiling and quasi-crystals. A non-local growth problem? In: *Aperiodicity and Order 2*, Jaric M, ed. New York: Academic Press

Penrose R (1989b): *The Emperor's New Mind*. Oxford: Oxford University Press

Pietgen HO, Richter PH (1986): *The Beauty of Fractals*. Berlin: Springer-Verlag

Price J (1991): Cell lineage and compartments in vertebrate neurogenesis. In: *Vision and Visual Dysfunction, Vol. 11*, Cronly-Dillon J, ed. London: Macmillan

Prigogine I, (1980): *From Being to Becoming*. San Francisco: W. H. Freeman

Purves D, Voyvodic JT, Magrassi L, Yawo H (1987): Nerve terminal remodeling visualized in living mice by repeated examination of the same neuron. *Science* 238: 1122–1126

Rotshenker S (1988): Multiple modes and sites for the induction of axonal growth. *TINS* 11: 363–368

Schneider G (1981): Early lesions and abnormal neuronal corrections. *TINS* 4: 187–192

Sharma SC (1975): Visual projection in surgically created 'compound' tectum in adult goldfish. *Brain Res* 93: 497–501.

Stuermer CAO (1991): The formation of topographically ordered connections during development and regeneration of the vertebrate visual system. In: *Vision and Visual Dysfunction*, Cronly-Dillon J, ed. London: Macmillan

Sperry RW (1976): Mental phenomena as causal determinants in brain function. In: *Conciousness and the Brain*, Globus G, Maxwell G, Savadnick I, eds. New York: Plenum Press

Thompson d'AW (1917): *On Growth and Form*. Cambridge: Cambridge University Press

Warner AE (1985): The role of gap junctions in amphibian development. *J Embryol Exp. Mophol Suppl* 89: 365–380

Wilson EO (1971): *The Inset Societies*. Cambridge: The Belknap Press of Harvard Univeristy

CHAPTER 16

Three Glimpses of Evolution

T.J. HORDER

Have we yet recognized the full complexity of evolution? Indeed, can we ever expect to do so? The aim of this chapter is to discuss the extent to which our methods for studying evolution may limit our perceptions of it and to consider whether any approach is capable of encompassing all the issues. The chapter focuses on an aspect of method which is particularly hard to stand back from. This is the area, beyond data gathering and experimental techniques as such, of relating the diversity of available data and inferring their meaning in relation to a full understanding of evolution: what one could call the tacit rationale, since the underlying arguments and assumptions are so rarely made explicit. Three such rationales are identified and characterized and then, in a final section, assessed in the light of their contribution to a possible single, coherent approach to evolution. The vertebrate eye is used to illustrate the arguments, for reasons that will become evident.

DARWINISM

Darwin's great achievement lay, not in the idea of evolution, which in itself was not new, but in the way in which he marshalled the evidence that made the case for the evolutionary origin of biological systems so overwhelming. Darwin's arguments all go back to one central inference from the body of facts available to him, which was essentially limited to the adult structures of known species as described by comparative anatomy and palaeontology[1]: that organisms combine elements of sameness and difference and that these signify evolutionary descent with modifi-

---

1. Darwin's brief references to embryology and Haeckel's original concept of recapitulation pre-dated both Mendel and an adequate descriptive or experimental study of embryology. Haeckel, following Darwin's lead, saw the remarkable similarities of the early embryonic stages of taxonomically related animal groups as corresponding to the adult of the common ancestor. This was an entirely understandable interpretation: successive terminal additions of adult forms during phylogenesis and their sequential repetitions in subsequent embryogenesis was the only way, in the absence of any others at the time, of envisaging the mechanism of heredity.

Formation and Regeneration of Nerve Connections
Sansar C. Sharma and James W. Fawcett, Editors
© 1993 Birkhäuser Boston

cation. Clearly his reasoning accounts magnificently in general terms for the random scattering of species we happen to know about, but is it powerful enough to make possible further, detailed understanding?

The recognition of evolutionary relationships hinges on the use of the concept of homology, i.e., the identification of features of organisms which can be used to measure descent from a common ancestor, while excluding convergence (e.g., "analogous" similarities of function whose independent evolutionary origins can be recognized by underlying differences in structure). It is necessary, therefore, to isolate components common to two or more organisms that show progressive modification, rather than being merely the same or so different that they cannot be related in any way. In other words, one is concerned with entities that have to be simultaneously similar and different. Any application of such evidence is, in the end, as good as the available rules for initially isolating such characters, for measuring change in relation to measures of the original state, and for combining the various measures of change (some of which may be mutually inconsistent) to define the relationships of the organisms as a whole. In practice, homology becomes a relative, and potentially arbitrary, matter, judged on a variety of criteria such as shape, size, or position.

Recent developments in taxonomy have served to demonstrate the absence of rational rules in the use of the concept of homology. Not only have a number of quite distinct taxonomic methodologies been advocated (phenetics, cladistics, etc.), but no basis for choosing between them has emerged. In an effort to avoid arbitrariness, these methods concentrate on discrete, well-defined characters, e.g., characters defined according to overt anatomical distinctions (such as specific bones or digit numbers) and scored in a binary fashion (e.g., as present or absent, ancestral or derived). To define the evolutionary relationships of species, one generates lists of shared and unshared characters. These are then pooled and relationships are measured on the basis of the relative numbers of unshared characters.[2] The objectivity is spurious, however, since the criteria for initial choice and definition of characters are not defined. Moreover, one has to consider the possibility that, in the attempt to make explicit what has traditionally been done almost intuitively, the modern methods of establishing homologies (and in turn taxonomies) may be ignoring aspects of the evidence, such as quantitatively varying characters and possible inbuilt measures of relative antiquity of morpho-

---

2. The cardinal difference between phenetics and cladistics lies in their use of shared and unshared characters; phenetics measures relatedness from overall similarity using all available characters, whereas cladistics uses exclusive (derived) characters only. Given the lack of rationale there has long been a tendency for taxonomy to become an end in itself, separate from phylogenetic reconstruction. With the modern methods it has been seen as a purely descriptive and pragmatic exercise aiming only for the best fit of the data into maximally discrete sets. Traditional taxonomic procedures may have been better than is realized because they unrestrictedly took into account the whole range of characters and used them in a flexible and quantitative way (Horder 1991).

logical features (as hinted at in the concept of the body plan[3]), which could link species in a more systematic and less statistical way. It is not surprising that the different taxonomic methods often suggest conflicting phylogenies, since the evolutionary inferences are only the consequences of the choice, number, description, and recombination of the characters used.

Thus in general, comparative anatomy and paleontology[4], can at best give rise to descriptions of evolution in terms of a tree of species in ranked order, and even then only with uncertain precision. The absence of any underlying rationale for analyzing morphological structure is understandable, given that we are dealing with random samples of the multidimensional and intrinsically variable products of evolution[5]; certainly we should not expect to be able to infer the underlying causal mechanisms. These limitations do not matter if we are only interested in relative ranking of species, but we are likely to want also to explain the gaps between known species (and to estimate their sizes and to predict intermediate forms), the specific character of the products of evolution, and particularly the extraordinary adaptedness of organisms. Darwin did, of course, provide a "mechanism" to explain the whole of evolution in the idea of natural selection. But this is too open-ended and nonspecific to provide any restrictive or predictive rules for relating morphologies.[6] Darwin (1859) summed up the crucial issues incomparably as follows:

---

3. The concept of body plan refers to an idealized or averaged set of morphological features shared characteristically by a broad group of related species: it is the defining feature of the larger scale taxa used in classification. It is an abstraction (no one member of the group may conform exactly to it and membership will be arbitrary in marginal cases) but it identifies more ancient characters retained from the common ancestor.

4. Absolute dating of evolutionary time is in itself not important, given that evolutionary change (as shown in stasis, extinctions, and punctuated equilibria) occurs at uneven rates. The geological record provides absolute dating, but for many well-known reasons, fossil data have limited resolution (Raup and Jablonski, 1986).

5. Although a number of evolutionary "laws" have been proposed to describe morphological trends, none has proved useful for taxonomy (Horder, 1991), including the problematic concepts of allometry and heterochrony (Horder, 1989). Limiting factors intrinsic to the organism have often been proposed, such as "internal" factors, Bauplan, constraint, genetic homeostasis, burden, ratchet, canalization (Scharloo, 1991), and so forth, but these are invariably nonpredictive and ill-defined (Raup and Jablonski, 1986). Fitch and Atchley (in Patterson, 1987) illustrate particularly clearly the difficulty of selecting morphological measures of evolutionary relationships, in a situation where accurate measures are available (Atchley and Fitch, 1991).

6. Selection is an entirely open-ended concept because its magnitude and direction (and therefore immediacy of evolutionary effect) are potentially unlimited, and because selective forces are in practice impossible to characterize fully (they are often nonspecifically measured as survival rates) since they depend on a reciprocal relation with the existing state of adaptedness of the specific organism as well as its whole ecological context. Distinctions between levels of selection are equally unclear if every component and property of an organism is open to selection. Despite the way in which embryos are protected from direct external selective forces, it can be argued that embryonic processes are particularly strongly selected because of the multitudinous consequences of developmental errors, even though selection of their effects may only occur at birth.

To suppose that the eye, with all its inimitable contrivances for adjusting the focus to different distances, for admitting different amounts of light, and for the correction of spherical and chromatic aberration, could have been formed by natural selection, seems, I freely confess, absurd in the highest possible degree.

Conclusion

Even today, morphological evidence can yield evolutionary inferences of only limited precision (i.e., in ranking species taxonomically) and little explanatory power (i.e., in predicting intermediates between known species). But the lack of methodological rigor in the interpretation of morphological data does not mean that we should question the very possibility of reconstructing phylogenetic paths, as has sometimes been implied. The various taxonomic methods have basic features in common and the broad phylogenetic inferences on which they could still agree are too successful in fitting and explaining the facts to be seriously doubted.

But all this is quite enough to explain why the origin of the vertebrate eye is an unsolved problem and presents evolutionists with something of a test case. The earliest vertebrates (agnathans as represented by the lamprey) have eyes indistinguishable in form and complexity from those of all later species. The few supposed ancestral chordates (like amphioxus) have nothing that can be recognized as remotely similar. The evidence and methods available (Walls, 1942) seem to leave us no alternatives: all we can say is that the eye, like Athena, appeared "full-grown and fully-armed from the brow of Zeus."

THE MODERN SYNTHESIS

It is easy to see why we have increasingly come to regard genetics as providing the missing explanations for the problems with evolution left over by Darwin. Classical genetics provided an all-embracing mechanism of both descent (in Mendelian transmission genetics) and modification (through mutation and genetic recombination). Species can now be defined in terms of gene pools and, together with speciation, assayed by the methods of population genetics. The Modern Evolutionary Synthesis of the 1930s and 1940s could plausibly argue a comprehensive and unified genetic basis for all evolutionary phenomena. Our confidence in these concepts has been massively boosted by knowing as much as we now do about their basis in DNA as the sole route of inheritance and as the precondition for the entire phenotype. However, by closing the door to Lamarckian inheritance, this theory makes the problem of adaptation all the more imponderable. Whereas Darwin could partly get round the problem by assuming direct molding by the external environment, we must now explain how discrete genetic units, inherited from previous generations and common to each cell in the organism, can anticipate and coordinate the adaptedness of the prospective, highly differentiated, adult state. It must be said that classical genetics, like the concept of selection, provides only a generalized form of explanation. What is missing is the detail concerning specific instances which is necessary to understand the patterns of actual evolutionary progression.

However, with the new era which has emerged as the result of the availability of molecular sequencing data, it could well be claimed that molecules now provide the most definitive possible way of describing evolution in detail and also the ideal characters for taxonomic purposes. When compared across species, lens proteins and opsins—it makes little difference whether DNA or amino acid sequences are considered—show a clocklike rate of change through evolutionary time and so offer an objective, quantifiable measure of degrees of homology. Rates of change differ widely for different proteins, but any would serve equally well if the only requirement is to arrange the given species in evolutionary rank order. However, this ideal is undermined by the fact that specific molecules may also vary their rates of change through time (Scherer, 1990). Increasingly it is possible to trace the ancestry of molecules back to progenitors known to have markedly different functions. For example, opsins are descended from second-messenger-activating proteins (Goldsmith, 1990), and one lens protein is a modified aldehyde reductase, as confirmed by its continued enzymatic activity (Harding, 1991). The crucial issue now becomes that of understanding how the varying rates of change of different molecules and their varying functions are coordinated through time, since it is a particular and appropriate combination of specific molecules that defines the species or the eye.

The above described variations in rates of change of molecular structure during evolution can, of course, be explained as due to variations in selection pressures both among molecules and at different times. But can we account any more precisely for the evolutionary forces driving particular molecular changes? Selection pressures are, obviously, determined by the state of adaptation of the organism[6]. Available methods are inadequate when it comes to defining the relationship between selection, adaptation and genetic change. Classic examples (such as industrial melanism in moths or selective breeding for wing patterns in *Drosophila* (Berry, 1977; Scharloo, 1991)) show that selection pressures and resulting genetic changes can in practice only be partially characterized; the latter normally involves multiple genes and has multiple, apparently incidental knock-on effects throughout the organism. The relationships can also be looked at by studying the effects of single mutations. The much confirmed finding is that any given mutation typically has diverse effects throughout the organism: single structures are polygenically determined (Clayton, 1985; Horder, 1989). Thus classical genetic methods point to the complexity and indirectness of the genetic programming of adapted adult structures. The more the final complexity of the adult is attributed directly to genes the more improbable becomes the number and specificity of genes required to explain it. To apply molecular methods to such a situation may well be inappropriate because such methods are simply unsuited to addressing the essential issues, such as control of gene expression and the integration of patterns of control through the collectivity of genes. The complexity of the interactions of the products of different DNA sequences is such that even a complete description of the genome may be very remote indeed from a description of the final nature of the organism! Knowing the steps in the evolutionary history of a lens protein from aldehyde reductase tells us a small part of the story and nothing about where it is expressed, let alone the conditions that led up to the evolution of the control of where it is expressed and how this was integrated with evolutionary changes throughout the rest of the genome.

Like morphological data, the molecular evidence is primarily structural in nature, and many of the same difficulties arise in defining homology at this level, e.g., selection of appropriate characters for comparison, choice of methods of comparison and—particularly given variable evolutionary rates of change—how to recombine the resulting lists of data and handle inconsistencies. The evolutionary process puts a premium on variation, and at the molecular level large-scale minor variation is easily achieved. It will manifest as selectively "neutral," random noise that will tend to obscure other, more progressive changes. This and a variety of other mechanisms (e.g., gene duplications leading to gene families, mutations, reverse mutations, deletions, transpositions, transfection, symbiotic associations, multiple and nonfunctional copies (like pseudogenes)) add new complexities to tracing homologies.[7] Unexpectedly, perhaps, it is beginning to appear that for the purpose of describing how particular combinations of characters are constrained during phylogeny, or of simply ranking species, molecular data may not necessarily lead to improvements in precision over morphological data.

Conclusion

The molecular data serve to highlight the limitations always inherent in the Modern Synthesis, which arise from the impossibility of achieving a complete causal account of any specific selection process or state of adaptedness in terms of genetic change.[8] Any such synthesis can be no more than a set of general and largely descriptive propositions because of the complexity of the interactive, causal chain intervening between genetic states and adult adaptedness, as well as the unknowable contingencies of past selective evolutionary forces. Even descriptively and taxonomically, any advantages the molecular data might

---

7. For a discussion of problems in defining homology at the molecular level, see Patterson (in Patterson, 1987) and Lalley and McKusick (1985). Atchley and Fitch (1991) show how dependent molecular taxonomy is on the molecules chosen and how valid measures of evolutionary relationships can be derived after averaging the varied rates of change of different molecules.

8. The Modern Synthesis has above all been characterized by its concentration on speciation and population genetics, and on the dynamics of gene pools rather than the specific character of the genes concerned. Such studies are in practice constrained by methodology to look at one isolated factor in a manifestly complex situation (typically geographical or selective variation of single genes or specified phenotypic markers). Phylogenesis was of little interest, paleontology dubious and macroevolution a potentially conflicting mechanism.

offer in terms of definable characters and objective measures of evolutionary change may be offset by the complexities of interpreting the huge data base. Any prospect of reconstructing unknown intermediate species or the histories of structures such as the eye from molecular data seems very remote indeed. In short, the molecular view of evolution has many of the characteristics and limitations of the morphological view and may be even more complex. It has made little impression on the Darwinian problems of selection, adaptation, or phylogenetic reconstruction.

IS THERE A THIRD APPROACH?

There remains a third source of evidence, and that is embryonic development. Figure 16.1 shows representative stages of the development of a typical vertebrate eye. Two features are particularly striking in that they could hardly have been anticipated or inferred from other sources of evidence. They are counterintuitive and demand explanation. First, there is the indirectness of the process: the origin of the eye from the neural tube tissue that will give rise to the central nervous system (CNS), the series of large-scale morphogenetic[9] displacements and rearrangements of the prospective eye tissues, the quite separate origin of the lens from the ectoderm, and so on. Second, there is the fact that a virtually identical set of events is seen in all vertebrates. Only at later stages do species-specific specializations emerge. Clearly, all embryological events must be the result of evolution and ultimately explicable by it. Granted that evolution would not be expected to create unnecessary complexity and that these transitory diversions during embryogenesis do not directly contribute to the selective advantage of the adult, an understanding of how developmental events relate to evolution seems likely to be a matter of some importance. The answer, put as simply as the issues will allow, is as follows.

The argument goes back to a feature of the evolutionary process which is so basic that its implications are easily overlooked: biological complexity must be acquired cumulatively. Given the interdependence of the components of organisms, new characters must be acquired on the basis, not only of their specific adaptive advantage, but also of their compatibility with the functioning of the entire set of components. As organisms become more complex, this consideration will mean that new acquisitions tend increasingly to be limited in degree. Furthermore, some previously acquired components will tend to become increasingly stable. For example, once DNA had become the mechanism for genetic coding, the point would be reached at which it, and the many structures and processes interdependently related to it (such as RNA, ribosomes, control of replication), would become an absolute precondition for any further evolution.

---

9. Morphogenesis is defined and its key role in embryonic pattern formation discussed in Horder (1989, 1991).

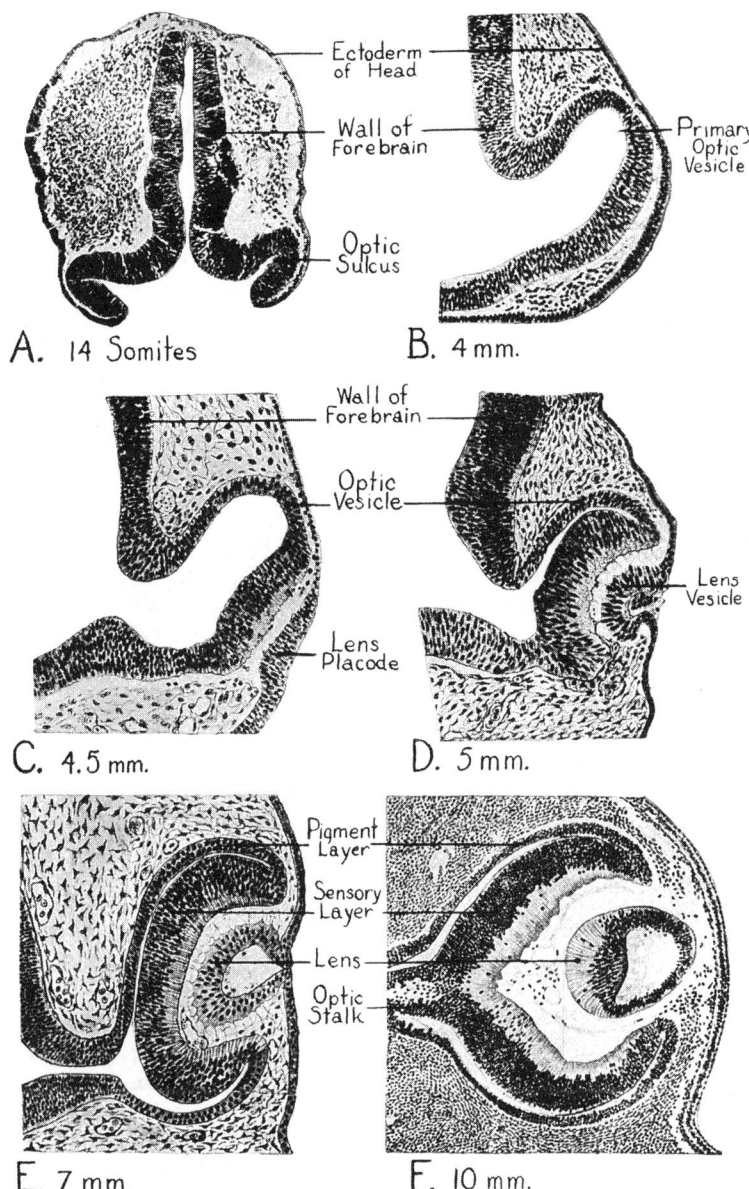

FIGURE 16.1. Early development of the human eye, showing stages in the continuous folding of the neural epithelium leading to the differentiation of the neural retina (sensory layer). Similarly the lens forms in continuity from the surface ectoderm. Choroid, sclera, and extraocular muscles differentiate in the surrounding mesenchyme later. The optic stalk becomes the optic nerve. In all vertebrates, including the earliest with eyes, these embryological stages are essentially identical. From Patten (1946).

One consequence is that organisms will contain characters of varying antiquities, and this is amply borne out in the varying stabilities of different types of molecules.

Once unicellular organisms had appeared, evolution faced a problem of special interest. Regardless of the inevitably obscure origins of the first metazoans, it is clear that for many fundamental reasons, the advantages of variation associated with sexual reproduction among them, the lines with potential for significant evolutionary advance retained the requirement for a periodic return to a single-cell (egg) stage. Metazoan evolution therefore could only advance by using mechanisms compatible with this starting point, i.e., the means of deriving groups of cells each expressing distinctive structural genes in a coordinated spatial pattern. The events and mechanisms of embryogenesis must mediate this function—the pattern-forming role of embryogenesis is reinforced by the fact that early embryonic cells are initially totipotent[10]—and must themselves have evolved through sufficiently small modifying steps that they were compatible with the survival of each intermediate organism in the evolutionary path.

The morphogenetic events in the embryogenesis of the eye clearly represent a continuous series of causally related effects. Given the structural continuity of epithelia, simple inspection is enough to show that each stage is consequential on the preceding stage, but the causality can easily be verified experimentally. If the neural tube is removed before eye outgrowth, for example, no eye is formed. These visible events directly demonstrate how the morphogenetic process leads to the positioning of the final, integrated pattern of differentiated cells making up the eye. Experimental evidence amply proves the dependence of differentiation on preceding events. For example, initially multipotent eye vesicle cells[10] become designated as neural retinal cells depending on the positions they arrive at during morphogenesis; approach of the eye cup induces the lens to form; and so on.[11] A causal chain such as this shows how embryological mechanisms solved the requirements of cumulative evolution; evolution of patterns of differentiation under the causal control of morphogenesis provides a way in which specialized

---

10. It is beyond dispute that, in vertebrates, early embryonic cells are totipotential, i.e. unrestricted regarding possible future specializations. The indirectness of embryogenesis is further evidence against pre-formed pattern in the egg (Horder, 1989). In some invertebrates egg cytoplasmic factors restrict cells fates, but only to a limited degree. The vertebrate condition is presumably the more advanced: free of maternal line contraints, it will allow greater evolutionary flexibility of adult pattern under the full control of the zygotic genome.

11. Throughout vertebrate embryogenesis, cell differentiation is always secondary to morphogenetic and mitotic events, which themselves orchestrate the changing ("inductive") contexts of cells that are the actual stimuli for specific differentiation. A great deal is known about the elaborate cascade of causal interactions between various tissue components of the eye (Coulombre, 1965), particularly lens induction (Jacobson and Sater, 1988) which is in part dependent on the eyecup. Derivation of the lens from the ectoderm may be secondary; the lens may initially have evolved from a common origin with the eye vesicle itself (Horder, 1986), as suggested by lens regeneration from the iris and retina, and development of a lens from the pineal optic vesicle.

structure can evolve through a series of nearly continuous steps.[12] The causal chain can only have been built up sequentially during evolution; the specializations of the final, differentiated eye could not have been selected for in ancestors unless they were expressed and at each stage their expression required already established morphogenetic and inductive preconditions. Any other interpretation leads to much less plausible implications. Evolution could hardly have created an eye independently of morphogenesis and then, secondarily, transferred entirely to the existing morphogenetic mechanisms. On this reasoning then, the morphogenetic sequence must parallel the evolutionary sequence. Early stages must reflect stabilized forms of an ancestral condition and later stages the gradually superimposed additions and variations acquired later in evolution. We can infer that the eye evolved, through a cumulative process of displacement and modified differentiation, from an initial origin as part of the neural tube, as shown in Figure 16.2.

This hypothesis can be tested against a considerable body of hard evidence. It is supported by a large assortment of facts, which themselves now gain a coherent explanation. Thus, for example, the neural retina itself can readily be seen as a variant of central nervous tissue, a tissue which had already evolved in chordates before the appearance of the eye. Morphogenetic movements are a prelude to organogenesis throughout the embryo, and variations in neural development throughout the CNS are associated with early modifications of shape of the neural tube similar to the eye. Visual receptors derived from cilia are common in invertebrate classes, and may even exist near the neural canal of amphioxus. Many non neural tissues contributing to eye structure correspond to tissues associated with the CNS in general and often pre-date the eye, so that, alongside the optic outgrowth itself, they could easily attain their pattern in the eye by displacement and minor variations in differentiation. For example, the sclera is a variant of the meninges, the extraocular muscles are variants of the segmental skeletal muscles of the head, and so on. These examples[13] are enough to illustrate how all aspects of eye structure can be accounted for as the result of a coordinated, continuous

---

12. Morphogenesis moves uncommitted cells into new, inductive contexts in the embryo, so controlling the pattern of resulting differentiation. Evolutionary changes in morphogenesis will allow changes in induction and differentiation, but only in the context of the continuity of morphogenesis. Eye evolution is unlikely to have involved sudden jumps because of the continuity of epithelial morphogenesis and its interlocking with many surrounding tissues. This "continuity principle" (Horder, 1983) applies to all adult structures since all cells are derived and localized in continuity—via mitosis and epithelial or individual cell movement—from the egg. In structures developed from diffuse, non-epithelial embryonic tissues (mesoderm, neural crest) apparently sudden evolutionary changes between discrete patterns (such as digit number) can be derived from continuously changing preconditions (Horder, 1989).

13. Further examples of supporting evidence are reviewed in Horder (1983, 1986). The origin of visual function in the ancestral equivalent of the thalamic region is supported by the original visual role of the pineal, the early developmental expression of crystallins (Clayton, 1985), and the experimental induction of retinal tissue within it (Sacerdote, 1971).

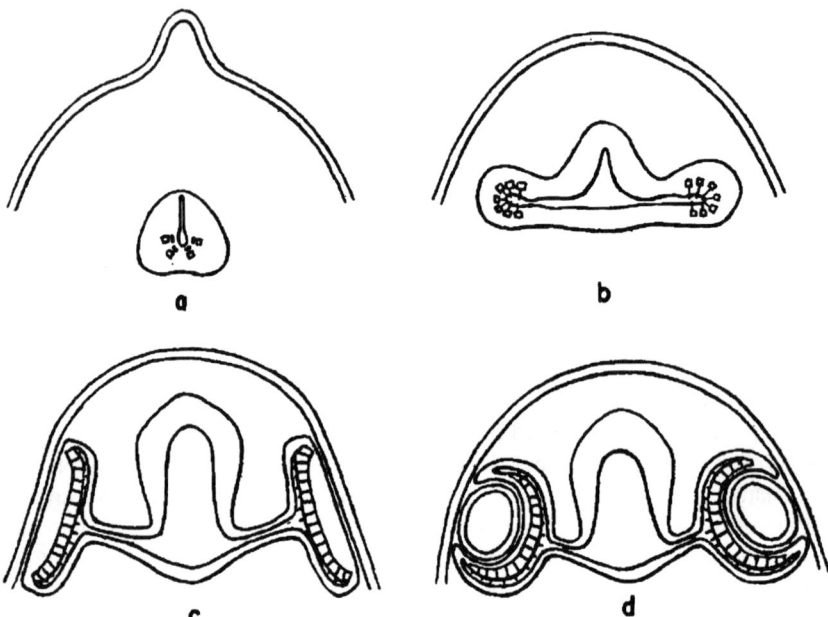

FIGURE 16.2. Four adult stages in the evolution of the vertebrate eye as originally proposed by Balfour and modified by Boveri. The scheme makes possible a potentially continuously gradation of intermediate steps, throughout which light receptors would have been placed at increasingly advantageous positions. At stage (a) light receptive cilia are present near the central cavity of the spinal cord. In stages (b–d) parts of the neural epithelium containing the visual receptors become everted and finally cupped. This scheme explains how rods (shown as protuberances into the cavity) come to be backward pointing in the retina and how a path is automatically provided (in the optic stalk) for optic fibers to reach the brain when they later grow out from the retinal ganglion cells. Adapted from Walls (1942).

evolutionary process involving minimal modification of preexisting states at each step. This principle applies equally to molecules. Lens proteins may originate as a group of molecules with initially diverse and widely expressed (Clayton, 1985) functions, needing in common the minimal changes required to become concentrated in lens cells with potential to change optical properties as a result: "The proteins we know and love as crystallins are hand-me-downs well used in some earlier role" (Harding and Crabbe, 1984).

The foregoing argument will, no doubt, bring to mind the veteran notion of recapitulation, a concept that is rightly regarded with considerable suspicion.[1] If the notion is taken to mean that the adult structure of ancestors is repeated during embryogenesis, this is easily discounted. The early stages of embryos are typically undifferentiated, so they cannot correspond to any adult state. Second, everything we know about evolution tells us that it occurs by modification, substitution, or deletion of ancestral characters. This must be true of embryonic phenomena, not just final adult specialization, because modified adult forms are only the result of their respective patterns of development. This implies that even the embryo stages of ancestors cannot be repeated in descendants, and, on the face of it, there is much against even this version of recapitulation; e.g. the earliest stages of development are conspicuous for their evolutionary variability (as in egg structure or in early adaptations to food supplies such as placentation or larval stages). However, the body plan common to all vertebrates bears witness to the conservatism of the somewhat later developmental events responsible for the initial laying out of overall morphological organization, such as gastrulation and neurulation. Indeed, the close similarities of embryos during these stages are striking.[14] Morphogenesis must have been among the earliest mechanisms to evolve in metazoans, an inevitable corollary of the origin of multicellularity from a single cell; it is ubiquitous in embryos (all animal embryos undergo a version of gastrulation, for example) and a general pre-condition for differentiation.[11] Once such mechanisms had evolved, any future change in adult structure could only be achieved by modification of morphogenesis. Because of the continuity of the causal chains involved, the repetition of embryological sequences and structures as they once occurred in ancestors is inevitable.

What we are seeing in eye morphogenesis is only recapitulation in a strictly limited sense.[15] The sequence of events in Figure 16.1 does not repeat ancestral adult stages as shown in Figure 16.2: such a claim would commit the Haeckelian error. It is ancestral embryological sequences that tend to be repeated in evolution. The remarkable similarities of embryogenesis among vertebrates is direct confirmation of this fact. But how closely and for how long in evolution any particular developmental events will repeat ancestral embryology will vary, depending particularly on how essential a given developmental sequence is to fundamental, body plan structural organization (Horder, 1991). It is the causal chain which has to be stabilized and repeated in successive generations, not a literal or complete repetition of ancestral embryonic structures or phenomena: the morphogenetic and

---

14. It has often been noted that embryos of related species resemble each other most closely midway through development, particularly at the phase of organogenesis. This is readily explained, since the effects of the earliest and maximally stabilized morphogenesis, e.g., gastrulation and neurulation, would then be most evident. Relatively greater variation at earlier stages would not be inconsistent with this stability, provided the earlier variation occurs (and evolves) in such a way as to leave the later morphogenesis largely unaffected: gastrulation is the effective starting point of all pattern formation.

15. Any attempt to depict evolution, pictorially or otherwise, confronts a complexity that can only be tackled by means of considerable idealization. Representation of a species cannot realistically include all individual variations among members of the species. Even more difficult to depict would be forms intermediate between known species. It is inherently improbable that known species are direct lineal relatives, and so true ancestral (and hence intermediate) forms have to be inferred by extrapolating from shared features of descendants.

differentiative cell mechanisms can vary in detail, provided they leave causal chains intact.[14] Thus, even at the embryological level, repetition of ancestral events is a relative matter; repetitions are increasingly modified as part of changes in the embryo as a whole or eventually lost through the evolutionary processes of substitution and deletion.

Conclusion

Thus, when understood in the light of broad features of the evolutionary process, the embryological evidence provides an appealingly simple solution to the problem of eye evolution.[16] Although the solution is inferential, it is derived from the interlocking of innumerable facts on the basis of rational criteria for establishing the relationships between the facts. The indirectness of embryogenesis is now entirely understandable, as are the evolutionary stabilities of early developmental events; these can be regarded as ancestral mechanisms and structures that have become invariant in descendants in the same way as the genetic code or the cell.

THE LIMITS TO OUR KNOWLEDGE OF EVOLUTION

Clearly we cannot directly observe or fully reconstruct evolution, and the extent to which we can modify or examine it experimentally is minuscule. This chapter has been concerned with the question of how close we can get to it by inference. In order to get as close as possible, our tacit rationale must match, in a sense, the nature and complexity of evolution itself.

The three rationales discussed above reflect the three different types of data on which they are primarily based, and their limitations reflect the restrictions of view that follow from any selective use of data. Of the three ways of looking at evolution, the morphological and molecular approaches have usually been considered independently. The two types of data are in many respects merely parallel descriptions of the same biological entities and it seems likely that, when molecular and morphological sets of characters are respectively pooled, as they have to be to produce a representative characterization of a given species, they will give much the same (and limited) taxonomic depictions of phylogeny. Embryology has up to now played at best a confused role, and more often than not has been ignored altogether.[17] But to regard the three approaches as alternatives at all is

---

16. The best discussion of eye evolution is still in Walls (1942). See also Horder (1983, 1986).
17. Many factors have conspired to cause a notable neglect of embryology from the time of the Modern Synthesis. These factors include its association with organicism (and even vitalism) and the implied incompatibility with reductionism; the reaction against Haeckelian recapitulation; the failure of classical experimental embryology to explain induction,

misleading. All three are ways of describing one organism: all three intergrade with each other[18] and must, in principle, be interrelated in strictly deterministic ways. The embryological description must contain much the same set of information as the others. Like the genetic account, embryogenesis concerns causal mechanisms and so, unlike adult data which represent the static end-results of those causes, it is explanatory of organisms. It corresponds, in effect, to a summary of the expression of the whole genome, including aspects not available—as we have seen—from either morphological or molecular data as such, including how genes are used in time and space, their functions and control,[19] and their interactions when acting collectively. Although not fundamental in the reductionist sense, the embryological evidence can be considered central, because only at this level is one unavoidably faced with the whole, unique, interacting combination of causes.

The embryological approach provides one distinct advantage for understanding of the evolution of higher organisms. The tight constraints imposed by the causal interconnectedness and continuities of embryogenesis imply that evolutionary progression can only occur through a series of near continuously changing adult morphologies.[12] For the same reason, it is less and less likely that any unsuspected mechanisms—genetic or otherwise—could mediate "saltatory" steps during evolution. The morphogenetic continuity of embryogenesis allows specific

---

e.g., in terms of a single stimulus (Jacobson and Sater, 1988) or biochemically; the absence of agreed general "models" or theories of embryogenesis (most of those proposed imply a one-to-one relation between the genome and the adult which cuts out the actual complexities of embryogenesis). On the latter point, the basic developmental phenomena, including the issues discussed in this chapter, are fully characterized experimentally and are not affected by contending theories (Horder, 1989).

18. The indirectness of embryogenesis shows how complex must be the relation between genes, molecular expression, adult anatomy, and final adaptations, and how dependent an understanding of all of these—particularly the complexity of the effects of mutations (Clayton, 1985; Horder, 1989)—must be on an understanding of embryology. Embryology is the appropriate and only practical level at which to understand fundamentally important cell and tissue interactions, such as morphogenesis, induction, and the responses of cells to functional stimuli (see note 19), even if these are ultimately explained by genes. These interactions determine the rules whereby genes in one cell are switched by genes active in others. Hence the control of morphological patterns of cell differentiation by the identical genetic information common to all cells; pattern can implicitly be programmed in a single integrated genome, but only in a form that can be read out by the complexities of embryogenesis. The genetic or molecular analysis of embryogenesis is inevitably so complex that one can question its usefulness. No one gene or molecule can be identified as responsible for a developmental switch in isolation, since all switches require both stimulus and receptivity. Since these involve interactions between whole, functioning cells or groups of cells, neither can easily be fully characterized.

19. The organization of tissue structure through direct response of developing cells to functional demands and interactions (Horder, 1983; Bock and Widdows, 1990) is a much neglected aspect of development. Not only do such mechanisms provide an immediate explanation for the fine tuning of structure to adaptational requirements, but they also minimize the number of genes needed to account for the final, detailed morphology.

hypotheses to be made about likely intermediate evolutionary stages between the morphologies of known species. In contrast to the limited information derivable from the molecular structure of lens proteins, the evidence on the morphogenesis of the eye leads directly to evolutionary inferences, around which all other types of evidence can be put into context.

The usefulness of embryological inferences is limited according to their closeness of fit to the available data. As we have seen in the example of the eye, there is a reciprocal relationship between the hard evidence (molecular, morphological and embryological) from extant species and embryological hypotheses linking them into an evolutionary sequence. The directness of the relation between embryological organization and adult patterns of differentiation varies.[12] In order to arrive at the best fit of data from all sources, the evidence must, moreover, be used in a flexible, iterative and recursive manner (Horder, 1991). How convincing such inferences are will depend on the size of the gaps between known species, but as the amount of data used is maximized—going beyond the eye to include all aspects of the anatomy of a species—the range of possible evolutionary sequences will be progressively narrowed. Despite the limitations, the embryologically-based approach can potentially transcend the particularities of specific species and a bridge can be made between the relatively direct recapitulatory evidence from embryology and the indirect evidence derived from comparative anatomy and molecular sequencing (Nei and Koehn, 1986).[20] It is very difficult, in fact, to see how evidence from these other levels can be interpreted evolutionarily or fitted together at all, without taking an embryological approach.

Darwin's worries about "organs of extreme perfection" are no longer quite so disturbing. With the conceptual framework described here, we are now well placed to explain how complex, adaptive organization can be arrived at by minute, successive evolutionary steps. Much can be explained by the genetic programming of cells to respond directly to functional requirements.[19] The embryological evidence provides opportunity for the optimization of the use of the whole body of available data. In place of the arbitrary, decompositional and binary procedures currently used in establishing homologies and phylogenies, embryology may make possible a rational choice of taxonomic characters (e.g. their systematic ranking in relation to the ancestral body plan and degrees of deviation from it) and may facilitate the flexible use of the totality of characters, including quantitative variations. It seems likely that the resolving power of taxonomy can thereby be improved, and extended to include quantitative measures of evolutionary distance between species, and increasingly reliable inferences about the nature of intermediates and of underlying common ancestors.

---

20. In contrast to embryologically-based causal chains, genetic or molecular-level sequences of events cannot be assumed to reflect evolutionary sequences, as the probable origin of RNA prior to DNA shows (Alberts, 1986).

Conclusion

It appears to me that we could now claim, with considerable confidence, to know the origin of the vertebrate eye.[21] This confidence is based on the nature of the rationale. Even though the conclusion is inferential, the underlying arguments stem from the most essential features of the evolutionary process. Most importantly, the arguments provide a means of relating the available data. The inference is supported at every step by a massive, internally consistent, interlocking network of observational evidence. This includes, ultimately, the whole imposing edifice of our biological knowledge, from molecular biology (e.g., the nature of genetic information and its consequences for the control of morphological differentiation) to the evidence in comparative anatomy and taxonomic grades for ancestral body plans reflecting evolutionarily stable, basic embryological mechanisms. Given its opportunistic as well as complex nature, we should not expect universal, provable laws in biology. Internal consistency is the most we can hope for, and its comprehensiveness is the most powerful argument of all for inferences about evolution.

Such confidence is in no way inconsistent with the fact that our knowledge of evolution is in the end bound to be limited in precision and completeness, regardless of any improvements that might come with the increasingly full and rational use of all the evidence as described above. These limits are inevitable—for reasons that this chapter has attempted to make plain—given the restricted nature of the available data, not to speak of the complexity of the unimaginably vast scale of the actual events. The one element which it is most in our power to change is the one of which we are probably least aware, the limit set by our own conceptual methods.

REFERENCES

Alberts BM (1986): The function of the hereditary materials: Biological catalyses reflect the cell's evolutionary history. *Am Zool* 26: 781–796

Atchley WR, Fitch WM (1991): Gene trees and the origin of inbred strains of mice. *Science* 254: 554–556

Berry RJ (1977): *Inheritance and Natural History*. London: Collins

Bock G, Widdows K (1990): *Myopia and the Control of Eye Growth*. Chicester: Wiley

Clayton RM (1985): Developmental genetics of the lens. In: *The Ocular Lens: Structure, Function and Pathology*, Maisel H, ed. New York: Dekker

Coulombre AJ (1965): The eye. In: *Organogenesis*, DeHaan RL, Ursprung H, eds. New York: Holt, Rinehart & Winston

---

21. The eye was obvious as the example of choice for this chapter, not only because of the challenges that it has long presented to evolutionists and embryologists, or because it surely represents a uniquely complex and precise assemblage of cell types, but also as a fitting recognition of my debt to Mike Gaze.

Darwin C (1859): *The Origin of Species by Means of Natural Selection*. London: Murray

Goldsmith TH (1990): Optimization, constraint, and history in the evolution of eyes. *Q Rev Biol* 65: 281–322

Harding JJ (1991): *Cataract: Biochemistry, Epidemiology and Pharmacology*. London: Chapman and Hall

Harding JJ, Crabbe MJC (1984): The lens: Development, proteins, metabolism and cataract. In: *The Eye, Vol 1b*, 3rd Edition, Davson H, ed. Orlando, FL; Academic Press

Horder TJ (1983): Embryological bases of evolution. In: *Development and Evolution*, Goodwin BC, Holder N, Wylie CC, eds. Cambridge: Cambridge University Press

Horder TJ (1986): Recapitulation reconsidered. *Plzen Lek Sborn Suppl* 51: 9–30

Horder TJ (1989): Syllabus for an embryological synthesis. In: *Complex Organismal Functions: Integration and Evolution in Vertebrates*, Wake DB, Roth G, eds. Chichester: Wiley

Horder TJ (1991): Molecular biology and evolution: Two perspectives. A review of concepts. In: *Developmental Patterning of the Vertebrate Limb*, Hinchliffe JR, Hurle JM, Summerbell D, eds. New York: Plenum

Jacobson AG, Sater AK (1988): Features of embryonic induction. *Development* 104: 341–359

Lalley PA, McKusick VA (1985): Report of the committee on comparative mapping. *Cytogenetics Cell Genet.* 40: 536–566

Nei M, Koehn RK (1986): *Evolution of Genes and Proteins*. Sunderland, MA: Sinauer

Patten BM (1946): *Human Embryology*. London: Churchill

Patterson C (1987): *Molecules and Morphology in Evolution: Conflict or Compromise?* Cambridge: Cambridge University Press

Raup DM, Jablonski D (1986): *Patterns and Processes in the History of Life*. Berlin: Springer

Sacerdote M (1971): Differentiation of ectopic retinal structures in the hypothalamic-hypophysial area in the adult crested newt bearing a permanent hypothalamic lesion. *Z Anat Entwickl Gesch* 134: 49–60

Scharloo W (1991): Canalization: Genetic and developmental aspects. *Ann Rev Ecol System* 22: 65–93

Scherer S (1990): The protein molecular clock: Time for a reevaluation. *Evol Biol* 24: 83–106

Walls LW (1942): *The Vertebrate Eye and Its Adaptive Radiation*. Bloomfield Hills: Cranbrook Press

CHAPTER 17

Models for the Formation of Ordered Retinotectal Connections

DAVID WILLSHAW

MEMORIES

I first met Mike Gaze in 1969 at Edinburgh. As a fledgling Ph.D. student, I was overawed by this eminent scientist, with educated haircut and notable bow-tie. To me it seemed that in the hands of Mike Gaze, the retinotectal mapping problem offered a set of beautifully unambiguous experimental results leading towards the possibility of constructing clean theories. My proper introduction to Mike's work came during a short collaboration with Martin Prestige, and I studied his papers avidly during my subsequent stay in Germany. I made several vain attempts to persuade him to take me on, before joining his group in 1977. Mike was always solicitous for my welfare, being keenly aware of the problems of persuading the MRC to employ someone who did not quite fit into the stereotype of an MRC scientist. Somehow he succeeded, and I worked in his group first at Mill Hill and then in Edinburgh, until 1988. "The leaven in the bread," I think he called me.

Anybody who has worked for Mike for any length of time will know that behind that stoically empirical exterior is a much less empirical view of life than he might admit. In this chapter I review a set of attempts to construct formal models for the formation of nerve connections in cases where Mike has been involved directly or where his influence has been felt and appreciated. The theoretical work that I describe has resulted in new candidate mechanisms for the establishment of ordered, retinotopic maps and of "striped" double retinotectal projections. This enterprise started in the 1970s, and is based on Mike's work on reevaluating the dogma of neuronal specificity.

THEORIES OF NEIGHBOR-MATCHING—RETINOTOPIC PROJECTIONS

One of the central themes that Mike explored in his book *The Formation of Nerve Connections* (Gaze, 1970) was the relevance of Sperry's hypothesis of neuronal specificity (Sperry, 1963) to the formation of retinotectal maps. There was at the time strong evidence that both retinal and tectal cells acquire immutable specificities that enable specific retinotectal connections to be made (Attardi and Sperry,

Formation and Regeneration of Nerve Connections
Sansar C. Sharma and James W. Fawcett, Editors
© 1993 Birkhäuser Boston

1963). However, there was also evidence to the contrary, from experiments involving surgical reduction of the optic tectum (Gaze and Sharma, 1970) and in particular from the projections made by surgically constructed *Xenopus* compound eyes (Gaze et al., 1963, 1965). Each half of a compound eye that is made out of two matching halves (such as two nasal halves) makes an ordered map over the entire tectum rather than to the "proper" half of the tectum, as would be predicted by neuronal specificity. The various possible interpretations of this result were examined by Mike and co-workers, and they describe the carefully crafted experiments carried out to test them.

All this led to the sowing of a certain amount of doubt that neuronal specificity was in fact the answer (Gaze, 1970).

My arguments have all been based on the assumption that the setting up and regeneration of ordered retinotectal connections is based on the mechanisms of neuronal specificity, which would require that each ganglion cell, during development, acquires a unique chemical label. Even though this may be so, it remains but one out of several possibilities. In particular, it would be worthwhile to investigate the applicability of self-ordering mechanisms in neural development and regeneration.

That retinotectal map-making may involve something other than the linking together of cells according to a fixed relationship between presynaptic and postsynaptic sheets was encapsulated by Gaze and Keating (1972) in their term "systems matching": that the ordered projection of retina onto tectum is always maintained, with the scale of the mapping being adjustable so that the entire retina projects to the entire tectum.

The question then was how this "systems matching" is brought about. Martin Prestige and I explored theoretically the idea of a competitive mechanism acting on fixed sets of labels and found it of limited application (Prestige and Willshaw, 1975). In 1974, Shin-Ho Chung (who was working in Mike's group at the time) spelled out a more focused proposal. In a paper in *Cell* (Chung, 1974), he suggested that the establishment of retinotectal connections could be controlled

by two sets of rules: one rigidly determining polarity *(of the maps)* . . . and the other stipulating that the spatial relationships of retinal elements be preserved on the tectal array. It has been suggested by J. Y. Lettvin (personal communication) that electrical activities in the optic nerve may be utilised by the nervous system in maintaining spatial contiguity between fibres. Two fibres arising from two adjoining retinal ganglion cells are likely to exhibit a temporal contiguity in their discharge patterns.

Models Employing Neural Activity

At roughly the same time, Christoph von der Malsburg and I were working along similar lines in Göttingen. We combined the ideas in Christoph's self-organizing model for the development of orientation-specific cells in visual cortex (von der Malsburg, 1973) with my interest in the retinotectal problem. The result was the neural activity model (Willshaw and von der Malsburg, 1976) for the formation of ordered nerve connections. This can be seen as a direct implementation of the Chung-Lettvin idea (although it was developed independently). Our model comprises a self-organizing mechanism that uses correlated neural activity for the local ordering of fibers on the tectum and a polarity mechanism for arranging the map of retinal fibers in the correct orientation.

The arrow model, published back-to-back with our paper, embodies the neighbor-matching idea developed by Mike and his colleagues (Hope et al., 1976). This employed a slightly different mechanism from ours. For each pair of fibers that find themselves at neighboring positions on the tectum, information about whether they are in the correct relative orientation in the map is used to decide whether they should switch positions or remain where they are. This is a neighborhood model, but not one where neighborhood information is clearly separated from polarity information. How this information about relative position of fibers in the map is made available was not discussed.

I cannot speak for the arrow model, but our activity model was not well received. There was at the time little evidence for correlated activity, and there seemed to be a general unwillingness to speculate. A more specific objection suggested that both proposals would have only a very short shelf life. Both models treated the tectum as a structure that carried only enough information to specify the polarity of the map, rather than the individual parts of the tectum carrying unique labels. This view seemed to be contradicted by the inference from tectal graft experiments that in some cases the reinnervating fibers can "sense" the area of tectum that they innervate, even when it is not in its usual position (Sharma and Gaze, 1971; Jacobson and Levine, 1975b). The possibility that the arrow model might have been disproved already by the experimental data was discussed, with characteristic evenhandedness, in the original paper (Hope et al., 1976). It was noted that, according to the arrow model, the result of rotating part of the tectum would be a map in which the part corresponding to the graft was also rotated; however, the maps resulting from interchanging two pieces of graft without rotation would remain normal. Mike and his colleagues didn't themselves take the arrow model further, although others did: principally Overton and Arbib in their Branch Arrow model, which combines notions of chemospecificity with the arrow mechanism (Overton and Arbib, 1982).

Marker Induction

The reaction of Christoph von der Malsburg and myself to this criticism was to search for a different implementation of the same type of neighborhood mechanism, this time through the action of a set of *markers* that were induced from the retina and labeled the tectum, thus making it possible for the tectum to carry some record of the projection. For explanatory purposes, we developed an analogy with a mythical commercial application involving the import of tea from India to Britain—hence the name "Tea Trade" model (von der Malsburg and Willshaw, 1977). The idea is that the neighborhood relations within the presynaptic sheet (retina) are expressed in terms of chemical markers, a set of markers being assigned to each presynaptic element. These markers are induced by the optic nerve fibers into the postsynaptic sheet (tectum), where they are used to guide the fibers to their places of termination.

It is assumed that (i) there exist sources of chemical markers fixed at certain positions in the presynaptic sheet and (ii) there are no such sources in the postsynaptic sheet. Due to processes of diffusion, eventually each presynaptic cell attains a unique collection of markers, which encodes the position of the cell relative to its neighbors. The presynaptic sheet then develops connections with the postsynaptic sheet. Initially these connections are arranged almost entirely at random. Through each synaptic contact, the collection of markers characteristic of the presynaptic cell of origin is induced into the postsynaptic sheet, where they diffuse as before. An iterative procedure operates to change the strength of each connection according to the instantaneous similarity between the collection of markers that the fiber carries and the collection of markers existing in the postsynaptic cell. Weak connections are abolished, while strong connections sprout to produce further connections on neighboring postsynaptic cells. To prevent synaptic strengths from growing without bounds, a presynaptic sum rule operates: the sum of the strengths of all synapses made by each presynaptic cell is normalized to a constant value.

The essence of this model is that a fixed set of markers is induced from retina to tectum. The difference between marker induction and other proposals can be illustrated in the case of the reinnervation of fish optic tectum following removal of half the retina to produce an ordered and expanded projection over the entire tectum. The alternative hypothesis of *regulation* (Meyer and Sperry, 1973) is that in this case the half-set of retinal markers would regenerate into a whole set to allow an expanded half-retinal projection on the tectum. According to marker induction, the retinal markers would not change. The axons from the regenerating half-retina would be guided by the existing tectal markers to their original sites and would then expand to cover the entire tectum, as the markers being induced through the newly reinnervating retinal fibers gradually came to swamp the markers that were initially present on the tectum. In the tectal graft experiments, the original set of markers are not overwritten so easily due to their abnormal disposition on the tectum. Thus translocated maps are obtained (although normal maps can also result, in concordance with other experimental findings; Jacobson and Levine, 1975a). Computer simulations with this model produced neighborhood-preserving mappings successfully and provided an account of various biological experiments investigating retinotectal projections in lower vertebrates (Willshaw and von der Malsburg, 1979). Mathematical analysis of the properties of the topographically ordered projections produced by the Tea Trade model is described by Häussler and von der Malsburg (1983).

FORMATION OF STRIPED PROJECTIONS

The other phenomenon that lends itself to the construction of formal models is the development of the striped projections formed on binocularly innervated cortex and on optic tectum in frogs and fishes. What is the mechanism for stripe formation? What determines the width of the stripes? As is well known, such projections occur naturally in mammalian visual cortex (Levay et al., 1975). However, they can be induced to form on the optic tecta of lower vertebrates, which offers greater opportunities for investigation (Constantine-Paton & Law, 1978). Hints of striped patterns were seen in adult goldfish when one tectum was removed to induce binocular innervation of the remaining tectum (Levine and Jacobson, 1975). The success of Constantine-Paton and Law (1978) in showing that doubly innervated tecta of three-eyed frogs are striped led James Fawcett and myself, working in Mike's lab at Mill Hill, to investigate whether the double projections made by compound eyes in *Xenopus* are also striped. As assessed by electrophysiological means, the projection on the tectum of each half of a compound eye is retinotopically ordered (Gaze et al., 1965). Using the horseradish peroxidase (HRP) technique for whole-mount preparations, as developed in Mike's lab by Vicky Stirling, made it possible to demonstrate that compound eye projections are also striped (Fawcett and Willshaw, 1982).

Although we were able to rule out mechanical segregation as the cause of stripes, we never followed up other possible causes conclusively. In an exhaustive series of experiments with Charles Straznicky, Mike had shown that the double projections made by compound eyes consisting of two halves of the same handedness exhibit point symmetry rather than mirror symmetry (Straznicky and Gaze, 1980). The obvious question to ask was whether these projections are also striped, which would rule out left–right differences between the two halves of the compound eye as a possible cause of stripes. Our preliminary results suggested that stripes were formed. There were also the double projections made by "isogenic" *Xenopus* compound eyes that were made by removal of a large portion of the eye rudiment at very early developmental stages. Mike had hinted in his book (Gaze, 1970) at the occurrence of such double projections (although he wondered whether batches of *Xenopus* had been mixed up). Joan Feldman and Neil MacDonald, working in Mike's lab, had further investigated this type of projection (Feldman and Gaze, 1975). Fawcett and I found evidence of stripes. However, all this is a matter of history since Ide et al. (1983) did the crucial experiment when they demonstrated that such isogenic compound eyes did form stripes.

As far as I am aware, positive evidence has yet to be obtained about (i) the mode and time course for the development of stripes in compound eyes and (ii) the role of neural activity.

Von der Malsburg was the first to produce a formal model for stripe formation (von der Malsburg and Willshaw, 1976). He adapted his earlier model for the formation of orientation-specific cells (von der Malsburg, 1973) to show how short-range correlated activity within one eye and anticorrelated activity between eyes could lead to the segregation of the projections made by two eyes on a target structure, to form a pattern of stripes. Swindale (1980) developed this idea into a more general model couched in terms of the positive and negative effects on the growth of one synapse by another. He showed mathematically that stripe width was determined by the range of the local correlations between the cells in one structure. He demonstrated how monocular deprivation would lead to a narrowing of stripe width, and suggested that the orientation of stripes has its origin in anisotropic growth of the tectum.

CURRENT PERSPECTIVES

Neural Activity Revisited

Although in 1976 the notion that the formation of nerve connections could involve a mechanism of correlated neural activity seemed somewhat absurd to neuroscientists, today the same idea sits much more comfortably in the neuroscientist's box of concepts. This may be because there is now more abundant evidence available for the existence and the effects of correlated neural activity (Arnett, 1978; Cook, 1991). Experimental work on the effects of neural activity is reviewed elsewhere in this volume and I do not discuss it here. From the theoretical perspective, recently there has been a revival of interest in how far the segregation of afferents in striped projections (or ocular dominance columns) formed on binocularly innervated optic tectum (or visual cortex) is influenced by activity. Notably, Miller and colleagues have devised a more complete and biologically realistic model than has been produced hitherto, with a more elaborate set of simulations (Miller et al., 1989), which can be viewed as a development of both von der Malsburg's and Swindale's ideas. The properties of this model have been subject to much analysis: in particular, the conditions under which target cells become either monocular or binocular have been established (Miller, 1990; Dayan and Goodhill, 1992).

Retinotopy and Stripe Formation

Producing computational models for the production of retinotopic maps or striped projections is now not such an exciting prospect as it once was. The current challenge is to consider whether a single mechanism is responsible for the formation of retinotopy and of stripes and, if so, what it is. The existence of striped projections in three-eyed frogs and in *Xenopus* with a compound eye, where such projections do not occur naturally, does suggest that the formation of stripes is an artifact of the system being used under atypical conditions, and therefore does not have a special mechanism reserved for it. Most existing models of retinopy cannot be extended to the formation of stripes without making unjustified *ad hoc* assumptions (von der Malsburg, 1979); existing models of stripe formation assume a preexisting retinotopy (Swindale, 1980; Miller et al., 1989). In my opinion, Mike's faithful workhorse, the compound-eye projection, which displays good stripes and retinopy, offers the best experimental model for investigating both phenemona together.

An intriguing possible mechanism is based on an algorithm called the elastic net (Durbin and Willshaw, 1987) for solving problems in combinatorial optimization, which was developed from a neural map-making algorithm. Shorn of its many ramifications, the basic problem in neural map-making is to discover the mechanism that connects the nerve cells in one structure to the cells in a second structure to produce the ordered pattern of connections (*map of the one onto the other*) observed experimentally. In most idealizations of the neurobiological mapping problem, both structures are regarded as two-dimensional. One could also think of constructing maps between structures that are not two-dimensional or between structures of different dimensionality. The Traveling Salesman Problem (Lawler et al., 1986), the problem of constructing the shortest circuit around a given set of points ("cities"), is an example of such a mapping. Like neural mappings, neighborhood relationships must be maintained: neighboring cities must be close together on the tour to be followed by the salesman. Richard Durbin and I reasoned that any algorithm designed for the neural task could be adapted for the computational problem, and we developed the elastic net algorithm from the Tea Trade model. As a computational algorithm, the elastic net has advantages over conventional algorithms, particularly in its implementation on parallel computers. In recent work, Geoff Goodhill and I have taken the reverse step of applying this computational algorithm to the problem of the formation of stripes and retinopy in the same projection. In the case of two retinae innervating a common target surface, the requirement is that patterns of stripes be formed, with the individual part projections made by each of the retinae being retinotopic. It turns out that according to the elastic net, such striped and retinotopic projections are formed (Goodhill and Willshaw, 1990); yet they are formed by the Tea Trade model (from which the elastic net was derived) only with the addition of certain severe assumptions (von der Malsburg, 1979). The problem now is to find a biologically feasible manifestation of the elastic net, and work on this is in progress.

CONCLUSION

There is currently growing interest in computational neuroscience—the use of formal models for the development of hypotheses in a way that makes all assumptions explicit and mutually consistent and that permits an exact evaluation of the limitations of the hypothesis under consideration. Mike Gaze numbers among the very few neuroscientists who on the one hand have kept their feet firmly on the experimental ground but on the other hand have encouraged the development of theoretical ideas. Those of us who have struggled to raise the profile of computational neuroscience above the parapet have Mike to thank. Long before it became fashionable for those working in an experimental laboratory to be on speaking terms with those with a theoretical bent, Mike was one of the few experimental scientists who were receptive to theoretical neurobiologists. His willingness to keep an open mind as to the other possible explanations of a phenomenon rather than to accept the conventional line of thought (*the deeply engrained worship of tidy dichotomies* (Gaze, 1970)) has been enriching and encouraging. I learned from Mike the necessity of checking one's theoretical notions continually (and persistently) against the experimental data, while valuing each piece of experimental data for no more than it is worth; the reward of writing clearly, however many redraftings this might take; and finally—but fundamentally—the need to establish that the phenomenon exists before attempting to account for it.

Acknowledgment

I thank Geoff Goodhill and Peter Dayan for their very helpful comments on earlier drafts of this paper.

REFERENCES

Arnett DW (1978): Statistical dependence between neighbouring retinal ganglion cells in goldfish. *Exp Brain Res* 32: 49–53

Attardi DG, Sperry RW (1963): Preferential selection of central pathways by regenerating optic fibres. *Exp Neurol* 7: 46–64

Chung SH (1974): In search of the rules for nerve connections. *Cell* 3: 210–205

Constantine-Paton M, Law MI (1978): Eye specific termination bands in tecta of three-eyed frogs. *Science* 202: 639–641

Cook JE (1991): Correlated activity in the CNS: A role on every time scale? *Trends Neurosci* 14: 397–401

Dayan PS, Goodhill GJ (1992): Perturbing Hebbian rules. In: *Advances in Neural Information Processing Systems Vol. 4,* Moody JE, Hanson SJ, Lippman RV, eds. San Mateo, CA: Morgan Kaufman: 19–26

Durbin R, Willshaw DJ (1987): An analogue approach to the traveling salesman problem using an elastic net method. *Nature* 326: 689–691

Fawcett JW, Willshaw DJ (1982): Compound eyes project stripes on the optic tectum in *Xenopus. Nature* 296: 350–352

Feldman JD, Gaze RM (1975): The development of half-eyes in *Xenopus* tadpoles. *J Comp Neurol* 162: 13–22

Gaze RM (1970): The formation of nerve connections. London: Academic Press

Gaze RM, Keating MJ (1972): The visual system and 'neuronal specificty.' *Nature* 237: 375–378

Gaze RM, Sharma SC (1970): Axial differences in the reinnervation of the goldfish optic tectum by regenerating optic nerve fibres. *Exp Brain Res* 10: 171–181

Gaze RM, Jacobson M, Székely S (1963): The retino-tectal projection in *Xenopus* with compound eyes. *J Physiol Lond* 165: 484–499

Gaze RM, Jacobson M, Székely S (1965): On the formation of connections by compound eyes in *Xenopus. J Physiol Lond* 176: 409–417

Goodhill GJ, Willshaw DJ (1990): Application of the elastic net algorithm to the formation of ocular dominance strips. *Network* 1: 41–59

Häussler AF, von der Malsburg C (1983): Development of retinotopic projecions: An analytical treatment. *J Theoret Neurobiol* 2: 47–73

Hope RA, Hammond BJ, Gaze RM (1976): The arrow model: Retinotectal specificity and map formation in the goldfish visual system. *Proc Roy Soc Lond B* 194: 447–466

Ide CF, Fraser SE, Meyer RL (1983): Eye dominance columns formed by an isogenic double-nasal frog eye. *Science* 221: 293–295

Jacobson M, Levine RL (1975a): Plasticity in the adult frog brain: Filling the visual scotoma after excision or translocation of parts of the optic tectum. *Brain Res* 88: 339–345

Jacobson M, Levine RL (1957b): Stability of implanted duplicate tectal positional markers serving as targets for optic axons in adult frogs. *Brain Res* 92: 468–471

Lawler EL, Lenstra JK, Rinnooy Kan AHG, Schmoys DB (1986): *The Traveling Salesman Problem*. Chichester: Wiley

Le Vay S, Hubel DH, Wiesel TN (1975): The pattern of ocular dominance columns in macaque visual cortex revealed by a reduced silver stains. *J Comp Neurol* 159: 559–576

Levine RL, Jacobson M (1974): Deployment of optic nerve fibres is determined by positional markers in the frog's tectum. *Exp Neurol* 43: 527–538

Levine RL, Jacobson M (1975): Discontinuous mapping of retina onto tectum innervated by both eyes. *Brain Res* 98: 172–176

Meyer RL, Sperry RW (1973): Tests for neuroplasticity in the anuran retinotectal system *Exp Neurol* 40: 525–539

Miller KD (1990): Correlation-based mechanisms of neural development. In: *Neuroscience and Connectionist Theory*, Gluck MA, Rumelhart DE, eds. Hillsborough, NJ: Lawrence Erlbaum: 267–353

Miller KD, Keller JB, Stryker MP (1989): Ocular dominance column development: Analysis and simulation. *Science* 245: 605–615

Overton KJ, Arbib MA (1982): The extended branch-arrow model of the formation of retino-tectal connections. *Biol Cyber* 45: 157–175

Prestige MC, Willshaw DJ (1975): On a role for competition in the formation of patterned neural connections. *Proc Roy Soc Lond B* 190: 77–98

Sharma SC, Gaze RM (1971): The retinotopic organisation of visual responses from tectal reimplants in adult goldfish. *Arch Ital Biol* 109: 357–366

Sperry RW (1963): Chemoaffinity in the orderly growth of nerve fibre patterns and connections. *Proc Natl Acad Sci USA* 50: 703–709

Straznicky C, Gaze RM (1980): Stable programming for map orientation in fused eye fragments in *Xenopus*. *J Embryol Exp Morphol* 55: 123–142

Swindale NV (1980): A model for the formation of ocular dominance stripes. *Proc Roy Soc Lond B* 208: 243–264

von der Malsburg C (1973): Self-organization of orientation sensitive cells in the striate cortex. *Kybernetik* 14: 85–100

von der Malsburg C (1979): Development of ocularity domains and growth behaviour of axon terminals. *Biol Cyber* 32: 49–62

von der Malsburg C, Willshaw DJ (1976): A mechanism for producing continuous neural mappings: Ocularity dominance stripes and ordered retino-tectal projections. *Exp Brain Res* 1 *(Suppl)*: 463–469

von der Malsburg C, Willshaw DJ (1977): How to label nerve cells so that they can interconnect in an ordered fashion. *Proc Natl Acad Sci USA* 74: 5176–5178

Willshaw DJ, von der Malsburg C (1976): How patterned neural connections can be set up by self-organization. *Proc Roy Soc B* 194: 431–445

Willshaw DJ, von der Malsburg C (1979): A marker induction mechanism for the establishment of ordered neural mappings; Its application to the retinotectal problem. *Phil Trans R Soc B* 287: 203–243

CHAPTER 18

Neural Specificity Revisited

Sansar C. Sharma

The sensory systems preserve the topography of stimuli by creating spatially ordered connections, or "maps" in the brain. The visual system has been a prototypical system for the study of mapping of spatial information onto appropriate areas of the brain. The retinal ganglion cell axons form terminal arbors in the contralateral optic tectum, and these arbors are ordered. Axons from a specific region of the retina project in a predictable manner to a specific region in the tectum. Sperry (1963) brought the problems of the formation of an orderly array of synaptic connections sharply into focus in his now-classic paper describing the chemoaffinity hypothesis which proposed fiber-to-target interactions and suggested that afferents were tagged with a biochemical label, synapsing only with neurons that carry a complementary label. Since Sperry's seminal paper, the concept of chemoaffinity has been expanded and additional theories of pattern formation have been proposed, including such examples of the following:

(a) Chemoaffinity may extend to include fiber-to-fiber as well as fiber-to-pathway interactions. Fiber-to-fiber affinity, for example, has been demonstrated in *in vitro* experiments on chick retina by the presence of a process of selective fasciculation of temporal retinal axons with each other (Bonhoeffer and Huf, 1985).

(b) Extrinsic factors such as substrate guidance may also provide important mechanisms underlying generation of topography; take for example the "blue prints" hypothesis of Singer et al. (1979). Comparable channels in the developing visual system have been described by Silver and Robb (1979), Krayanak and Goldberg (1981), Navascues et al. (1987), and Cima and Grant (1982).

(c) Preformed molecularly labeled pathways may exist in the absence of channels. For example, the extracellular matrix component laminin is found in the inner limiting membrane of the chick retina (McLoon, 1984) and encourages axonal outgrowth. In addition, retinal axons of 6-day chick embryos grow extensively on laminin in culture, whereas older axons, at embryonic day 11, do not (Cohen et al., 1986). Transient expression of laminin is noted in the developing rat and chick optic axons (Cohen et al., 1987). In adult goldfish, following optic nerve section, laminin is widely distributed; however, in the rat,

Formation and Regeneration of Nerve Connections
Sansar C. Sharma and James W. Fawcett, Editors
© 1993 Birkhäuser Boston

such is not the case. Another glycoprotein, N-CAM, demonstrates a localized distribution along the chick visual pathway and is presumed to be an important developmental molecular marker (Silver and Rutishauser, 1984). Preformed molecular pathways may be necessary for initial fiber outgrowth. However, such pathway markers may not regulate final synaptic events as suggested by studies showing that optic fibers entering the brain at the wrong position nevertheless grow directly to the correct tectal target region (Harris, 1986).

(d) Fiber–glia interactions may play a role in pathway selection. Maggs and Scholes (1986) showed that, in a cichlid fish, astroglia in the optic nerve expressed vimentin-like peptide, whereas astroglia in the optic tract expressed only glia fibrillary acidic protein (GFAP). The chiasm thus marks a point of transition in the optic pathway in glial cell populations. Our studies of the goldfish visual system have also shown a change in glial cell characteristics at the level of the chiasm. The response of optic fibers to different glial cell populations may play a major role in determining fiber behavior.

(e) Baier and Bonhoeffer (1992) recently showed that growth cones of temporal half-retinal ganglion cell axons in chick retina can respond preferentially to a concentration of gradients of the posterior tectal cell membrane component. These authors elegantly showed that temporal retinal axons face an "uphill" concentration gradient of a chemical distributed on the caudal tectal cells. Posterior tectal membranes repel growth cones from the temporal retinal axons in an *in vitro* assay system and the observed repulsion depended upon the steepness ("uphill") of the gradient. This inhibitory growth pattern of temporal retinal axon growth cones by the caudal tectal cell membrane may contribute *in vivo* to a mechanism whereby temporal axons are targeted to appropriate anterior tectum.

Following this brief description of mechanisms presently under investigation in various laboratories, it might be helpful to review how much we have learned in the past two decades regarding development of the retinotectal connections. I shall first describe the development of the retina. Some of the relevant problems have been discussed in Chapters 4, 12 and 13 and will be avoided here.

The migration of retinal axons is guided by factors present on the neurons on the glia cells (for example, see Rakic, 1985). The molecular nature of such factors may include the dynamic changes observed in cadherin (a cell surface glycoprotein) expression during the early differentiation of the retinal ganglion cells (Matsunaga et al., 1988). Cadherin may be responsible for placing early differentiated retinal ganglion cells along the inner margin of the retina.

These recently emerged retinal axons grow in contact with laminin containing basal lamina. Laminin, however, does not impose directionality on growing axons (McKenna and Raper, 1988). Recent evidence by Brittis et al. (1992) suggests that chondroitin sulfate, a glycosaminoglycan, serves as a regulator in guiding growing axons in the retina. Chondroitin sulfate appears early in the optic vesicle and at the optic center of the retina, and its concentration is substantially reduced around the optic fissure. These authors suggested that axons grow only in the direction devoid of chondroitin sulfate. Furthermore, chondroitin sulfate appears to recede centrifugally in the retina thereby perhaps allowing retina to differentiate in centrope-ripheral sequence and, hence, additionally orientating the axons in the proper direction.

The progression of growing axons in the optic nerve may be facilitated by the laminin present in the glia (Cohen et al., 1987). Neugebauer et al. (1988, 1991) and Neugebauer and Reichardt (1991) have shown that retinal neurite outgrowth on astrocytes is promoted by N–cadherin, NCAM and integrins. Chapter 4 describes in detail the fiber–fiber interactions in the growing optic axons in the nerve. Emergence of order in the optic nerve and tract (also discussed in previous chapters throughout this book) is essential in facilitating the orderly deployment of axons on the tectum. The axonal fasciculation in the nerve is maintained by NCAM (Thanos et al., 1984) and this order is disrupted when anti-NCAM antibodies are applied to the eye.

The questions of why fibers cross at the optic chiasm and why some later arriving axons (in rodents) go to the ipsilateral tectum are unclear. Nothing is known about cues that are present at the chiasm allowing this sorting of axons.

Speculation also exists as to the mechanism controlling the directionalities of growing axons in the optic tract. Are they intrinsic to the pathway, as suggested by Harris (1989), or do growing axons follow preformed pathways (as has been discussed in Chapter 3)? Another possibility, that growing axons in the tract are guided by target derived chemo-attractants, can not be totally eliminated.

Additionally, channeling of growing axons into a pathway may be facilitated by certain growth-promoting molecules. The possibility of tectum derived trophic factor seems untenable, as the tectum, at least in *Xenopus*, is totally undifferenti-ated at the time the first optic fibers reach the tectum (Gaze and Grant, 1992). In an exciting paper, Gaze and Grant (1992) provided evidence that at the time of arrival of the first optic axons in the tectum in *Xenopus*, no post-mitotic cells are present in the tectal precursor region. The first tectal cells appear around stage 41–45. These authors suggested that the retinotopic ordering in the optic pathway present from stages 37/38, may be responsible for the initial distribution of optic axons in the caudal diencephalic nuclei and not the optic tectum. The earliest arriving fibers have dorsoventral retinal fiber order. The ventro-temporal retinal fibers either arrive late or are influenced by the tectum for their ordering. It is clear that the order of the optic axons in the optic tract is established first and synapses are first formed in the diencephalic nuclei and not on the tectum as tectal cells have not yet differentiated. It is therefore plausible that optic fibers, after reaching the caudal diencephalic nuclei, wait there for the tectal cells to become post-mitotic and differentiate into neurons. It is only after tectal cell differentiation begins that optic fibers enter the tectum. It is fascinating to note that Bonhoeffer and his colleagues, who have elegantly shown inhibitory effects on retinal growth cones in choice assay systems and the ability of axonal growth cones to detect and respond to a gradient of tectal derived membranes, prepared their retinal explants from embryonic day 6 in the chick eye and the tectal membrane preparation from embryonic day 9 tecta. This difference in timing suggests that day 6 tectal cells, still in the process of differentiation in chick embryo, perhaps have not yet developed the membrane glycoprotein involved in the process of axonal growth cone inhibition. Future experiments should be directed to resolving the problem whether arrival of the optic axons on the tectum is dependent upon the differentiation of the tectal cells and that early arriving axons on the tectum are responsible for inducing the differentiation of tectal cells. The fact that axonal growth cones have preferences for appropriate tectal target areas provides forceful and compelling evidence for axonal guidance by target-derived factors.

The question of forming a topographically ordered map on the tectum seems to have complex set(s) of rules. The distribution of appropriate quadrentic retinal axonal terminals have been postulated to be governed by target derived trophic or inhibitory factors (Bonhoeffer and Huf, 1985). The initially formed axonal arbors are diffused and are brought into restricted foci by sprouting and rearrangement of terminals which may be activity dependent (for details see Chapters 5, 8 and 14). In addition to the sensitivity of temporal axons to a concentration gradient of "guiding components" derived from caudal half tectum in chick (Baier and Bonhoeffer, 1992), other molecules which are expressed in a discontinuous fashion both in the retina and the tectum have been isolated (Trisler et al., 1981; Trisler and Collins, 1987; McLoon, 1991; Rabacchi et al., 1990; McCaffery et al., 1990; Constantine-Paton et al., 1986). Different molecules which are seemingly involved in forming the gradients in the retina have also been isolated. Trisler et al. (1981) showed the presence of a "TOP" molecule and its effect on retinal development. It is present in a gradient fashion on the dorsoventral axis of the developing retina. The TOP molecule has been shown to be involved in the synaptic maturation in the retina. The role this molecule plays in generating retinotophic topography remains unclear.

Similarly, another specific protein isolated by Rabacchi et al. (1990) shows graded distribution in dorsal versus ventral retina. How this protein, with its possible linkage to laminin receptor, deploys itself in setting up a gradient is unclear. The first example in which a protein distinguished between nasal and temporal retina was shown by McLoon (1991). The molecule was termed "TRAP". The distribution of TRAP in the tectum is unknown. None of the molecules whose distribution in the retina or perhaps in the tectum is graded, provides testimony to the evidence of orthogonally graded systems in the tectum. Great advances have been made regarding the molecular aspects of the problem of neural specificity in the developing retinotectal system in the past 10 years. Future research should be directed towards understanding how axonal growth is affected and, most importantly, how single axon of a specific type find the correct class of a neuron in the tectum and form the synapses.

It is clear from the above that intensive investigations of the developing visual system have generated a plethora of theories; yet we are far from understanding the details of any one theory and we do not yet understand whether or how different classes of mechanisms interact to establish an orderly map.

Further efforts to discern the mechanisms of map formation have focused on the adult goldfish, whose visual system exhibits extensive plasticity in its pattern of retinotectal connectivity. The size disparity experiments have provided a useful paradigm for investigating how regenerating optic axons establish orderly synaptic connects with the target, although the proposed underlying mechanisms of connectivity are still controversial (Sharma, 1986; Edwards et al., 1985; Marotte, 1983; Hayes and Meyers, 1988; Murray and Sharma, 1992).

Much of the controversy in this field exists because no anatomical methods have been developed that would permit a complete and precise determination of what ganglion cell axons do when they grow to the tectum either during normal development or during optic nerve regeneration. Numerous anatomical techniques have been employed (for review, see Sharma and Romeskie, 1984), but none allows visualization of a single complete terminal arborization showing also the locus of ganglion cell origin in the retina together with its electrophysiological characteristics. A complete picture of the connectivity process requires combined anatomical and electrophysiological characterization of a ganglion cell and its arborization pattern as well as characterization of cells in the tectum on whose dendrites the synapses are formed.

In the tectum of the goldfish, cell types receptive to the primary visual input have recently been described (Guthrie and Sharma, 1991). HRP methods, as utilized so far by various authors, provide scarce arbors, and those that are shown are probably incomplete (Fujisawa, 1987). An excellent effort has been made by Schmidt et al. (1988 and see Chapter 14) to show HRP labeled optic axon arbors during regeneration. In this study, early regenerating arbors were up to five times wider in extent as compared to normal; during the late regeneration period the arbor size returned to normal. These transitory changes in arbor size suggest restriction of arbor during the course of regeneration and correspond in time to the period in which activity has been implicated in sharpening the visual maps. Sorely lacking in such work is information regarding the retinal ganglion cell types responsible for various kinds of arbors. Additionally, only arbors in the SFGS layer of the tectum were described; optic axons terminating in deeper layers of the tectum were never addressed.

To date there are three known optic fiber termination zones in goldfish: SFGS, SGC and SPV (Sharma, 1972; Springer and Gaffney, 1981). The electrophysiological characteristics of these three terminal areas were shown earlier by Jacobson and Gaze (1964). In relation to the structure of the optic tectum in teleosts, particularly in the goldfish, a few limited studies exist which used the Golgi method (Meek and Schellart, 1978; Sharma and Romeskie, 1979). A recent study by Kageyama and Meyer (1988) utilizes cytochrome oxide (C.O.) labeling of the goldfish tectum and suggests a new scheme of tectal lamination based upon oxidative metabolic organization. Optic input laminae had the highest C.O. activity, with four distnt lamina outlined in the SFGS of the tectum. Their suggestion that these sublaminae may be innervated by different classes of ganglion cells once again begs the question as to what specific ganglion cells innervate what sublaminae.

The identifications of optic axon arbors in the goldfish tectum also suggest three types of retinal ganglion cells (Schmidt et al., 1988). However, another study describes four types of ganglion cells in the goldfish (Hitchcock and Easter, 1986).

In general, three types of cells are commonly accepted in the mammalian retina (for example, Leventhal and Hirsch, 1982), and three types of retinal ganglion cells have been described in amphibia (Sakaguchi and Murphy, 1985) and catfish (Dunn-Meynell and Sharma, 1986). Only one study correlates a single retinal ganglion cell type with the terminal arbor and its tectal lamination (Stirling and Merrill, 1987), and no such study exists to date in any fish. Extensive studies have been done on the mammalian visual system, in which X, Y and W retinal cells and their morphologies have been correlated with their lamination in the lateral geniculate nucleus (for review, see Sherman and Spear et al., 1982, and Sur et al., 1987). Similar developmental studies in the cat visual system have revealed the entire terminal fields of a single identifiable retinal axon (Friedlander et al., 1985). In addition, synaptic circuitry of identified X, Y and W cells in the cat LGN have been delineated (Wilson et al., 1976). These later studies provide evidence of differences in X, Y and W cells in the LGN which are responsible for specific visual pathways. Since X, Y and W type cells in the retina of goldfish exist and because the visual topography in lower vertebrates is restored after surgical manipulations it would be extremely useful to study whether specific visual pathways exist for separate topographies.

The adult goldfish visual system has lent itself to a vast variety of experimental manipulations which have provided significant insight into the mechanisms responsible for the formation of the retinotectal system. Since the optic nerve in adult goldfish is capable of regeneration, the goldfish has been used as a model for biochemical analysis of nerve regeneration, plasticity in retinotectal connections, and psychophysical experimentation. As described above, studies in the lower vertebrate visual system have been limited in the finer details of what constitute retinotectal system topography. Totally lacking are the studies which correlate the observed categories of ganglion cell types in the eye with the lamination and shape of their arbors in the tectum. Future experiments should be directed: 1) to discern which retinal ganglion cell axons connect where in the tectum, 2) on what cell type(s) they connect, and 3) the fine nature of this pattern of connectivity. Once this information is discerned, we can begin to ask more specific questions regarding detailed anatomical substrates of observed neural plasticity in the adult visual system and the re-establishment of connections following optic nerve regeneration.

The concept of neuronal specificity in the retinotectal system is comparatively easy to comprehend when one studies either quadrant, i.e., retinal representation onto the tectum or point-to-point electrophysically mapped retinotectal connections. However, when one looks at detailed electromicroscopic pictures of retinotectal connections, at least in fish and amphibians, the complexity of the number of synapses onto a single cell, creates a monumental picture.

From the above discussion, it is apparant that most experimental designs have limitations as only one single cue for the pattern formation is tested in each experiment. No single experiment to date describes all the details of the pattern formation. As described earlier in the first section, certain molecules seems necessary for initial elongation of axons and they appear in specific sequence. It is possible that the visual system uses multiple mechanisms in series and parallel to congregate its connections. Exclusion of one single factor does not rupture the system. The concept that positional cues present on the tectum are *sorely* responsible for the navigation of optic axon is equally redundant as optic fibers have been shown to innervate nontectal and nonvisual areas where the fibers maintain retinotopic order (Sharma, 1981). Hence, the ordering of patterns during development and regeneration in the visual system must depend on multilevel factors acting in concert. Hopefully, future experimental designs will be made to test more than one factor(s) responsible for the pattern formation in the retintotectal system.

From relatively few studies, it appears that an extensive divergence of information exists in retinotectal connections. It is easy to picture axonal terminal arbor of a single ganglion cell forming synapses onto dozens of tectal cells. Similarly, as optic axonal arbors overlap onto the tectum, it is predictable that one tectal cell may receive information from more than one axonal terminal. This complex situation led George Székély (1990) to recently claim, perhaps in exasperation, that "in a system with this extent of divergence and convergence, any king of *neuronal specificity* which determines specific neuronal interconnections would be meaningless."

As complex and hopeless at times as the concept of "neuronal specificity" appears to be, it has served the scientific community well by stimulating generations of investigators. From my personal perspective, it has been a very provocative and challenging and at times rewarding concept. When asked by someone a few years back in Philadelphia how neuronal specificity works, Mike Gaze replied "when I started in this field the ideas and experiments were simple to discern. In the last two decades I have learned a lot about the details of this system. However, I haven't the foggiest idea how does neuronal specificity work."

It is a credit to Mike Gaze that so many of his students and collaborators, whose contributions make this volume, have made important contributions to the field of retinotectal development and plasticity of adult connections. My hesitation at the end is the fear that Mike will detect any and all ambiguities in this assay.

REFERENCES

Baier H and Bonhoeffer F (1992): Axon guidance by gradients of target-derived component. *Science* 255: 472–475

Bonhoeffer F and Huf J (1985): Position dependent properties of retinal axons and their growth cones. *Nature* 315: 409–410

Brittis PA, Canning DR, Silver J (1992): Chondroitin sulfate as a regulator of neuronal patterning in the retina. *Science* 255: 733–736

Cima C, Grant P (1982): Development of the optic nerve in *Xenopus laevis* I. Early development and organization. *J Embryol Exp Morph* 72: 225–249

Cohen J, Burne JF, McKinley C, Winter J (1987): The role of laminin and laminin/fibronectin receptor complex in the outgrowth of retinal ganglion cell axons. *Develop Biol* 122: 407–418

Cohen J, Burne JF, Winter J, Bartlet PF (1986): Retinal ganglion cells lose response to laminin with maturation. *Nature* 322: 456–467

Constantine-Paton M, Blum AS, Mendez-Otero R, Barnstable CJ (1986): A cell surface molecule distributed in a dorsoventral gradient in the perinatal rat retina. *Nature* 324: 459–462

Dunn-Meynell AA, Sharma SC (1986): The visual system of channel catfish (Ictaluruspunctatus) 1. Retinal ganglion cell morphology. *J Comp Neurol* 247: 32–55

Edwards MA, Sharma SC, Murray M (1985): Selective retinal innervation of a surgically created tectal island in goldfish. Light microscopic analysis. *J Comp Neurol* 232: 372–385

Friedlander MJ, Martin KAC, Vahle-Hinze C (1985): The structure of the terminal arborizations of physiologically identified retinal ganglion cells y-axons in the kitten. *J Physiol* 359: 293–313

Fujisawa H (1987): Mode of growth of retinal axons within the tectum of *Xenopus* tadpoles, and implications in the ordered neuronal connections between the retina and the tectum. *J Comp Neurol* 260: 127–139

Gaze RM, Grant P (1992): Development of the tectum and diencephalon in relation to the time of arrival of the earliest optic fibers in *Xenopus*. *Anat Embryol* 185: 599–612

Guthrie DM, Sharma SC (1991): Visual responses of morphologically identified tectal cells in the goldfish. *Vision Res* 31: 507–524

Harris WA (1986): Homing behavior of axons in the embryonic vertebrate brain. *Nature* 320: 266–269

Harris WA (1989): Local positional cues in the neuroepithelium guide retinal axons in embryonic *Xenopus* brain. *Nature* 339: 218–221

Hayes WP, Meyers R (1988): Optic synapse number but not density is constrained during regeneration into surgically-halved tectum in goldfish: HRP-EM evidence that optic fibers compete for a fixed number of postsynaptic sites on tectum. *J Comp Neurol* 274: 539–559

Hitchcock PF, Easter SS (1986): Retinal ganglion cells in goldfish: A quantitative classification into four morphological types, and a quantitative study of the development of one of them. *J Neurosci* 6: 1037–1050

Jacobson M, Gaze RM (1964): Types of visual response from single units in the optic tectum and optic nerve of the goldfish. *Quart J Exp Physiol* 49: 199–209

Kageyama GH, Meyer RL (1988): Histochemical localization of cytochrome oxidase in the retina and optic tectum of normal goldfish: A combined cytochrome oxidase-horseradish peroxidase study. *J Comp Neurol* 270: 354–371

Krayanak S, Goldberg S (1981): Oriented extracellular channels and axonal guidance in the embryonic chick retina. *Develop Biol* 84: 41–50

Leventhal AG, Hirsch HVB (1982): Effects of visual deprivations upon the morphology of retinal ganglion cells projecting to the dorsal lateral geniculate nucleus of the cat. *J Neurosci* 3: 332–344

Maggs A, Scholes J (1986): Glial domains and nerve fiber patterns in the fish retinotectal pathway. *J Neurosci* 6: 424–438

Marotte LR (1983): Increase in synaptic sites in goldfish tectum after partial tectal ablation. *Neuroscience Lett* 36: 261–266

Matsunga M, Hatta K, Takeichi M (1988): Role of N-Cadherin cell adhesion molecule in the histogenesis of the neural retina. *Neuron* 1: 289–295

McCaffery P, Neve RL, Drager UC (1990): A dorso-ventral asymmetry in the embryonic retina defined by protein conformation. *Proc Natl Acad Sci USA* 87: 8570–8574

McKenna MP, Raper JA (1988): Growth cone behavior on gradient of substratum bound laminin. *Develop Biol* 130: 232–236

McLoon S (1984): Evidence for shifting connections during development of the chick retino-tectal projection. *J Neurosci* 5: 2570–2580

McLoon S (1991): A monoclonal antibody that distinguishes between temporal and nasal axons. *J Neurosci* 11: 1470–1477

Meek J, Schellart NA (1978): A Golgi study of goldfish optic tectum. *J Comp Neurol* 182: 89–122

Murray M, Sharma SC (1992): Target regulation of synaptic number in the expanded retinotectal projection of goldfish: The half-retinal preparation. *Restorat Neurol Neurosci* 4: 97–105

Navascues J, Rodriquez-Gallando L, Garcia-Martinez V, Alvarez IS, Martin-Partido G (1987): Extra axonal environment and fibre directionality in the early development of the chick embryo optic chiasm: A light and scanning electron-microscopic study. *J Neurocytol* 16: 299–310

Neugebauer KM, Reichardt LF (1991): Cell surface regulation of B_1-integrin activity on developing retinal neurons. *Nature* 350: 68–71

Neugebauer KM, Emett CJ, Venstrom KA, Reichardt LF (1991): Vitronectin and thrombospondin promote retinal neurite outgrowth: Developmental regulation and the role of integrins. *Neuron* 6: 345–358

Neugebauer KM, Tomaselli KJ, Lillien J, Reichardt LF (1988): N–Cadherin, NCAM and integrins promote retinal neurite outgrowth on astrocytes *in vitro*. *J Cell Biol* 107: 1177–1187

Rabacchi S, Neve RL, Drager UC (1990): A positional marker for the dorsal embryonic retina is homologous to the high affinity laminin receptor. *Development* 109: 521–531

Rakic P (1985): In: *The Cell in Contact*, Edelman GM, Thiery JP, eds. New York: Wiley

Romeskie M, Sharma SC (1979): The goldfish optic tectum: A Golgi study. *Neuroscience* 4: 625–642

Sakaguchi D, Murphy R (1985): Map formation in the developing *Xenopus* retino-tectal system: An examination of ganglion cell terminal arborizations. *J Neurosci* 5: 3228–3245

Schmidt JT, Turcotte JC, Buzzard M, Tieman DG (1988): Staining of regenerated optic arbors in goldfish tectum. Progressive changes in immature arbors and a comparison of mature regenerated arbors with normal arbors. *J Comp Neurol* 269: 565–591

Sharma SC (1972): The retinal projection in adult goldfish: An experimental study. *Brain Res* 39: 213–223

Sharma SC (1981): Retinal projections in a non-visual area following bilateral tectal ablation in goldfish. *Nature* 291: 66–67

Sharma SC (1986): The effects of surgical manipulations of the optic tectum. In: *Process of Recovery from Neural Trauma*, Gilad G, Kreutzberg, Giorio A, eds. *Exp Brain Res Suppl* 13: 258–267

Sharma SC, Romeskie M (1984): Plasticity in the retino-tectal connections in goldfish. In: *Comparative Neurology of the Optic Tectum*, Vanegas H, ed. New York: Plenum Press

Sherman SM, Spear PD (1982): Organization of the visual pathways in normal and visually deprived cats. *Physiol Rev* 62: 738–855

Silver J, Robb RM (1979): Studies on the development of the eye cup and optic nerve in normal mice and in mutants with congenital optic nerve aplasia. *Develop Biol* 68: 175–190

Silver J, Rutishauser U (1984): Guidance of optic axons in vivo by a performed adhesive pathway on neuroepithelial endfeet. *Develop Biol* 106: 485–499

Singer M, Nordlander RH, Egar M (1979): Embryogenesis and regeneration in the spinal cord of the newt. The blueprint hypothesis of neuronal pathway patterning. *J Comp Neurol* 185: 1–22

Sperry RW (1963): Chemoaffinity in orderly growth of nerve fibers and connections. *Proc Natl Acad Sci USA* 50: 703–710

Springer AD, Gaffney JS (1981): Retinal projections in the goldfish. A cobaltous-lysine study. *J Comp Neurol* 203: 401–424

Stirling VR, Merrill EG (1987): Functional morphology of frog retinal ganglion cells and their central projections: The dimming detectors. *J Comp Neurol* 258: 477–495

Sur M, Esquerra M, Garraghty PE, Knitzer MF, Sherman SM (1987): Morphology of physiologically identified retinogeniculate x- and y-axons in the cat. *J Neurophysiol* 58: 1–32

Székély G (1990): Problems of the neuronal specificity concept in the development of neural organization. *Concepts Neurosci* 1(2): 165–197

Thanos S, Bonhoeffer F, Rutishauser U (1984): Fiber–fiber interactions and tectal cues influence the development of chicken retinotectal projections. *Proc Natl Acad Sci USA* 81: 1906–1910

Trisler D, Collins F (1987): Corresponding spatial gradients of TOP molecules in the developing retina and optic tectum. *Science* 237: 1208–1209

Trisler D, Schneider MD, Nirenberg M (1981): A topographic gradient of molecules in retina can be used to identify neuron position. *Proc Natl Acad Sci USA* 78: 2145–2149

Wilson PD, Rowe MH, Stone J (1976): Properties of relay cells in the cat's lateral geniculate nucleus: A comparison of w-cells with x- and y-cells. *J Neurophysiol* 34: 1193–1209

Keyword Index

This index was established according to the keywords supplied by the authors. Page numbers refer to the beginning of the chapter.